Einstein, Picasso

Einstein, Picasso

Space, Time, and the Beauty That Causes Havoc

Arthur I. Miller

A Member of the
Perseus Books Group

Published by Basic Books,
A Member of the Perseus Books Group

Designed by Brent Wilcox
Set in 11-point Adobe Caslon

Library of Congress Cataloging-in-Publication Data
Miller, Arthur I.
Einstein, Picasso : space, time, and the beauty that causes havoc / Arthur I. Miller.
p. cm.
Includes bibliographical references and index.
ISBN 0-465-01860-2
1. Art and science—History—20th century. 2. Picasso, Pablo, 1881–1973. 3. Artists—France—Biography. 4. Einstein, Albert, 1879–1955. 5. Physicists—United States—Biography. I. Title.

N72.S3 M55 2001
709'.2—dc21
[B]

00-065130

02 03 04 / 10 9 8 7 6 5 4 3 2 1

To Mike Brady

Contents

Acknowledgments

One of the pleasures of interdisciplinary research is meeting like-minded people. While working on this book I had the good fortune to discuss my ideas with art historians who were extremely generous with their time in criticizing my aims and setting out tough-minded issues for me to tackle, forcing me to clarify my own views on art and science. In particular I am grateful to Chris Green, Mike Leja and, of course, Linda Dalrymple Henderson, who is the groundbreaker in studying the interplay between cubist art and science.

Without assistance from the staff of the magnificent Musée Picasso, this book would have been extremely difficult to write. I thank particularly Sylvie Fresnault and Yvonne Sudour, and especially Anne Baldassari, whose pioneering research on Picasso as photographer was of great value to me.

I thank Jonathan Betts, Curator of Horology, The Old Royal Observatory, Greenwich, of the National Maritime Museum, Greenwich, England, for his comments on the setting up of time zones; Barbara Bürki and Adolf Meichle of the Einstein Hause, Bern, for their help in forming a picture of Einstein's home life in 1905; and Jürgen Renn of the Max Planck Institut für Wissenschaftgeschichte, Berlin, for his splendid hospitality. Howard Gardner's incisive comments on the psychology of creativity were extremely helpful.

My good friend Mike Brady read the entire manuscript and made invaluable comments that added immeasurably to the book's final form. Among others who read and commented on chapters are Dorothy Edinburgh, Marilia Duffles, Jill Shaw Ruddick and Sarah Star.

I am deeply appreciative of the sabbatical leave for the academic year 1999/2000 generously granted by Derek Roberts in his capacity as Provost of University College London. I need not emphasize the value of uninterrupted work on a complex project.

The exquisite care that my editor William Frucht took with my manuscript made this a better book. Working with Bill is always an educational experience and a great pleasure.

Writing any book requires pulling together the intellectual and emotional areas of one's life. This was certainly the case here, in more ways than one. Mike Brady, Leon Fine, Bill Frucht, Katie Lane and Semir Zeki always asked how things were going, and I deeply appreciate their intellectual and emotional support.

Arthur I. Miller
London, 2000

1

Two Worlds as One

Everything is possible, everything is realizable, in all and everywhere.

—André Salmon

Albert Einstein and Pablo Picasso, exemplars of genius, inspiration for generations of artists and scientists, are icons of the twentieth century. Modern science is Einstein and modern art, Picasso. How this came about is one of the great sagas in the history of Western thought.

While it is a truism that one can always find amazing coincidences between any two people, in the case of Einstein and Picasso the similarities in their personal lives, working lives and creativity are uncanny and documentable. The parallels between the two during their period of greatest creativity—the first decade and a half of the twentieth century—show us much more than the common points of their own thinking. They also offer glimpses into the nature of artistic and scientific creativity and of how research was carried out at the *common* frontier of art and science.

In those exhilarating days at the beginning of the last century, when everything seemed possible and realizable everywhere, Einstein and Picasso made no distinction between their personal and working lives. From a single cauldron emerged ideas that set into motion everything we call modern. I am more interested in *why* Einstein and Picasso made their discoveries than in *how* they went about developing insights. The psychologist of art Rudolf Arnheim has written: "How then are we to discover what takes place when a work of art is created? We can listen to what the artist reports about himself."[1] This holds for scientists, too.

Pablo Picasso's biographer John Richardson quotes a comment by Dora Maar, one of the artist's more perceptive mistresses. Although she was speaking of Picasso's postcubist days, her remarks apply best to the period in which he discovered cubism. "There were five factors," Maar said, "that determined his way of life and likewise his style: the woman with whom he was in love; the poet, or poets, who served as a catalyst; the place where he lived; the circle of friends who provided the admiration and understanding of which he never had enough; and the dog who was his inseparable companion."[2] These factors began to come together in May 1904, when Picasso moved into 13 rue Ravignan, a dilapidated tenement in the Montmartre district affectionately known as the Bateau Lavoir. By midsummer of 1907 he had produced *Les Demoiselles d'Avignon,* the painting that brought art into the twentieth century. Whereas many of Picasso's artist and literary friends had already achieved some notoriety in their own right and greatly respected his artistic acumen, none of them was prepared for what emerged from his atelier that July.

Except for the dog, Albert Einstein's situation was similar in spring 1905, when he and Mileva moved into a cramped third-floor walk-up at 49 Kramgasse, in the old city center of Bern, Switzerland. Einstein's close friends in Bern were obscure civil servants like him, and certainly none of them had the remotest clue to what he would soon produce.

The general line of argumentation among art historians is that the roots of cubism are in Paul Cézanne and primitive art. This view discounts completely how astounding developments in science, mathematics and technology contributed to the very definition of "avant-garde."[3] It has long been known that the roots of science were never totally within science itself. Why then should the roots of the most influential art movement of the twentieth century lie totally within art? By widening our viewpoint of the origins of Picasso's *Demoiselles* to include science, mathematics and technology, we gain deeper insight into Picasso's monumental struggles.

Parallel biographies are thus a means to explore the intellectual climate at the beginning of the twentieth century, an era of genius unmatched since the Renaissance. The best works produced in that era will be forever among those that define the high road of civilization. Relativity and *Les Demoiselles* represent the responses of two people—Einstein and Picasso, although geographically and culturally separated—to the dramatic changes sweeping across Europe like a tidal wave.[4]

At the epicenter of these enormous transformations was the debate about representation versus abstraction. In art, there was a strong countermovement to the figuration and perspective that had held center stage ever since the Renaissance, which surfaced most forcefully in the postimpressionism of Paul Cézanne. New developments in technology such as airplanes, wireless telegraphy and automobiles were altering everyone's conception of space and time. The multiple images in the pioneering cinematography of Eadward Muybridge and Étienne-Jules Marey permitted change with time to be portrayed either on successive frames of film or on a single frame, in addition to depicting different perspectives on serial frames. In science the discovery of X rays seemed to render inside and outside ambiguous, the opaque became transparent and the distinction between two and three dimensions was blurred. Radioactivity, with its apparently limitless amounts of energy, seemed to prove that space is full of alpha, beta, gamma and X rays flying everywhere and opening up everything. Even more abstractly, mathematicians mused over exotic new geometries that could be represented in dimensions greater than three. People were especially fascinated by the idea of four-dimensional space, with its implication of motion in space *or* time.

All of this was discussed in newspapers, magazines and cafés, as well as in elegant and accessible philosophical writings by people like Henri Bergson and the great French polymath Henri Poincaré. These developments and what they meant were debated among the tight group of friends known as *la bande à Picasso* who met in Picasso's atelier, on whose door hung the sign *Rendezvous des poètes.* The group comprised poets, devotees of the occult and avant-garde literary fantasists such as Alfred Jarry, who had published parables on non-Euclidean geometry, the fourth dimension and time travel. Coincidentally, in Bern, Switzerland, a comparable study group debating similar themes called itself the "Olympia Academy." They met in similar, if less flamboyantly romantic, bohemian poverty. Each group took all knowledge as its province and orbited a central sun: in Paris, Picasso; in Bern, Einstein.

Ideas were everywhere and so was the desire for change. Alongside the developments in mathematics, science and technology was the discovery of the conceptual quality of African objets d'art. All of these ideas helped Picasso to free himself from earlier modes of thinking. Everyone involved in cubism considered it a highly intellectual adventure with the specific goal of reducing forms to geometry. Picasso's exploration of space in his groundbreaking *Les Demoiselles d'Avignon* employed notions of four-dimensional space described to him by Maurice Princet, an insurance actuary interested in advanced mathematics and a member of *la bande à Picasso.*

The two men were introduced in 1905 by Princet's notoriously unfaithful mistress Alice Géry, who at one time had been involved with Picasso. Although never a central figure in Picasso's group, Princet was frequently seen with them in cafés, participated in their hashish sessions and visited the Bateau Lavoir during critical times in spring 1907, when Picasso was struggling with *Les Demoiselles d'Avignon.* Picasso listened to his discourses on non-Euclidean geometry and the fourth dimension, which Princet gleaned mostly from Poincaré's widely read book *La Science et l'hypothèse.* Whereas Bergson wrote poetically about time and simultaneity, and Jarry in ways that were fantastically subversive, it was Poincaré, via Princet, who delivered the goods about simultaneity and non-Euclidean geometry.

Poincaré is a common denominator in our story. In 1904, Einstein read the excellent German translation of *La Science et l'hypothèse* and likewise was inspired by the sweep of its mathematics, philosophy and science. Just as his suggestive play with higher dimensions was among the factors that spurred Picasso's discovery of geometry as the language of the new art, Poincaré's insights on time and simultaneity were inspirational to Einstein's discovery of relativity.

In the intellectual atmosphere of 1905 it is not surprising that Einstein and Picasso began exploring new notions of space and time almost coincidentally. The main lesson of Einstein's 1905 relativity theory is that in thinking about these subjects, we cannot trust our senses. Picasso and Einstein believed that art and science are means for exploring worlds beyond perceptions, beyond appearances. Direct viewing deceives, as Einstein knew by 1905 in physics, and Picasso by 1907 in art. Just as relativity theory overthrew the absolute status of space and time, the cubism of Georges Braque and Picasso dethroned perspective in art.[5]

Einstein's approach to space and time was not primarily mathematical. Notions of aesthetics were essential to his discovery in 1905 of relativity and a new representation for light, and then in 1907 of a means to widen relativity theory to include gravity. Nor were Picasso's studies of space totally artistic in the narrow sense of this term, as his interest in scientific developments reveals. Picasso's new aesthetic for the *Demoiselles* was the reduction of forms to geometry.

Cézanne's influence on Picasso was complex, inasmuch as it was less significant for the *Demoiselles* than for later developments. Of great importance was Cézanne's bold new manner of producing spatial ambiguity, which he accomplished by merging foreground and background in a way that fused planes and integrated objects and space. This is called *passage*.[6] Cézanne went on to further organize his paintings so as to create several perspective points, which change as you view the painting from different angles. This required of Cézanne at least an intuitive understanding of spatial relations that verged on the geometric. Picasso referred to Cézanne as his "one and only master."[7]

Einstein's Cézanne was the great Dutch physicist H. A. Lorentz, of whom Einstein wrote, "I admire this man like no other; I might say, I love him."[8] Although Cézanne made the great leap to free art from a single perspective point, he remained rooted in the nineteenth century. Similarly Lorentz had almost formulated a proper theory of electromagnetic phenomena, and yet could not bring himself to interpret it as predicting the relativity of space and time. Einstein and Picasso, on the other hand, because they sought realities beyond appearances, each accomplished something entirely new.

Nor were the two men's personal working styles dissimilar. Both came to terms early on with the loneliness of the creative effort. As Einstein wrote some years later, "I live in the solitude which is painful in youth, but delicious in the years of maturity."[9] Picasso recalled the "unbelievable solitude"[10] he felt when working on *Les Demoiselles d'Avignon.*

Both men emphasized that despite their apparently revolutionary advances, they were actually extending the work of past masters. *Les Demoiselles* contains vestiges of Cézanne, El Greco, Gauguin and Ingres, among others, with the addition of conceptual aspects of primitive art properly represented with geometry. Likewise in Einstein's relativity theory we note legacies of Lorentz, Ernst Mach, David Hume, Immanuel Kant and Poincaré, to name but a few of his scientific and philosophical ancestors. Technology too played a role in Picasso's development of cubism, as we see from his adroit use of photographs as models for paintings and his interest in cinematography. Similarly did technology provide key input into Einstein's thinking toward relativity in 1905, particularly the design of electrical dynamos and practical problems of wireless telegraphy.

Yet at first, both their seminal works were terribly misunderstood. That Einstein had accomplished something new in 1905 would not be apparent to everyone until 1911. If relativity was appreciated at all before that, it was mostly for the wrong reasons. Let us not forget that Einstein sat in the Swiss Federal Patent Office from 1902 until 1909 and received his first academic position on the basis of research results that had nothing to do with relativity theory. The initial reaction to the *Demoiselles* by three members of *la bande à Picasso* was an embarrassed silence, and in a subsequent viewing Braque was scandalized. In the fall of 1907 Picasso put the painting aside and did not exhibit it again until 1916. It was not widely recognized as anything revolutionary until the early 1920s. Just as only Einstein understood his 1905 paper on electrodynamics as a major conceptual advance, so did Picasso with the *Demoiselles.*

Picasso's and Einstein's personal lives bear similarities and differences that, to some extent, reflect their intellectual and social milieus. Recently discovered love letters between Einstein and his college girlfriend, Mileva Marić, reveal a side of him as yet unexplored. By 1909, Mileva, Einstein's wife since 1902, was in a position of disfavor not unlike that of Picasso's then-mistress Fernande Olivier. Like Picasso with Fernande, Einstein had learned to harness Mileva's moods to his vision, and his passions provided some of the dynamics for his greatest creations.

By 1911 many artists were familiar with X rays, radioactivity and Poincaré's writings on geometry. All of this influenced their practice of art and was instrumental in producing early offshoots of cubism that were formulated specifically to diverge from cubism's "figurativeness," as this term had been reinterpreted. An early representative of this trend was Wassily Kandinsky, who produced the first entirely nonfigurative painting in 1910. He was among artists who were especially interested in the mass-energy equivalence, X rays and radioactivity, which they took as proof that, ultimately, everything is amorphous. While art was moving toward a highly abstractive phase, physics underwent a parallel movement after the geometrization of space and time in Einstein's general relativity theory of 1915, and then even more dramatically in the 1920s with the development of quantum theory. Yet pure abstraction was a Rubicon that Picasso never crossed, and Einstein never agreed with the high abstractions of quantum theory. Each man ultimately lost contact with the implications of his own revolution.

Instead of referring to an "interplay" between art and science, we must begin to speak of ideas that were developed in common by artists and scientists. The age-old quest of both art and science has been to seek new representations of phenomena beyond appearances. This effort becomes focused at the nascent moment of creativity, when boundaries dissolve between disciplines and notions of aesthetics became paramount. Coming to grips with this phenomenon requires delving into the nature of creative thinking.

For the purpose of parallel biographies of Einstein and Picasso, I have divided their stories up into six chapters, three each. To set the stage for their anni mirabiles in 1905 and 1907 respectively, Chapters 2 and 3 discuss their formative years, which include their education, the social, scientific and intellectual milieus within which they lived and which they attempted to break with, their lives as young men including their female confidants and lovers and the closed circles of male friends with which they surrounded themselves.

Chapters 4 and 5 continue the life of Picasso into the second decade of the twentieth century. Chapter 4 focuses on the scientific, technological and mathematical elements of the avant-garde that affected his discovery of the proper representation for *Les Demoiselles d'Avignon.* I will look into Picasso's work habits, cultural tastes and personal life and the tensions that provided the dynamics for his creative efforts. Science, technology and mathematics turn out to be important tiles in this mosaic.

Late in 1907 Picasso met Georges Braque. Chapter 5 investigates their joint efforts toward a developed cubism. Recently, Anne Baldassari of the Musée Picasso, Paris, has located over 5,000 photographic documents in Picasso's archives, of which roughly one hundred predate the 1920s. These photographs, which date back to 1901, demonstrate Picasso's skill in taking pictures as well as manipulating negatives and prints. In this way Picasso discovered a new space of reference, pictorial space, which he used not only for paintings but to test prototypes of new visual approaches such as *papier collé* and *collage*. This chapter highlights his adroit and highly creative use of photography.

Chapter 6 explores how Einstein discovered the special relativity theory in 1905. One common point that emerges is the important role Poincaré played for both men. Not unconnected is the impact on Einstein of the technological element of the avant-garde, which includes setting clocks using electromagnetic signals and issues concerning electrical dynamos. Einstein had a second, less profound annus mirabilis in 1907, when he widened the 1905 relativity theory to include gravity. This is discussed in Chapter 7.

The tools to understand these parallel biographies come from cognitive scientific theories. Among them are results on how information held in memory is processed during unconscious thought and Gestalt psychological concepts. This approach is explored in Chapter 8, which also serves the important role of summing up and drawing conclusions.

The search for parallels leads inevitably to the general issue of parallelisms in how art and science developed in the twentieth century. The common trend toward abstraction and new forms of visual imagery turns out not to be serendipitous. That art and science should have progressed in a parallel manner in the twentieth century is abundantly clear from the intellectual struggles of Einstein and Picasso. As Gertrude Stein put it, in words that hold for Einstein, too: "The things that Picasso could see were the things which had their own reality, reality not of things seen but of things that exist."[11]

I wrote *Einstein, Picasso* for lovers of art and science practiced at their most fundamental and exciting level, for aficionados of thinking across disciplines and generally for readers interested in the drama of high creativity. We wonder about the moment when everything comes together to produce incredible insights. How does this happen? How do thoughts emerge that go beyond the information at hand? Answering these questions demands a multidisciplinary mode of thinking and analysis that is becoming progressively more important as lines between disciplines become blurred. It is my hope that this book will further inspire this method of twenty-first-century thought, demonstrated so spectacularly and set as a cornerstone of high creativity by Einstein and Picasso.

2

A Good-Looking Bootblack

We had no other preoccupation but what we were doing and . . . saw nobody but each other. Apollinaire, Max Jacob, Salmon. . . . Think of it, what an aristocracy!

—Pablo Picasso

In May 1904 the twenty-three-year-old Pablo Picasso arrived in Paris from Barcelona for his fourth stay. He moved into a recently vacated studio in Montmartre on 13 rue Ravignan, an odd-shaped ramshackle building dubbed the Bateau Lavoir, or "laundry boat," by Picasso's close friend, the poet Max Jacob. After three discouraging trips to Paris, this time Picasso was determined to stay and make a name for himself in a city that was the center of the art world, and the avant-garde itself. He would accomplish this beyond his dreams.

THE YOUNG ARTIST'S EDUCATION

Pablo Ruiz Picasso was born 25 October, 1881, at Málaga, a small town in southern Spain that had seen better times. His father, don José Ruiz Blasco, was a painter, teacher of art and curator of the Municipal Museum. His mother, María López Picasso, was a conventional Spanish wife with the customary aura of piety. Picasso's biographer John Richardson describes don José as "having the inclination to be an artist but not the gifts."[1] Don José soon felt this defect to the depths of his soul.

From his earliest days Pablo had premonitions of genius. Visiting an exhibition of children's art organized by the British Council in 1946, he remarked that "as a child I would never have been able to participate in a show of this kind: at the age of twelve I drew like Raphael."[2] He was an incredibly gifted child, able to draw before he could speak. His mother reported that Pablo's first words were "piz, piz," Spanish baby language for *lapiz,* or pencil.[3] A typical party game with the young boy was to have him draw, say, a horse, starting from the tail, then starting from the nose, and so on.

As is abundantly clear from drawings done at Corunna, where the family lived during 1891–1895, Picasso's assessment of his early ability was not exaggerated.[4] At the time he was a student in his father's art class at the School of Fine Arts. Seeing the boy's work don José was elated, yet despairing at his son overtaking him. Picasso's relationship with his father was tangled but fundamentally oedipal. In 1898, in Madrid, Picasso was remembered as saying that "in art one must kill one's father."[5] This competitiveness, love and embarrassment for don José are among the factors at the heart of Picasso's creativity. He later fantasized that his father gave up painting after seeing the young Pablo's works at Corunna.[6]

His teenage years contained more than the usual turbulence. In July 1895, the thirteen-year-old Picasso visited Madrid and saw the Prado. This first exposure to great art shook his self-confidence. At the time the family was in transit to Barcelona, where don José took up an appointment at the School of Fine Arts.

The high quality of Pablo's artwork for the entrance exam led to his being permitted to skip preliminary courses and enter advanced drawing classes with students five or six years older, who quickly accepted him as an extraordinary artistic talent.[7] In 1896 and 1897 he completed his first two substantial paintings, *First Communion* (1896)[8] and *Science and Charity* (1897).[9]

During 1897–1898 Picasso studied at the prestigious San Fernando in Madrid—sort of. He cut his art classes at every opportunity and railed against his teachers as not "having a grain of common sense." He wrote his friend Joachim Bas after two months at the school, "They just go on and on, as I suspected they would, about the same old things."[10] Spain, the letter continued, was the wrong place to study art; in Munich, "painting is studied seriously without regard for dogmatic notions of pointillism and so on." Evidently no one in the academy had anything to offer the young prodigy. In 1936, in reply to his then-secretary and old friend Jaime

Sabartés, who inquired whether Picasso ever regularly attended classes in Madrid, Picasso replied, "I should say not! What for?"[11]

Picasso's loathing for the Madrid Academy, and his lack of a workable studio and proper living quarters, led to depression and illness. Once back in Barcelona, he gratefully accepted an invitation from his friend Manuel Pallarès to spend some months working at the family's farm in the wild and rugged countryside of Horta de Ebro, high in the hills of Catalonia. Here, the seventeen-year-old Pablo Ruiz underwent a spiritual transformation. From June 1898 to February 1899, Pallarès and Pablo worked in the fields like peasants, lived in mountain grottos where they painted frescos and ran naked in the countryside. This experience restored Pablo's confidence in himself and in his creative abilities as an artist.

So confident did Pablo become that upon returning to Barcelona in February 1899, he directly confronted his father by dropping Ruiz and taking instead his mother's last name, Picasso. Then, counter to his father's wishes, Picasso did not reenter the formal Art Academy but an unstructured one, where he taught himself advanced drawing techniques. This move away from structure and academicism reflected Picasso's involvement in the intellectual and artistic activities at the Els Quatre Gats, a Barcelona tavern modeled on a Montmartre café, and more popularly known as Els 4 Gats. The goal of the young bohemians that gathered there was to integrate developments from the European avant-garde into Catalan culture. Picasso's brief association with Els 4 Gats—February 1899 to September 1900—prepared him well for his Parisian future.

In this provincial hotbed of modernist debate, he found himself discussing Friedrich Nietzsche, Henrik Ibsen, Oscar Wilde and Arthur Schopenhauer, all read in Spanish translation. Nietzsche's effect on young intellectual Spaniards, like their fellows elsewhere in Europe, cannot be overestimated.[12] His call for explosive developments in art, for unhindered self-expression and for the conception of the artist "as heroic, defiant, and full of eruptive sexual energy overthrowing accepted styles," in addition to a Will to Power charged with sexual energy, struck a resonant chord in Picasso.[13] Roger Shattuck writes of Nietzsche's dramatic pronouncement of the death of God in *Thus Spoke Zarathustra* as clearing the field of "supernatural encumbrances [so that] the ancient sin of *hubris* disappeared." God is a creation of one's own mind, and so the mind is the highest level of existence.[14] Such ideas, on the threshold of the new century, were inspiration to produce new and strikingly different art and literary forms.

In February 1900, Picasso had his first serious exhibition, at Els 4 Gats. One of the paintings he showed there is *The Last Moments*, subsequently included among the Spanish paintings shown at the Exposition Universelle in Paris, to celebrate the onset of the new century.

THREE PARISIAN SÉJOURS

Arriving in Paris for the first time in October 1900, accompanied by his close friend Carles Casagemas, Picasso became immediately immersed in Montmartre sexual intrigues. The two were introduced to three models, Germaine, Odette and Antoinette. Casagemas fell in love with Germaine at first sight. Picasso later boasted about sleeping with all three.[15] By this time Picasso was rather blasé about sex. At age fourteen his friend Pallarès had introduced Picasso to the Barri Xino, the red-light district of Barcelona, which Pablo then frequented almost daily.

Consequently the Montmartoise habit of maintaining open relationships suited Picasso just fine. For him, comparison would always remain the most alluring facet of sex, whereas fidelity was well-nigh incomprehensible. For his friend Casagemas, however, Montmartre was disastrous. Germaine, the lover to whom he was committed, was a woman of voracious sexual appetites, a situation that did not augur well for a man suffering from bouts of impotence compounded by an addiction to alcohol and drugs. Owing to Casagemas's deteriorating state of mind, they returned to Barcelona on 1 January, 1901. Casagemas committed suicide soon after.

Picasso would return to Paris in May 1901 to present a show at the gallery of the dealer Ambroise Vollard, as arranged by an influential friend in Madrid. Picasso prepared the exhibit in just three weeks, producing an average of three pictures a day, most of them in a colorful springlike pre-fauve mode. Of the sixty-four works, over half were sold. But Picasso chose not to continue in that style. Profoundly affected by Casagemas's suicide, toward the end of 1901 he slipped into his Blue Period. The subjects included scenes of his dead friend and of Saint Lazare, a Parisian prison that housed prostitutes with venereal disease, which Picasso had visited several times. The Blue Period paintings would sell hardly at all until 1905, when Picasso's new patron Gertrude Stein began buying them. In January of 1902 he returned to Barcelona virtually penniless.[16]

Perhaps the most important consequence of the Vollard exhibition was that Picasso met Max Jacob. Even though Picasso spoke almost no French and Jacob no Spanish, the attraction between the two men was immediate and they "shook hands with enthusiasm," as Jacob recalled some quarter of a century later.[17] Jacob, an art critic and poet five years older than Picasso, became his first close Parisian friend, from whom he would begin to learn sophisticated French and who would form his literary tastes. But Jacob's first important service to Picasso was to keep him alive during his third stay in Paris from October 1902 to January 1903.

Picasso's goals were to continue his study of French art as well as to sell paintings. Although he succeeded on the first point, he failed in the latter, and this stay was a miserable one. He was flat broke, living in incredible squalor, until Jacob offered to share his room. There being a single bed, Picasso slept in it during the day when Jacob was at work. Neither did Jacob have much money, and from time to time they both went without food. This was the low point of Picasso's youth. Once again he returned to Spain a commercial failure. Yet he was taken with Paris and determined to come back.

MONTMARTRE AND THE BATEAU LAVOIR

By 1904, the spirit and thrust of the avant-garde was focused with explosive force, in the two most highly intellectual parts of Paris, Montparnasse and Montmartre. The explosion would take place in Montmartre with Picasso as the lightning stroke that catalyzed an incredible flowering of art and literature.

This was the Paris of *la belle époque.* The fashionable salon had declined, its place taken by the more democratic café that served as a vital scene for the exchange of ideas—on art, literature, science, mathematics and technology—and a fertile atmosphere for the growth of new ones. Paris had developed into a theatre with a distinctive intellectual verve. Life itself became a performance.

At the end of the nineteenth century Montmartre comprised a hill, or Butte de Montmartre, that was still in transition from a rural to an urban area. At the base were the major boulevards, de Clichy and Rouchechouart, as well as Place Pigalle. For the most part the Montmartoise were laborers with a scattering of bourgeoisie and a liberal sprinkling of artists and writers attracted by the cheaper rents. On spring and summer weekends, Parisians flocked to Montmartre to sit in the outdoor bistros, particularly on the boulevards, which were lined with them. The crowds were especially intense during the various festivals. Despite all this gaiety, crime was rampant: Violent gangs of so-called *apaches* roamed the quarter robbing the unwary and well-to-do.

An especially popular bistro on Boulevard Rouchechouart was Le Chat Noir, whose owners had the genial idea of encouraging artists and writers to gather there and give readings. This was good for business. The pleasant atmosphere even attracted writers from the Left Bank. Le Chat Noir was the model for such successful Left Bank cafés as Closerie des Lilas in Montparnasse and Des Deux Magots on Boulevard Saint Germain. The serious artists and writers frequented small dingy cafés on the Butte, such as Le Lapin Agile with its colorful owner Frédé Gérard, whom Picasso had met on an earlier trip to Paris in 1901. At that time Frédé owned a sordid *troquet*, or small bar, on rue Ravignan, called Le Zut. In Le Chat Noir, Picasso at last found a hangout to match Els 4 Gats from his Barcelona days.

He also spent many summer evenings on the terrace of Le Lapin Agile. On weeknights the clientele was sparse and one could converse in agreeable tones. It was a different story on the weekends when upwards of one hundred people pressed themselves into the café's two small rooms. The air was thick with smoke, along with the odors of absinthe, beer, cheap wine, cheap cuisine and human stench, this being an epoch when taking a bath was considered abnormal. Jean-Paul Crespelle sums up the weekend crowd: "There was a little for everyone, a veritable enchantment of Parisian avant-garde society: employees, petites bourgeois, kept women and their lovers, pretentious young ladies looking for an evening of adventure," in addition, of course, to the usual artists and writers.[18] To keep the noise down somewhat, Frédé strummed his guitar and sang in a mezzo voice. During 1905–1910, Le Lapin Agile was the favorite café of avant-garde poets and artists.

The principal reason why restaurants on the Butte were so numerous was the miserable conditions in which artists lived and worked. Their quarters usually lacked adequate cooking facilities. Picasso's atelier in 13 rue Ravignan was typical. The building itself was constructed in 1860 as a piano factory. In 1889, in an attempt to lure artists back to Montmartre and raise the quarter's image, the factory was divided up into ateliers. Not being very imaginative, the architect simply compartmentalized the levels with wooden partitions, creating a labyrinth of cabins connected by a bewildering array of staircases.

Since the building was essentially on the peak of the Butte, the stacking up of levels gave the impression of a ramshackle ziggurat or perhaps a gigantic shack. Although Picasso's studio was one floor below the entrance on Place Ravignan, it was on street level because the building was on a hill. Air shafts penetrating the structure to provide ventilation for the studios presented somewhat of a hazard. On one much-talked-about win-

FIGURE 2.1
Picasso on the Place Ravignan, 1904.

ter evening, a German tenant attempting to clear snow from his atelier window fell into one of the shafts and broke his neck. Sometimes the concierge, Madame Coudray, helped the artists in time of dire need by cooking them soup and providing wake-up calls.

Max Jacob dubbed the building the Bateau Lavoir, after its resemblance to the laundry boats along the Seine. It was also known as la Maison du Trappeur after the trappers' cabins in Alaska. Before Picasso moved in, the Bateau Lavoir had a reputation as a hangout for anarchists, who were forcefully evicted by the police in the 1890s.[19]

The seediness of the area and its not infrequent violence kept the rents quite low. Consequently, there was no shortage of artists and writers in residence. Picasso paid about fifteen francs per month, for which he got little more than four dingy walls and a roof. Like many Parisian flats at the time, the Bateau Lavoir ateliers had no electricity or gas. Such amenities were not installed until the 1930s. Neither did the units offer running water or toilet facilities. Tenants either filled their bucket from a common tap on the first floor or queued up at the fountain on Place Ravignan. The only toilet in the building was situated next to the water tap and had a perpetually broken door. This lack of facilities did not disturb Picasso, who saw it as a way to meet neighbors, especially women, and in particular Fernande Olivier, who lived close by on Picasso's level.

PICASSO IN LOVE

Fernande had been at the Bateau Lavoir for a short time before meeting Picasso. She was living with a sculptor in the Montmartoise fashion. Between often bumping into Picasso during the day and noticing the constant stream of Spanish visitors to his atelier, she wondered, "Whenever does he work?"[20] Later Fernande discovered that Picasso preferred painting at night, to avoid interruption.

Picasso had noticed Fernande around the Bateau Lavoir as well as at his favorite art supply store on Boulevard de Clichy. She in turn had first spotted him on Place Ravignan, filling his water pitcher and chatting with his Catalan buddy Ricardo Canals. Their love affair began in August of 1904 after a classic pickup. Dashing into the Bateau Lavoir at the start of a violent thunderstorm, Fernande was suddenly confronted by Picasso, who emerged from his atelier carrying one of his cats. "He held it out to me, laughing and blocking my path. I laughed too, and he took me to see his studio," recalled Fernande twenty-nine years later. Their occasional affair increased in intensity during the next year, as each gradually discontinued other liaisons. At the time he met Fernande, Picasso's mistress Madeline was pregnant. He was involved with two other women as well, Frédé's stepdaughter and Alice Géry, the perpetually unfaithful mistress of Maurice Princet, of whom we will have much to say. Picasso's increasing involvement with Fernande was a factor in Madeline's decision to obtain an abortion.

In September 1905 Fernande moved into Picasso's atelier, and they remained together until 1912. For the first year or so they were monogamous, and then reverted to somewhat muted Montmartoise relationships. During their most passionate period Fernande played the part of Picasso's muse; after that she might be charitably described as providing negative reinforcement. He succeeded in spite of her.

La belle Fernande was a beauty. With a mass of auburn hair, green almond-shaped eyes and a voluptuous body, she was capable of turning heads and of flirtations that were more than playful. She had an air of

FIGURE 2.2 Joan Vidal Ventosa, *Portrait of Fernande Olivier, Pablo Picasso and Ramón Reventós.* Barcelona, 1906. Ramón Reventós was a good friend of Picasso.

lightness and elegance, an open and curious mind, strong determination and some artistic talent.

She was born 6 June, 1881, out of wedlock, and her real name was Amélie Lang. Raised by an adoptive family, at the age of eighteen Amélie was forced into marriage with a brutal shop assistant who had seduced her. After a year she ran away and there ensued a series of modeling jobs and affairs. As models and other women on the fringes of society customarily did, she took a *nom de guerre*—in fact a succession of them, ending up for unknown reasons with "Fernande Olivier." By the time of Picasso's whirlwind courtship of her in August 1904, Fernande could boast of a pedigree in the arts that included associations with Othen Friesz and Raoul Dufy.[21] Her published memoir, *Picasso and His Friends,* remains a valuable source of information, despite its sometimes spotty dating. Even Picasso grudgingly allowed that it caught the mood of the time.[22]

Fernande's first reaction upon entering Picasso's atelier was, "My God, in what chaos."[23] There was paint everywhere. The chipped gray paint on the walls had further chippings from canvases. The bathtub was a receptacle for piles of drawings, newspapers and books. The furniture consisted of a rustic wobbly chair to which Picasso usually attached his dog Frika, a combination police dog and Breton spaniel. A pedestal table of Napoleon III vintage in black wood was used for the toilette and for meals. A rusted frying pan served for cooking and pissing. In a drawer Picasso kept a white mouse whose musklike odor provided a peculiar counterpoint to the mélange of dog odors, paints and turpentines. Picasso adored pets and during his stay in the Bateau Lavoir amassed three Siamese cats, a turtle and a female monkey, in addition to Frika. The finishing touches on the decor were large numbers of old tins for preserves, in which he kept his brushes. The light was intentionally cold, owing to the windows having been painted blue to obtain a constant lighting. Fifty years later, Picasso's principal dealer, Daniel-Henry Kahnweiler, recalled of Picasso's atelier that "it was trying."[24] Crespelle pungently summarized the situation in the Bateau Lavoir: "In fact, it was no more trying than most of the other ateliers."[25]

There was also an entrée, or smaller room, almost entirely filled by a divan that the intimate called the *chambre de la bonne.* By 1906 Picasso had transformed it into a shrine for Fernande. Besides the divan, there was a crate draped with the red sash that Fernande had worn at the time of their first meeting, a portrait of her sketched with pen and two vases won in a lottery at a trade fair on Boulevard de Clichy.

Privacy was at a premium in the Bateau Lavoir because of the thin inner walls. Fernande recalls that "it was glacial in winter and like a Turkish bath in summer."[26] If there was no fuel delivery in winter for the stove, the couple stayed in bed. During the summer Picasso painted naked, except for a scarf around his waist. He intentionally left the door to his atelier open for air circulation and to draw admiring comments from passing women.

Until he met Fernande, Picasso's relationships with women were for the most part with whores, punctuated with a succession of models in Paris, and the occasional brief romance. Overwhelmed by his emotions for Fernande, Picasso became extremely protective of her. Fernande's flirtatious ways gave Picasso good cause for jealousy. To prevent advances from other bohemians, after Fernande moved into his atelier Picasso kept her as if in a seraglio. He locked her in and ran all errands himself.[27]

Sometimes Picasso's jealousy erupted into public displays of violence. Once Fernande heard some shouting in the street about a shooting at Le Lapin Agile. She managed to get out of the atelier and went over to have a look for herself. It turned out that Picasso had also gone to investigate. Spotting Fernande he grabbed her, slapped her around and then dragged her back to the Bateau Lavoir. They reconciled, as they did after other furious fights, when Fernande walked out, with Picasso pursuing her down the rue Ravignan begging forgiveness. All in all, Fernande was the ideal mistress for him at that time. Neither cold nor hunger bothered them. She passed her days stretched out on the divan, drinking tea, reading novels and smoking Turkish cigarettes.

Before 1907, when Picasso's paintings began selling well, he and Fernande frequented cafés that offered credit or were incredibly inexpensive. In this they were no different from most other writers and artists on the Butte. Le Lapin Agile offered the best of both worlds. Dinner at the less expensive restaurants was around ninety centimes, while the more expensive ones charged around two francs fifty. For ninety centimes one could eat a substantial meal of beefsteak, *frites, tart aux pommes* and an espresso. It was the quality that varied. Fernande had tastes that often went beyond the basic fare of Le Lapin Agile. Picasso, on the other hand, was content with a piece of chorizo and a tomato, in the company of friends with whom he could discuss painting.

Those who accepted the hospitality offered at bistros on the Butte not infrequently ran up such large bills that owners found themselves unable to turn them away lest they never pay. This generosity sometimes drove cafés out of business. There were times that Picasso's close Catalan friend and earliest supporter in Paris, Paco Durrio, had to bring the couple food because they had no money at all. Another friend, the artist Maurice de Vlaminck, recalled overhearing Fernande order a cutlet for the dog Frika in a bistro that offered credit. Vlaminck was certain that they had no money to buy even scraps of meat from the neighborhood butcher.[28] Another artist friend from the Butte, Kees Van Dongen, remembered instances from around 1905 when he and Picasso would steal bottles of milk and croissants from the doorsteps of Montmartre apartments.[29] Life was not easy on rue Ravignan. As Francis Carco recalls, "We breathed there an atmosphere of poverty, of abandonment, of austerity and black misery."[30]

Things began to change for Picasso as he moved from the moroseness of the Blue Period into the lively circus scenes, Harlequins and *saltimbanques* of the Rose Period, a transition pretty much coincident with his falling in love with Fernande. This is also, however, about the time he met the writers Guillaume Apollinaire and André Salmon.

GUILLAUME APOLLINAIRE AND ANDRÉ SALMON

Picasso became acquainted with Apollinaire and Salmon in October 1904. Max Jacob nostalgically recalled his own animated introduction to

Apollinaire by Picasso, also in October 1904 and with Salmon present, at Austin's Railway Restaurant, near Gare Saint-Lazare: "The three of us left together and Guillaume carried us off for a stroll which never came to an end . . . . Here began the best days of my life."[31] Picasso's magnetism and genius were immediately apparent. By 1905, Jacob and Salmon had taken up studios in the Bateau Lavoir and Apollinaire had moved close by. Together with Jacob, they formed the nucleus of *la bande à Picasso,* which met almost daily at the Bateau Lavoir for discussions that ranged across literature, politics, philosophy, mathematics, technology, science and whatever else was vital to the avant-garde. The members of *la bande à Picasso* revolved about Picasso like planets around a sun. Through them Picasso widened his group of friends beyond the Catalan circles in Paris into the vibrant world of French literature and culture. In every sense of the term, *la bande à Picasso* constituted Picasso's "think tank."[32] It was also a closed social club with its own argot and ceremonies, and a language and cynical tone that were not for the fainthearted.

We glimpse some of their conversations in a letter of 7 July, 1906, from Jacob in Paris to Picasso at Gósol in Spain.[33] In the confrontational style of the Bateau Lavoir, Jacob writes of his criticisms of symbolism and his views on horoscopes, the history of the French Republic, the state of metaphysics in England, Germany and France, and the materialistic mindset of students in the Latin Quarter—"prats" who are not experiencing the passions and joys of life. Salmon recalled how *la bande à Picasso* would cynically criticize the "isms" of painting with their game "à faire Degas [pretend to be Degas]," in which each took turns "criticizing" Picasso's paintings as if they were Degas or Renoir, with Jacob doing hilarious impersonations.[34]

Salmon was Picasso's age and at the time of their meeting a poet and literary journalist. Already a member of the literary avant-garde, he had been working with Apollinaire since 1903 on the staff of several new wave journals. In 1905 Salmon moved into the Bateau Lavoir and for the next two or three years saw more of Picasso than the others.

Among the young French literati of the days, Apollinaire was one of the leaders of the revolt against the symbolist "school of 1895," whose principal figures were Paul Verlaine, Arthur Rimbaud, Stéphane Mallarmé, Jean Moréas and Stuart Merrill. He played a major role in widening Picasso's intellectual horizons. The facts surrounding Apollinaire's lineage and the intervening twenty-two years of his life before he settled permanently in Paris in 1902 are so complex as to have fascinated even Picasso. We know little for sure other than that Apollinaire demonstrated early on a precocity and talent for writing. By 1903 he was at the forefront of the Parisian literary scene.

In 1905 Apollinaire and Salmon introduced Picasso to the intense Tuesday night soirées of the new wave journal *Vers et Prose*, held at the Closerie des Lilas in Montparnasse. With no money for transportation, the trio walked halfway across Paris and back. On the return trip Picasso scoured trashcans for food for his pets. The artists and writers whom Picasso met at the *Vers et Prose* were truly of international stature, and some would later play their role in Picasso's life. Among these were Maurice Raynal and Henri-Pierre Roché. In 1906 Roché introduced Picasso to Gertrude and Leo Stein, whose patronage would prove of great importance.[35]

It is no exaggeration to say that Apollinaire was the lord of the café society of Montparnasse.[36] With a fiefdom ranging from Boulevard Montparnasse with its Closerie des Lilas to Boulevard Saint Germain and Des Deux Magots, he was "the impresario of the avant-garde."[37] His ambition was to construct "a special language linking poets and artists."[38] Apollinaire, Salmon and other poets did not restrict themselves to the Left Bank. They also trekked to Montmartre and particularly to Picasso's atelier, where poets and artists mutually inspired each other. Around 1905, Picasso hung a sign on his door that read *Rendezvous des poètes*. The Bateau Lavoir became the new headquarters of the avant-garde.

Scholars of this period agree that the myriad connections between artists and painters could not have been coincidental. At the time, "painters and poets influenced each other *tour à tour*."[39] Both groups held the view that something dramatically new was about to occur in art. Not too many years later, in 1922, Salmon recalled of those heady days: "Tout est possible, tout est réalisable en tout, partout et avec tout [everything is possible, everything is realizable, in all and everywhere]."[40] Such inspiration and optimism cannot be overestimated.

Picasso and his group of friends lived in an era of dramatic change that occurs rarely in Western history. Great shifts were occurring in art, literature and science, with even bigger ones expected. These young men believed that they were living in an heroic age where anything was possible. They needed no accolades from society. They were impoverished and had nothing to lose. They shared everything, including knowledge, and strove to produce art and literature that would match the incredible achievements in science, mathematics and technology.

When Apollinaire and Picasso met, Picasso knew little about French literature except what he had learned from Jacob, and Apollinaire knew little about art except for what he had gleaned from brief conversations with André Derain and Vlaminck. Yet Apollinaire and Picasso each immediately recognized the other as a kindred soul. Although Picasso's spoken French was not of a high level, Kahnweiler recalls that "even if there were some years when he spoke little French, he was absolutely able to judge, to taste immediately the beauty of a poem."[41] Jacob noted this as early as 1901. Apollinaire encouraged Picasso to accentuate the poetic dimensions of his art, in this way helping Picasso to liberate himself from absolute rules, "to listen to the propositions of his heart,"[42] and generally to widen his horizons. Picasso's subsequent *oeuvres* gave Apollinaire the clue toward a common language of poet and artist that emerged in Apollinaire's "calligrammes" or "ideogrammes." By virtue of his own imagination, culture and intelligence, Apollinaire opened avenues of Picasso's thinking that were essential for his artistic breakthroughs in 1907. He provided intellectual support and confidence to Picasso and other young artists of the Butte.

Picasso, in turn, encouraged Apollinaire to become an art critic. Apollinaire's artistic knowledge, however, was another story. As Braque said in an interview of 1954: "He was incapable of recognising a Rubens from a Rembrandt."[43] Although Apollinaire never became an expert on art, he was a tireless propagandizer and unconditional admirer of the new emerging art, especially during cubism's early days when the reviews were nothing but scathing. This may in fact have been Picasso's hidden reason for encouraging Apollinaire to write art criticism.

The constant bond between the two men was their mutual support regarding the search for new forms of representation in art and literature—particularly a common language for the two fields, and the high regard in which they held the creative process. In 1905, in his first article written about Picasso, Apollinaire emphasizes his friend's "perseverance in the pursuit of beauty," that is, their common pursuit of a new aesthetic.[44]

What were Apollinaire, Picasso and Salmon up against in their search for new forms in art and literature?

PHILOSOPHICAL AND LITERARY TRENDS

Nineteenth-century science had seen cycles of worldviews that alternated between realism and romanticism. So, to some extent, did literature.[45] The end of the century witnessed a reaction against the dominant realism or naturalism represented by the literature and theatre of Émile Zola,

Henrik Ibsen, August Strindberg and André Antoin. They focused on the particularities of everyday life. Plots and stage sets became increasingly elaborate in order to be as faithful to actual life as possible, while dreams, illusions and legends were excluded in favor of philosophical and moral themes. Naturalist art included, for instance, the work of John Constable, whose cloud paintings were meticulously dated and timed as if they were scientific data.

All of this was entwined with positivism, the predominant philosophical outlook in philosophy and science. The positivistic view was first spun in 1830 by the French philosopher Auguste Comte, who advocated progress toward a science cleansed of theology and metaphysics. These ideas were elaborated in the 1880s by the fortyish Viennese philosopher-scientist Ernst Mach, whose brand of positivism stressed that only phenomena reducible to sense perceptions (or laboratory data) could be considered physically real: What you see is what you get. Imagination could play no role: Anything beyond appearances was mere illusion. (I will have more to say about Mach in Chapter 3.)

Positivism fit very well with the materialism of the day and its unsavory association with rampant industrialism and literary and theatrical commercialism. The avant-garde felt the world was drowning in mediocrity. There was "a sense prevalent among intellectuals of alienation and exclusion from the forefront of public life, coupled with a political disillusionment which was exacerbated by the scandals and corruption of contemporary political life."[46] This was the *fin-de-siècle* mood.

The philosophical reaction against Comte's and Mach's positivism was an idealist revival championed in part by the distinguished French philosopher Henri Bergson. Bergson's idealism emphasized élan vital and faith in a creativity that cannot be explained by science, as well as a rethinking of the relation between mind and reality. The fashionable belief in the occult and the wide popularity of séances among the intelligentsia were partly a reaction to positivism's doctrine of the nonexistence of the invisible and the ineffable.

La bande à Picasso could not have failed to have heard about Bergson through his widely advertised lectures and from Max Jacob, who read him as a young philosophy student.[47] Central to Bergson's philosophy is the concept of "duration," according to which what we know as reality is what in retrospect we have experienced as the sum total of a *continuous* uninterrupted flow of sensations. Duration is a dynamic process that permits us to reflect all at once—simultaneously—on the inner unconscious experience that constitutes our memory and so is the source of all that we know: "Pure intellect is a contraction . . . of a more extensive power [which is a] vague intuition" that is exercised in the deep unconscious.[48]

To Bergson the purest perception of the world can be obtained only by rejecting an exclusively materialist interpretation. Only the artist can reach this apogee of thought because "art has no other object than to dissipate the practically useful symbols . . . in order to bring us face to face with reality itself."[49]

For Bergson the true self is unconscious and nonlogical and can express itself only through intuition. Bergson denied to science any possibility of understanding physical reality because scientific symbols and units, as intellectual constructs, are not reflective of the continuous manner in which the individual experiences time.[50] Certain philosophers found Bergson's views vague and anti-intellectual. Yet it was his poetic vagueness and emphasis on art and creativity that impressed antisymbolist poets such as Apollinaire, Salmon and Jacob, and through them Picasso.

The literary movement known as symbolism styled itself for the most part on a philosophical idealism driven onto the path of mysticism. Thus it cut itself off from life entirely and so was doubly counter to positivism. The *fin-de-siècle* decadence that emerged was resisted particularly in Barcelona, where the intellectuals favored "a Nietzschean energy and defiance of the bourgeois rather than the lilies and languors of Swinburne and Burne-Jones or the pessimistic irony of Laforgue."[51] In Barcelona this movement was more strenuous than in Paris: Active anarchist movements in Catalonia resulted in almost daily bombings. At Els 4 Gats, Picasso had been in the eye of the storm.

Apollinaire and his young colleagues sought a new form of literature: one that contained some of the fantasy of symbolism and yet did not turn its back on the world; that did not exclude intellectual and literary elements; that did not reduce itself to an outpouring of sentimentalism on the one hand or an exact copy of nature on the other; that glimpsed a world beyond appearances. Just as the symbolists turned to music for many of their themes,[52] the new wave would turn to art. The fauvism of Henri Matisse was somewhat in the right direction, as Apollinaire wrote in a 1907 issue of *La Phalange*: "The eloquence of your work arises primarily from the combination of colours and lines. It is what constitutes the art of the painter and not, as some superficial minds still believe, the simple reproduction of the object."[53] Matisse, twelve years younger than

Picasso, was the only painter that Picasso ever judged to be a rival and eventually considered an equal.[54]

By the time he met Picasso in October 1904, Apollinaire had already been experimenting with "an accommodation between a form of lyricism anchored in reality, whether urban or rural, and the symbolist notion inherited from Mallarmé of the poem as an enigma."[55]

In science, by contrast, the mood was quite different.

SOME SCIENTIFIC AND TECHNOLOGICAL TRENDS

Three momentous discoveries at the very end of the century lifted science out of its own *fin-de-siècle* doldrums: the discoveries of X rays in 1895, radioactivity in 1896 and the electron in 1897. Scientists were forced to take seriously the idea that these effects might be caused by entities beyond sense perceptions.

X rays, in particular, struck the public's fancy. The immediate philosophical-scientific message is that what you see is *not* what you get: There are limits to human perception. This relativity of knowledge fuelled antipositivist critiques. Space was no longer empty. Instead rays were flying everywhere: alpha rays from radioactive emissions, beta rays, which are another name for electrons, gamma rays, which would eventually be recognized as a species of light, as would X rays. The very name "X ray" denoted that scientists were not exactly sure what they are. The notion of X-ray vision, which had titillated the imaginations of writers since the 1890s, now seemed about to be made real. Cartoonists had a field day.[56]

Recall that Picasso's first trip to Paris, in 1900, was to attend the Exposition Universelle at which his painting *The Last Moments* was exhibited. The Exposition's Palace of Electricity was set up as testimony to the incredible industrial progress made in the last few decades. Within the past generation, everyone's experience of time and space had been altered by such technological innovations as the telephone, wireless telegraphy, X rays, bicycles, movies, the automobile, the dirigible and then the airplane.[57] Given his interest in photography and in any experimentation with images, Picasso could not have failed to have a look at the X-ray photographs and equipment on exhibit.

SOME REPORTS OF SCIENCE IN NEWSPAPERS AND JOURNALS

Picasso and his literary friends would also, of course, have read about X rays and other technological developments in the newspapers. In the *Paris-Journal* of 10 May, 1905, an article entitled "Choses de l'Invisible [Invisible Things]" begins, "There are more things in Heaven and Earth Horatio than we believe communally. . . . And our century seems effectively to be the epoch whence the invisible, the occult, relegated to the level of chimeras by positivists triumphantly, seems to be revealed to us. . . . It is the discovery of that extraordinary spy, the X-ray." Other articles of this period also cited X-ray photographs as revealing an invisible reality, just like supposed spirit photographs.[58] Such articles had special appeal for writers such as Jacob, who dabbled in the occult. Artists and writers began to feel that like the scientists perhaps they, too, could reveal invisible realities.

The 31 January, 1906, issue of *Paris-Journal* quotes German physicist Wilhelm Röntgen himself, in an article entitled "Les Rayons X [X Rays]": "I have found rays that permit one to see the invisible, to see things inaccessible to your eyes." "This time," the article comments, "the miracle was real." It then continues in a more somber vein by advising readers on the dark side of X rays: Their ability to kill rat embryos and to make rats and rabbits sterile, "Can they be used to eradicate the human species too?" The writer warns the public to be careful with radiographs because "already a physician died who experimented on himself with X-rays."

Le Temps of 31 December, 1906, gave the two most important scientific discoveries of the year as the transmission of pictures over distance by telephone and heavier-than-air flight. Other dailies, such as *Le Matin* ("Les Photographies miraculeuses," 8 Feb., 1907) and *L'Intransigeant,* carried detailed articles on long-distance transmission of photographs. This would bring an entirely new look to newspapers, which up to then were dependent on artists. *Le Journal* of 9 January, 1906, reported on Wilbur Wright's visit to Paris in order to convince the French of his feat; he flies 38 kilometers in 38 minutes. *Le Temps* of 11 November, 1906, said of a dirigible flight from Paris to London: "Le Manche a disparu [The English Channel has disappeared]."

Le Matin of 8 February, 1907, trumpeted the avant-garde and its associated technological progress: "Men of today are no longer surprised. They have seen come true so many miraculous solutions: telephones, wireless telegraphy, dirigibles. All these surprises of human genius have accustomed us to expect the unexpected, to smile at the impossible, because we have the certitude of seeing it realised." The phrase "La Conquête de l'Air" occurs often in articles extolling the glory, romance and danger in air travel.

Like most everyone interested in current trends in literature, Apollinaire, Jacob and Salmon followed the bimonthly literary journal *Mercure de France,* which also carried quasi-scientific articles. I say "quasi" because these articles were written mainly by literary fantasists such as Alfred Jarry and a well-known reviver of the Rosicrucian order, Joséphin Péladan, known also as the "Sâr." This revival, like the interest in occultism and Theosophy, went hand in hand with the upsurge of symbolism and its goal of finding release from the everyday world. The February 1904 issue carries Péladan's article "Le Radium et l'Hyperphysique," which tries to connect X rays and supernatural phenomena. He writes that "hyperphysics has for its object the study of supernatural phenomena."[59] As a prelude, however, the explanations of X rays, cathode rays and Ernest Rutherford's work on radioactivity are not bad. Although some considered Péladan a rather preposterous figure, Apollinaire, Jacob and Jarry, with their attraction to the occult, were somewhat influenced by his ideas concerning the Apocalypse, androgen and the mystical nature of sex. In turn, they no doubt transmitted these ideas to Picasso.[60] In the same issue of *Mercure de France* we also find Albert Prieur's review of Louis Fabre's book, *L'Esprit Scientifique*, which explores the issue of scientific relativism in some detail.[61] Another issue contains an article by Marcel Réja that includes a discussion of time travel.[62]

In still another issue, a review of Gustav LeBon's 1905 book *L'Evolution de la Matière* focuses on the claim that any sort of radiation is a result of atomic disassociation.[63] According to LeBon, atoms are not permanent, but are continuously being transformed into energy. In the end everything is amorphous. These views reflect the philosophy of LeBon's good friend Bergson: Both men stressed continuity and process over tangible materiality.[64] LeBon's book became a best-seller.[65]

Another notable book review in *Mercure de France* is by Louis Weber of the great French polymath Henri Poincaré's *La Science et l'hypothèse*, published in 1902.[66] Weber discusses how, in a superficial sense, science can appear to be an infallible logic system whose truth is beyond any doubt. Yet by taking note of the role played by hypotheses and their dependence on interlocking assumptions, Poincaré shows how unstable scientific theories actually are. Poincaré's probing, writes the reviewer, reveals as well that our choice of any hypothesis among a theoretically infinite number that can explain any set of data, is based on "convenience." Consequently, there is no reason to believe in the uniqueness of scientific theories, nor even of the real existence of unseen entities postulated by theories such as atoms. *La Science et l'hypothèse* was also a best-seller and, as we will see, it played an important role for Picasso.

All of these newspapers and literary journals would have been read by various members of *la bande à Picasso.* Their friend Alfred Jarry, who was in a better position than they to understand the science, exerted a strong influence on their literary endeavors and would soon influence Picasso's art as well.

ALFRED JARRY

Alfred Jarry was the very personification of the avant-garde as a way of life. He was an intellectual agent provocateur who specialized in demolishing bourgeois literary and social conventions.

Jarry was born 8 September, 1873, in Laval in Brittany. At the Lycée in Rennes he was a brilliant student, excelling without much effort while demonstrating a talent for being the worst sort of troublemaker. Scatological humor was his particular delight. He was graduated with an exemplary record in Greek, Latin, German and drawing. More relevant to Jarry's future literary career, however, was his hopelessly incompetent physics teacher, Professor Hébert. Hébert's class in "my science of physics" was sheer pandemonium, with explanations hopelessly inept and demonstrations always going awry. Jarry participated with great gusto in the general uproar and in concocting plays about the unfortunate professor. Some of this material would later form the basis of his *Ubu* plays, dating from 1896, in addition to his posthumously published play, *Gestes et opinions du docteur Faustroll, pataphysicien.*

Jarry had his heart set on an engineering degree at the École Polytechnique, but decided instead to attend the Lycée Henri IV in Paris in order to prepare for entrance to the École Normale. While majoring in sophisticated pranks, Jarry also found the time to learn about Nietzsche and to hear, from Henri Bergson himself, the beginnings of a new sort of philosophy with a decidedly antipositivist edge, as well as about the primacy of the imagination, an aspect of symbolist literature that would always be central to his work.[67] In time Jarry abandoned formal education to strike out into Paris as an *homme de lettres*. Beginning in about 1895, aided by an intentionally enormous daily intake of alcohol and drugs, Jarry transformed himself into a person whose explosively antitraditional artistic and literary goals were inseparable from the man: He lived his literary creations.

Although Jarry had made somewhat of a name for himself by 1896, it was in that year that he burst onto the French literary scene with his play *Ubu Roi*. Père Ubu is a one-man demolition squad. With full Nietzschean nihilism, Jarry aimed to uproot the pompous French theater of the time and through absurdity to poke fun at middle-class culture generally. The first word of *Ubu Roi* is Jarry's variation on the famous *mot de Cambronne*, "Merdre!" Such a word had never before been uttered on the French stage (at least not in so prominent a spot). Making matters worse was the emphatic pronunciation Jarry asked for, with the addition of the sonorous "r"—"MerdRe." At the first performance it took some fifteen or twenty minutes to restore order before the play could resume. The play's second word was "Merdre!"[68]

Typically Jarry could be seen riding his bicycle clad in a bicycle racer's outfit, with a carbine over his shoulder, one or two Browning revolvers on his belt and possibly a fishing rod to catch a meal from the Seine. The revolvers were legendary. In 1897 he once became so exasperated at the stuttering of a Belgian comic that Jarry shot him—with a blank cartridge, of course. Another time at Le Lapin Agile, Jarry blasted away at three Germans whose questions about aesthetic theories exasperated him.

Among the few constants of Jarry's personality were his inconsistency and his absolute lack of moderation. No convention escaped flouting: When he felt like it, he ate his meals backwards, starting from dessert. He sometimes wore a paper shirt with the tie drawn in ink.[69]

These traits were not idiosyncrasies or mere excesses for show, but Jarry's statement about the unity of life, literature and art, a unity that, in his view, required completely redefining one's conceptual framework. Such views made an indelible impression on young writers such as Apollinaire and Salmon, whom Jarry met in 1903, and through them on Picasso. Jarry's message was that they must rid themselves of any constraints on their thinking. His influence reinforced their own lyrical and artistic fantasies, even if they were not prepared to go Jarry's own route. In 1907 Picasso began to respond artistically to Jarry's message.

Whereas the destructive Père Ubu was a parody of a bumbling savant with his "science of physics," Dr. Faustroll reflected Jarry's own search for another avenue, a "pataphysics," to probe worlds beyond our perceptions. *Gestes et opinions du docteur Faustroll, pataphysicien* is a journey through imaginary worlds of art, literature, philosophy and science.[70] Jarry spends the most time on *l'imagination scientifique*. Some years later "he declared categorically that he conceived of no other form of imagination."[71]

Jarry's exposition of pataphysics makes it clear that the scientific and artistic dimensions are the ones he most sought to delineate. "Pataphysics will be, above all, the science of the particular, despite the common opinion that the only science is that of the general. Pataphysics will examine the laws governing exceptions."[72] "*Pataphysics is the science of imaginary solutions, which symbolically attributes to the lineaments of objects the properties described by their virtuality.*"[73] As follows from Jarry's belief in the consistency of being inconsistent, in life as on the page, everything is an exception and so anything can be an imaginary solution to any problem, be it scientific, artistic or literary. Furthermore, all of these imaginary solutions are admissible, and so all worlds are possible ones. Intellectual freedom is complete.

Jarry immediately gives an example of what he means by an object's "virtuality." The question, What is the shape of a clock? permits a large number of replies, depending on how you view it, that is, its virtuality. Any one of these solutions is possible. In other words, there is no absolute "shape of the clock." This idea would influence Picasso in developing an art in which several views of an object are set down simultaneously.

As to the propriety of such "imaginary solutions," Jarry goes on in *Gestes* to discuss what life would seem like to someone shrunk to the size of a mite walking on a cabbage leaf covered with dew. Certainly the view one could formulate from sense perceptions would differ drastically from our daily world. Such examples of the relativity of knowledge illustrate Jarry's stance against traditional science and its assumption that ours is the only possible universe.

In his first years in Paris, Picasso's French was not good enough to let him read Jarry's published works. But Jarry's good friend Apollinaire informed Picasso all about them, and in such a way that Picasso could understand what Jarry was getting at: One must free one's mind to imagine all imaginary worlds, to look beyond appearances. Nor could Apollinaire overlook a dominant theme of Jarry's work, the high place that Jarry accorded to the artist. Following Nietzsche's pronouncement of the death of God, the resulting vacuum is filled by the creative artist, the only person capable of searching for the absolute with the infinite creativity needed to conjure worlds of the imagination.[74] Seeking the absolute was of great importance to Jarry, for whom all relative truths were, at bottom, lies.

In the search for the absolute, Jarry suggested, further extensions of art required geometry. As he wrote in his 1901 *Almanach illustré du Père Ubu*, a conversation between Père Ubu and his conscience:

> Conscience: Père Ubu, you only utter stupidities. What are, to change the subject, your latest stupidities on the subject of painting?
>
> Père Ubu: I no longer do paintings, I am of the persuasion of Saint Jerome who said to his students: "Distrust Titian! Be on guard with Correggio!" I who even have more general views, say: "Be on guard with painting!" I have ceased to aid Mr. Bougrereau with my advice, I make geometry.[75]

Disputes over whether Picasso ever met Jarry are red herrings.[76] He knew about Jarry through Apollinaire as well as via Jacob, Raynal and Salmon. Jarry's dedication to exploding bourgeois norms, his regard for art and interest in science and the high place he accorded to artists could only have impressed Picasso. Jarry's charge to both artists and writers was to move away from realism toward imaginary worlds. This message directly affected Apollinaire, as it would Picasso.

Picasso went out of his way to emulate Jarry's lifestyle, though without the excesses. For example, by 1907 Picasso also possessed a Browning revolver loaded with blanks, which he would fire at admirers inquiring about the meaning of his paintings, his theory of aesthetics, or anyone daring to insult Cézanne's memory.[77] Like Jarry, Picasso used his Browning as a pataphysical weapon, in a sense playing Père Ubu *au natural*,[78] disposing of bourgeois boors, morons and philistines. Père Ubu's statement to his conscience about discarding old forms of painting by turning to geometry could only have been inspirational to Picasso during his work on *Les Demoiselles d'Avignon.* Geometry turned out to be the language of the dramatically new art that Picasso sought so passionately beginning in 1907. Before that, however, he required two key periods of transformation.

TRANSFORMATION: 1904–1905

The first is Picasso's transition from the Blue to the Rose Period, during the winter of 1904 and 1905. Everything was in place for a dramatic change, to paraphrase Dora Maar:[79] the woman with whom he was in love, Fernande Olivier; the poet or poets who served as catalysts, chiefly Apollinaire, Jacob and Salmon; the place where he lived, the Bateau Lavoir; the circle of friends who provided the admiration and understanding of which he never had enough, *la bande à Picasso*; and the dog who was his inseparable companion, Frika.

The combination of Picasso's settled personal life and the intellectual stimulus of *la bande à Picasso* provided the spark. Frequenting the Cirque Médrano, Picasso became fond of the clowns and acrobats, figures who in their character and activities merged such traditional sixteenth-century figures as Harlequins and *saltimbanques*, outsiders from society. Fernande recalled Picasso spending time with them at the circus's bar. To her exasperation, he even invited one home for dinner.[80] By this time Picasso had styled himself as the genial outsider, and so these figures were all the more attractive to him.

To poets such as Apollinaire, the *saltimbanques* were a metaphor for "artistic creation, a magical process, seen as divine, not at all subject to laws of nature and even less to social conventions."[81] Apollinaire wanted to inculcate this metaphor into the mind of the artist. His poem *Les Saltimbanques*, completed in November 1905, greatly affected Picasso's thinking toward his *The Family of Saltimbanques*,[82] whose figures can be identified as members of *la bande à Picasso*, with Picasso himself as Harlequin.[83] These Harlequin paintings depict Harlequin dressed in the traditional bright colors and delineated patterns that removed him from the direct world of appearances, connecting him instead with a mysterious world beyond it.[84] In *Saltimbanques* Apollinaire gave this figure a dominant position and referred to him as "Harlequin Trismegistus," the name given to Hermes, timeless and mysterious god of alchemical secrets.[85] Such was the poet's assessment of Picasso's attraction and creative powers, even in 1905.[86]

After completing *Saltimbanques*, at the very end of 1905 Picasso effectively concluded his Rose Period by killing off Harlequin in *Death of Harlequin*.[87]

FURTHER TRANSFORMATIONS: 1905–1906

Scholars have conjectured additional, complementary, reasons for Picasso's turning from the heavily sentimental Blue Period to the Rose Period, also called Picasso's first neoclassical period. In about 1904, much talk began to circulate at cafés such as Closerie des Lilas about a renewal of the "Mediterranean tradition." Essentially this was a countermovement by southern Europeans against the decadent *fin-de-siècle* mood propagated by German painters and the Nietzschean cult of *le moi* (the self). Classical art was being "rediscovered" and for Picasso trips to the Louvre became essential. He also began to look more closely at Gauguin's work, to which he had been introduced in 1901 by his friend Paco Durrio, who had once lived with Gauguin. To Picasso and other young artists, Gauguin represented a form of primitivism more literary and philosophical than aesthetic. His life was an implicit criticism of modern civilization, comparing it to a more idyllic lifestyle that had been almost obliterated by harsh French colonialism.[88] In response to these varied intellectual and artistic currents Picasso abandoned his Blue Period and embraced a lighter, more abstract, freer style.

The first indications of this conceptual shift are evident in sketches and paintings Picasso made in Holland during June and July of 1905. They are flat studies with no perspective point of women depicted as giantesses—"schoolgirls like guardsmen," Picasso is reported to have said.[89] This change of style from the Rose Period paintings of Harlequins and *saltimbanques,* which were beginning to sell, can be perhaps related to Picasso's "mistrust of his own virtuosity," which included the incredible speed with which he produced works.[90] In 1901, despite the success of the pre-fauve, Lautrec-like paintings bought by Vollard, Picasso had shifted to his unpopular Blue Period paintings. Now, in the face of his moderately successful Rose Period, he was taking chances with a more conceptual style, paintings that are "without subject, silences."[91]

At the time of his trip to Holland, Fernande had not yet moved in with Picasso. She resented his not taking her along and posed nude to arouse his jealousy, no doubt mixing in some affairs as well. On his return Picasso showed her the nudes he had painted, with their unspoken implications. This led to a severe falling-out, followed by a passionate reconciliation. Fernande moved into Picasso's studio in late summer, 1905.[92]

COLLECTORS AND DEALERS

Another input into Picasso's stylistic transformation was his introduction, by Roché, to Leo and Gertrude Stein. In October 1905, Roché arranged for Leo to visit the Bateau Lavoir. Leo, amazed at what he saw, was completely taken by Picasso: "He spoke little and seemed neither remote nor intimate—just completely there. . . . He seemed more real than most people while doing nothing about it."[93] A few days later, Fernande and Picasso dined at the Steins' flat at rue de Fleurus, on the Left Bank. Gertrude was struck by the "good looking bootblack."[94] Picasso was equally taken by her and they began to see each other fairly regularly. Perhaps they got along so well because they shared a poor command of French, the only language they had in common. Picasso had the gift to get along with people of whose language he knew little or nothing.[95] He and Gertrude became like brothers, and Picasso often called her "pard," American slang that he picked up from the Westerns he enjoyed so much.[96]

Fascinated with him, Gertrude asked Picasso to paint her portrait. She claims that there were some ninety sittings at the Bateau Lavoir. On Saturday evenings Fernande and Picasso would walk her back to the Left Bank, where they would all have dinner at rue de Fleurus. These dinners were the origin of her salon at which Picasso met, among others, Henri Matisse in March 1906. With his bad French, Picasso was often frustrated at following discussions, especially with Matisse: "Matisse talks and talks, I can't talk, so I just said *oui, oui, oui*. But it's damned nonsense all the same."[97] He brought along members of *la bande à Picasso* such as Apollinaire, Salmon, Jacob and Princet for support.[98] From the Steins, particularly Gertrude, he received an education in art history and current aesthetic theories. Their patronage also improved his finances. But no less importantly, Picasso saw his paintings hung beside those of El Greco, Gauguin, Renoir, Cézanne and Matisse.

Most artists on the Butte were literally at the mercy of dealers, most of whom unashamedly took advantage of them. For example, the dealer known as *le père* Soulié never gave an artist more than one hundred francs a painting, a relative windfall. More often he bargained the artist down to such a low level that the artist was bound to refuse. At lunchtime or dinnertime, Soulié would reappear at the poor artist's atelier and once again press the low price, which the artist, now hungry, would often accept.[99] Whenever Picasso made money he immediately bought art supplies, leaving himself still almost broke.

The poor sales of his Blue Period paintings, exhibited from 25 February to 25 March 1905 at the Galeries Serrurier, made Picasso decide never to exhibit in Paris again. This attitude had taken a while to develop and only became final after the artist met such collectors as the Russian Sergei Shchukin and the Steins. Why not sell directly to the collectors themselves?

Picasso did not change his mind again about dealers until July 1907, when he met Daniel-Henry Kahnweiler.[100] Kahnweiler's advantage over other dealers was that he genuinely not only liked painting but painters, too. He established a reputation of driving a hard deal, while being scrupulously fair and fiercely loyal to his artists.[101] Starting in 1907 Picasso had no more material problems and no more need to solicit the dealers of Montmartre in order to pay for his dinner. This material independence is without doubt another tile in the mosaic of Picasso's creativity.[102]

Three other episodes enter into Picasso's thought during his first great creative year, 1907: the major Paris salons that he attended scrupulously over the years—the Salon des Indépendants and the Salon d'Automne; his discovery of Iberian sculpture at the Louvre some time before May 1906; and his trip to Gósol with Fernande in the summer of 1906.

THE SALONS OF 1905 AND 1906

The Salon des Indépendants, created in the latter part of the nineteenth century, was intended specifically as a countermovement to the academic system of the Salon des Beaux-Arts. Its exhibitions were enormous unjuried shows. In an attempt to raise the overall quality of these shows, in 1903 the entirely separate Salon d'Automne was created with distinguished jurors to choose the best in contemporary art.[103]

The Salon d'Automne exhibition of 1905 was the first of the series to make an indelible impression on Picasso. First there was the scandal of the Cage aux Fauves, the room where Matisse and such followers as Derain and Vlaminck exhibited. They used bright, highly contrasting colors employed almost arbitrarily, giving an overall impression of violence of human forms and faces. The fauvists' direct challenge to all previous art forms led the critic Louis Vauxcelles to refer to them as "wild men" and to ask whether this was really art.[104]

Also at this exhibition were a few paintings by Cézanne, which may not have attracted Picasso at the time. Similarly with Seurat's paintings. He admired Manet's technique. But what struck him most of all, besides the fauves, was the Ingres retrospective, particularly Ingres's *Le Bain Turc,*

FIGURE 2.3
El Greco's
Apocalyptic Vision,
1608–1614.

which had been hidden away in a private collection for almost forty years. The discovery of a seraglio scene by a painter known for austere portraits created the expected stir. The close grouping of the nude women in *Le Bain Turc*, and the stance of the two with arms arched over their heads, influenced Picasso's composition of *Les Demoiselles d'Avignon*. So did El Greco's *Apocalyptic Vision*, which Picasso came across by 1906.

At the 1906 Salon des Indépendants, Matisse exhibited his *Le Bonheur de Vivre*, regarded by critics as the most advanced painting of its time: the very epitome of everything that was current and adventurous in art, a painting against which Picasso's *saltimbanques* and Harlequins paled. He took *Le Bonheur de Vivre* and its reception as a major challenge. Gertrude fanned the jealousy between the two men at her Saturday evening soirées. Nevertheless Picasso and Matisse met regularly, commented upon each other's work and even exchanged paintings. As we will see in Chapter 4, the relation became strained to the breaking point after Matisse saw *Les Demoiselles d'Avignon*.

Richardson emphasizes that by fall 1906 Picasso's competitiveness began to come to the fore. He took up boxing, he "went into training; he could no longer allow Matisse's supremacy to go unchallenged."[105] At this

FIGURE 2.4
Pablo Picasso, *The Harem.*
Gósol, 1906.

point Picasso decided he needed a rest from the pressures of Paris and time to reflect. But before leaving for Gósol he made a discovery that would lead him toward a new style, at the exhibition in May 1906 of recently discovered primitive Iberian sculpture at the Louvre.

PRIMITIVISM AND GÓSOL

The masklike stone faces shown at the Louvre impressed Picasso because the artist had been interested not in a naturalistic representation but a conceptual one. This exhibit also struck Matisse and Derain, who likewise felt the need to return to "primitive" sources of art.[106] The Louvre experience led Picasso to reinterpret Gauguin. He would fold Gauguin's soft primitivism and romantic image of paradise into the "rawness" in spirit of the Iberian sculpture.

Picasso assimilated the Louvre experience over the summer he spent at Gósol with Fernande. This was a happy time for them. Fernande recalled that in Spain "he became a different and altogether nicer person."[107] They passed through Barcelona, where the photograph in Figure 2.2 was taken by Picasso's friend Joan Vidal Ventosa in his atelier as the couple were departing for Gósol.

At Gósol Picasso "discovered" Fernande's nude body and made many sketches of her in the midst of her toilette. His sketch *The Harem* depicts Fernande-like houris grouped closely and combing their hair (Figure 2.4). In the foreground is a muscular Iberian-like naked man lounging on the floor while grasping a *porrón,* with the remains of a picnic lunch of sausage and bread lying on the ground. The connection in theme and composition with Ingres's *Bain Turc* is clear. It is Picasso's first attempt at this sort of figure grouping.

Another discovery at Gósol was their landlord Josep Fontdevila, whose face, in Picasso's hands, was gradually transformed into a deathlike mask that would pervade Picasso's studies in portraiture, eventually in old age becoming Picasso's own. Picasso's sketchbooks from Gósol "indicate his experimental work of simplifying and schematising the female face and body, of focusing on the reduction of face to mask."[108]

On returning to Paris in late August or early September, Picasso immediately set about finishing Gertrude Stein's portrait. The story has been told many times of how at the end of about ninety sittings Picasso whited out her face and put the portrait aside. Now he could complete it and did so, legend has it, in a flash:[109] Gertrude's face is that of an Iberian relief, a stonelike mask. This is perhaps the earliest example of what Apollinaire caught so well when he wrote six years later about the development of cubism "as an art of painting new ensembles with elements borrowed not from the reality of vision, but from the reality of conception."[110] In order to further his discovery of conception over perception, Picasso embarked on a series of paintings of stonelike nudes. His style became more impersonal, less sentimental. Like his literary friends Apollinaire, Jacob and Salmon, Picasso had made the transition away from symbolism toward a new representation of reality.

This brings us to the end of 1906. Picasso is producing stonelike giantesses such as the women in *Two Nudes*.[111] In his *Self-Portrait with a Palette* the artist portrays himself with a stonelike self-confidence that directly challenges Matisse's role as leader of the avant-garde.[112] The challenge is not only in the subject's aggressive posture: the minimalist coloring, the clenched fist and the conceptual presentation of face as mask all propose a forceful alternative to Matisse's fauve art. As Richardson writes, "Matisse wanted to soothe, comfort and delight, whereas Picasso wanted to challenge, excite and shock."[113] This Nietzschean self-portrait threw down the gauntlet. Leo Stein, recalling Picasso's extreme competitive mood at this time, quoted him as saying that "the strong should go ahead and take what they want."[114]

Life with Fernande was beginning to deteriorate. For the first time in his life Picasso was absolutely absorbed in his work. The honeymoon in Gósol was over. Picasso withdrew into himself, convinced that he was on the brink of producing a great work of art, and met interruptions with extreme anger. Fernande resented his deep involvement and perhaps also Picasso's representations of her, in which she was gradually changing from a lithe, beautiful woman to a stony Gertrude Stein look-alike. Picasso was already on the path, described decades later by Françoise Gilot, of transforming his mistresses from "goddesses to doormats."[115] By the end of 1906 the tension between them was at a fever pitch and growing worse. In September 1907 they tried a brief separation.

At the nascent moment of creativity, artists cannot separate themselves from the world around them. The influence of this highly charged sexual situation cannot be discounted in Picasso's masterpiece of 1907, *Les Demoiselles*.

3

The Kind of Male Beauty That Caused Such Havoc

If I have the good fortune to pass my examinations successfully I shall go to the Zurich Polytechnic. I shall remain there for four years in order to study mathematics and physics. I dream of becoming an instructor in those branches of the natural sciences, specialising in the theoretical portion. . . . There is a certain independence in the scientific profession that greatly pleases me.

—Albert Einstein,
Aarau, 18 September, 1896

With youthful optimism, a seventeen-year-old Albert Einstein thus mused about his future. The occasion was a French language examination toward graduation from a preparatory school in the Swiss canton of Aargau. His prospects were not at all assured. The year before, in 1895, Einstein had failed the entrance examination for the Swiss Polytechnic in Zurich and was advised to brush up on certain subjects before trying again. Before that, in 1894, the boy had dropped out of the Luitpold Gymnasium in Munich without a diploma, to the relief of his exasperated teachers. In 1900 he graduated from the Swiss Polytechnic but was unable to find steady work for two years, owing primarily to personality conflicts with eminent professors, as well as mediocre grades. Finally, in 1902, a good friend's father arranged for a position at the Swiss Federal Patent Office in Bern.

Then, in 1905, this middle-level civil servant published four papers that completely overturned the world of physics and made an indelible impact on the culture of the twentieth century. The fourth paper introduced the equation that has become a synecdoche for physics, cosmology, and indeed all of modern science, $E = mc^2$.

We are particularly interested in the third paper, the so-called special relativity paper, which in its new concepts of space and time closely paralleled, in scientific terms, Picasso's visual explorations.

THE YOUNG SCIENTIST'S EDUCATION

Like Picasso, Einstein spent his childhood within a closely knit family.[1] His father, Hermann Einstein, showed some youthful ability in mathematics, but the family's financial problems forced him to give up any scientific aspirations, and he became a businessman. Einstein's mother, Pauline (née Koch), was ten years younger than Hermann. They were a devoted couple who strove to raise their children in an enlightened and cultured household. Hermann was fond of reading Heine and Schiller aloud, while Pauline encouraged their musical abilities.

At the time of Albert's birth, on 14 March, 1879, in Ulm, Germany, Hermann, age thirty-two, owned a featherbed business of questionable success. A year later he became a partner in his younger brother Jakob's mechanical engineering firm in Munich that specialized in gas boilers. In 1882 they became convinced of the great future in electrical enterprises, to which they extended their activities. With Hermann running the business side, Jakob, a qualified engineer, took care of the scientific end.

Young Albert was so late in starting to speak that those close to him feared he would never learn. "But the fears were unfounded," because at about age two and a half, his sister Maja's arrival triggered the boy's speech.[2] The warmth of the Einstein family and their nonreligious attitude of free thought were apparently good for Albert's self-esteem. At age three or four the boy could make his own way through Munich's busiest streets. "Self-reliance was already ingrained in his character," recalled Maja.[3] Although Albert did not lack for friends, he preferred solitary pursuits such as working on puzzles and building multistoried houses from cards. Einstein entered public primary school at age seven and was considered only moderately intelligent because he took such a long time to solve problems. He preferred to examine them slowly and systematically instead of replying automatically as teachers expected. Patience, persistence and organization, traits so essential to his future work, emerged early. But they were misunderstood by his teachers.

The misunderstanding worsened at the Luitpold Gymnasium in Munich, which Einstein entered at age ten. He later wrote of his early education, "the teachers in the elementary school appeared to me like drill sergeants and the teachers at the Gymnasium like lieutenants."[4] In this Prussian atmosphere, teachers were to be feared and their word not to be questioned. Rote learning was the order of the day.

The Gymnasium curriculum emphasized languages first, then history and biology, with the physical sciences and mathematics given lowest priority. A good memory was essential to advancement. This was a problem for Einstein because his "principal weakness was a bad memory, especially a bad memory for words and texts."[5] In science this "bad memory" was partly an asset because he preferred to work from first principles rather than memorized results. For example, when confronted with a problem involving gravitation, Einstein began with the basic axioms of Newton's theory and worked from there, instead of trying to remember already derived results. This is extremely difficult and is the distinct feature of those desiring to understand more than they know.

Dire situations arose for Einstein at the Gymnasium. A teacher of Greek told him that owing to his poor performance he would never amount to anything—which, in Greek studies, he never did. Other teachers were equally blunt, one going so far as to say that his mere presence was destroying the "respect which a teacher needs from his class."[6] These teachers apparently sensed in Albert more than the usual resistance to authority, and reacted accordingly, in a manner that was harsh even for that era. Einstein's response was to withdraw from the "merely personal" into worlds beyond appearances.[7]

Einstein's first inkling of such worlds had occurred at age four or five. Ill in bed, his father brought home a compass for his entertainment.[8] Einstein was amazed: No matter which way he turned the compass, its needle pointed persistently in the same direction. This was completely foreign to his everyday experience, in which such an effect could occur only if the needle were physically held in place. The boy concluded that "something deeply hidden had to be behind things."[9] Later Einstein referred to a phenomenon that dramatically conflicts with our everyday expectations as a "wonder."[10]

He made another attempt to free himself from the merely personal at about age twelve. Despite growing up in an irreligious household, Einstein went through a phase of deep religious immersion. He prepared for his bar mitzvah and was eager to become an active member of the local Jewish community. But this was not to be, owing in large part to a struggling young medical student, Max Talmud, who took lunch at the Einstein home once a week, as was the custom in Jewish households in Munich. While Talmud encouraged the boy's deep religiosity, he also fostered Einstein's budding curiosity about science and philosophy. Talmud introduced him to popular science books such as Aaron Bernstein's *People's Books on Natural Science*, a work of five or six volumes that Einstein recalled having "read with breathless attention."[11] Bernstein's books underwent numerous editions in the second half of the nineteenth century and were widely read among the emancipated Reform Jewish community. These books are Einstein's earliest introduction to atomism, a topic that Bernstein developed as one that could uncover relationships between different scientific disciplines.[12] Einstein would keep Bernstein's approach in mind.

As a gift, Talmud gave the twelve-year-old Einstein a text on Euclidean geometry, which Einstein called the "holy geometry booklet."[13] The subject struck Einstein as another "wonder" because its counterintuitive assertions could "be proved with such certainty that any doubt appeared to be out of the question."[14]

Uncle Jakob posed difficult algebra and geometry problems to Albert, who, when he persevered through to the solution "was overcome with great happiness [and became] aware of the direction in which his talents were leading him."[15] He asked his uncle, his father and Talmud for more mathematical texts, which he then worked through methodically. Einstein's remarkable powers of concentration were beginning to appear.

When he was thirteen Talmud introduced Einstein to Immanuel Kant's *Critique of Pure Reason.* Talmud's conversations on science and philosophy complemented Uncle Jakob's mathematical lessons; both took the boy beyond the Gymnasium curriculum.

At his mother's suggestion, at age five Einstein had also begun violin lessons. He found the essential musical drills so trying that he threw a chair at his first music teacher, who ran out of the house in tears. Then, at age thirteen, along with Kant's writings, Einstein discovered Mozart's sonatas.[16] Although they were beyond his technical competence, he practiced them repeatedly, although not systematically. At this time he also gave up formal lessons and independently honed his natural talent, which was not unimpressive. Music became an integral part of his life, often serving as a means to gather his thoughts on a scientific problem. That he was more than proficient we know from an inspector's report on a music examination Einstein took at the cantonal school in Aarau on 31 March, 1896. The inspector singled out "one student, by name of Einstein," out of seventeen tested, who "even sparkled by rendering an adagio from a Beethoven sonata with deep insight."[17]

Thus while Uncle Jakob supplied mathematical grist for Einstein's mill, Max Talmud broadened his horizons to include issues of science and philosophy. This potent mix exploded the boy's "religious paradise of youth."[18] Einstein realized that "Out yonder there was this huge world, which exists independently of us human beings and which stands before us like a great, eternal riddle, at least partially accessible to our inspection and thinking. The contemplation of this world beckoned like a liberation."[19] Einstein had found his calling: solve the riddle.

Hermann and Jakob Einstein's failing business brought an abrupt end to these youthful musings on *de rerum natura*. Hermann's lack of understanding of business realities, combined with Uncle Jakob's grand plans for expansion, led to severe problems. In March 1894 they moved their electrical firm to Pavia, Italy, and in summer 1894 the family joined them in Milan, except for Albert. The plan was for Albert to remain in Munich and complete his last year at the Luitpold Gymnasium. But this went totally against the young man's grain. Meanwhile, several prominent faculty members had voiced increasing displeasure with a student whose rebelliousness was, in large part, rooted in an attempt to think differently. Added to the miserable situation at school and his homesickness was impending service in another institution that Albert had detested ever since childhood, the Prussian military.[20]

Einstein acted decisively. Without informing his parents, he obtained a medical certificate that he was on the verge of a nervous breakdown and had to leave school. Then he prevailed upon a sympathetic mathematics teacher to give him a certificate attesting to his advanced knowledge of the subject, which would qualify him for entry into university-level courses. Einstein was not the average high-school dropout: every step was well orchestrated.[21]

Late in December 1894, six months before his expected graduation, Einstein arrived in Milan, to the utter amazement of his parents. Einstein's father urged him to "forget this 'philosophical nonsense'" and enroll in an engineering curriculum to prepare for a career that offered financial security.[22] But lack of a Gymnasium diploma limited his choice of universities. Luckily the prestigious Swiss Polytechnic, in Zurich, did not require one. Einstein reassured his parents that he would immediately settle down to prepare independently for the entrance examination. He also began the process of renouncing his German citizenship, which was finalized in January 1896. He remained stateless until he became a Swiss citizen in 1901.

WANDERLUST AND EARLY PHYSICS

Until fall of 1895 Einstein traveled through Northern Italy. Like Goethe some hundred years before, he found that the sunny Italian landscape freed him from the Sturm und Drang of his Munich years. And, like the teenage Picasso at Horta, Einstein became surer of himself and of his powers. In the summer of 1895 he sent his maternal uncle, Caesar Koch, his first scientific essay, with the grand title "Examination of the State of the Ether in the Magnetic Field."[23] Although even in retrospect this essay shows no signs of great genius, it demonstrates Einstein's familiarity with advanced topics in electromagnetic theory and his love of physics. At sixteen he had the perseverance of an autodidact.

Caesar Koch, a merchant in Brussels, probably could not understand a word of it. With this in mind, Einstein wrote in the cover letter that he would "not be the least offended if you do not read the stuff at all."[24] Most likely Einstein sent it to Uncle Caesar so that word would get back to his family that he was studying. At this time Albert also did some occasional consulting for the Einstein brothers' electrical firm and succeeded in solving certain difficult problems in machine design.[25] Uncle Jakob is reported to have said to a colleague at the factory, "where I and my assistant engineer have racked our brains for days, this young fellow comes along and solves the whole business in a mere quarter-hour. He'll go far one day."[26]

But not yet. The examination for the Swiss Polytechnic began on 8 October, 1895. As expected, Einstein did extremely well on the mathematics and physics parts but inadequately in languages and history, subjects that require rote learning.[27] On the basis of his scores in the physical sciences, the eminent Professor Heinrich Friedrich Weber proposed that Einstein attend his lectures for second-year students. But Einstein decided to take the advice of the Polytechnic's director and make up his deficiencies at the cantonal school in the town of Aarau, thirty miles west of Zurich, in the canton of Aargau. He entered the school at the end of October 1895; it turned out to be just what he needed. The move gave him time to reflect on his (largely self-acquired) knowledge as well as to hone his social skills.

EINSTEIN AT AARAU

The cantonal school Einstein attended was founded by followers of the eighteenth-century Swiss educational reformer Johann Heinrich Pestalozzi, who emphasized the innate power of *Anschauung*. The German term *Anschauung* is philosophically loaded and can be understood as "intuition" or as the "visual imagery" abstracted from phenomena we have witnessed with our senses. It plays an important role in the philosophy of Pestalozzi's contemporary Immanuel Kant, for whom *Anschauung* is the highest form of visual imagery.[28] In his principal book, *How Gertrude Teaches Her Children,* published in 1801, Pestalozzi wrote, "I must point out that the ABC of *Anschauung* is the essential and the only true means of teaching how to judge the shape of all things correctly."[29] The cantonal school taught according to Pestalozzi's belief that "conceptual thinking is built on *Anschauung*." There was no forced feeding of knowledge at the cantonal school. Rather, emphasis was on independent thought and "young people saw in the teacher not a figure of authority, but alongside the scholar, a man of distinct personality. [Albert's] time in Aarau was thus very instructive for him in many ways and one of the best periods of his life."[30] Conceptual thinking would become Einstein's specialty.

At Aarau Einstein boarded with the Winteler family. Jost Winteler and his wife Pauline quickly became, and remained, "Papa" and "Mama."[31] Jost, a scholar with impressive credentials who taught Greek and history, exposed Einstein to liberal political views that were in striking contrast to those from the Gymnasium.[32]

On his arrival at Aarau Einstein was, to some extent, a man of the world. Although a bit younger than his classmates, he had the reputation of someone who had rebelled against the Prussian Gymnasium regime and had traveled on his own throughout northern Italy. A fellow student at Aarau, Hans Byland, remembered Einstein somewhat romantically:

FIGURE 3.1 Einstein's graduating class from the Canton School in the town of Aarau. Einstein is seated in the first row at the left.

> The grey felt hat pushed back over the silky mass of dark hair, he strode along with vigour and assurance, at the rapid—I am tempted to say sweeping—tempo of the restless mind that carried a world within it. Nothing escaped the acute gaze of the large sun-bright eyes. Whoever approached him was captivated by his superior personality. A mocking trait around the fleshy mouth with its protruding lower lip did not encourage the Philistine to tangle with him. Unconfined by conventional restrictions, he confronted the world spirit as a laughing philosopher, and his witty sarcasm mercilessly castigated all vanity and artificiality.[33]

Although Byland's recollection comes almost thirty years after the fact, a group photograph from Aarau supports it. Despite the need to be absolutely still while being photographed, Einstein's demeanor is that of a young man at ease with himself, and in a state of deep contemplation. He is also the only one with collar open and tie undone.

Einstein applied himself seriously when the subject matter interested him. His development as a scientist benefited greatly from the emphasis at Aarau on the power of visual understanding, of *Anschauung*, rather than

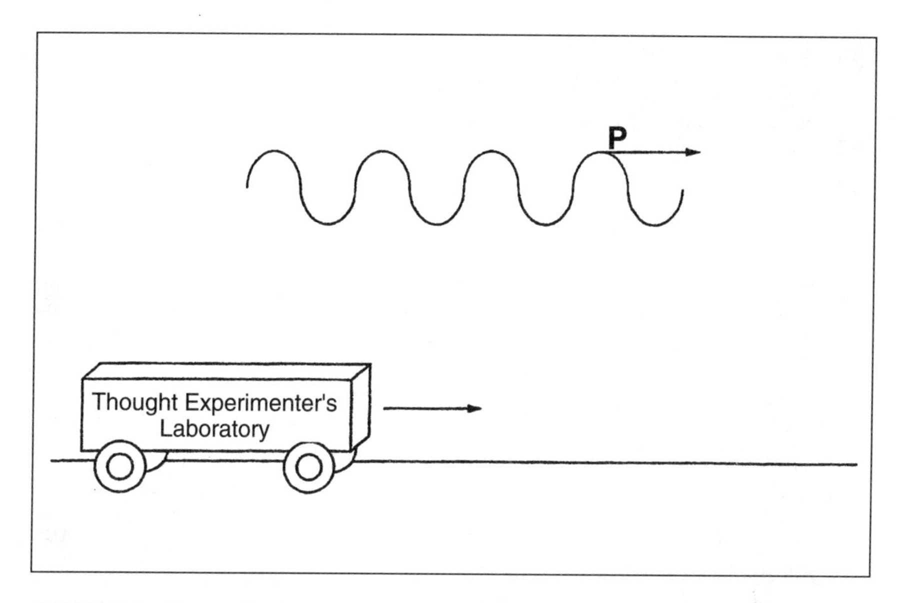

FIGURE 3.2 Einstein's thought experimenter tries to catch up with a point on a light wave labeled *P*, which moves at the velocity of light as measured by an observer on the ground.

on memorization and rote learning.[34] Einstein combined this mode of thinking with his knowledge of physics into a "thought experiment" that he kept in the back of his mind for ten years until, in 1905, he realized that it contained the "germ of the special relativity theory."[35] I will return to this experiment in Chapter 6. It is a nice example of *Anschauung* in which Einstein imagines the case of someone chasing a point on a light wave (Figure 3.2). The visual representation, or *Anschauung*, of a light wave is a sweeping abstraction from the properties of water waves.

Another highlight of his time at Aarau was his romance with Marie Winteler, Papa and Mama Winteler's eighteen-year-old daughter. Their relationship had all the trappings of youthful first love, to which Marie's musical talent added a dash of artistic passion. In a letter Einstein wrote to Marie on 21 April, 1896, from Pavia, while visiting his parents, he addresses her as "Beloved sweetheart!" and tells Marie how much he misses her "dear little eyes" and "dainty little hands," how indispensable she is to him, and how he now knows the "meaning of homesickness and pining" for "my dear little sunshine." The relationship was serious enough for Einstein to write how much his mother has "taken you to her heart." In fact, Einstein's mother added a postscript, "Without having read this letter, I send you cordial greetings! Pauline Einstein."[36] Writing to Einstein at the Swiss Polytechnic, Marie pleads with her "dear dear darling" not to work so hard and promises that she will come to Zurich to see where "my darling lives."[37] She concludes with "Good night darling."

Like many high school relationships, Albert and Marie's did not survive his going off to university. As Einstein wrote to Mama Winteler ("Dear mommy") in May 1897: "I am writing to you . . . in order to cut short an inner struggle whose outcome is, in fact, already firmly settled in my mind. . . . Strenuous intellectual work and looking at God's Nature are the reconciling, fortifying, yet relentlessly strict angels that shall lead me through all of life's troubles. . . . One creates a small world for oneself."[38] So he would do throughout his life. Among the very few reasons Einstein was willing to emerge from his "small world" was for a woman. Women found him physically attractive, an attraction to which his carefree manner, his distinctively passionate style of violin playing, and later on, of course, his legendary intelligence, added immensely. As a woman friend of his second wife, Elsa, put it: "He had the kind of male beauty that especially at the beginning of the century, caused such havoc."[39] Yet when pursuit threatened to change into a serious relationship, Einstein scurried back to his "small world."

Some years later Marie described her relationship with Einstein as one of "ideal love."[40] Perhaps. Einstein was still sending his dirty laundry back to Marie even after he had met and fallen in love with Mileva Marić, another Swiss Polytechnic student; Marie promptly mailed it back to him, clean.[41] Einstein revealed everything to his new girlfriend. In a letter written toward the end of September 1899 he mentions that he will be making frequent trips to Aarau since his sister Maja had just enrolled as a student in the town, but "you don't have to be afraid at all [even though] the critical little daughter is coming home, with whom I fell so terribly in love 4 years ago."[42]

EINSTEIN AS A UNIVERSITY STUDENT

In September 1896 Einstein took the secondary school leaving exam, or *Matura,* which permitted him to enter the Swiss Polytechnic.[43] Out of a possible 6 points, he scored highest in algebra and geometry, with a 5–6 in physics, and his lowest score in French. Regarding the physics part, the examiner noted Einstein arrived late and finished early.[44] He scored an overall $5\frac{1}{3}$, the highest in his class.

In October 1896 Einstein entered the Swiss Polytechnic Institute in Zurich, renowned for its teaching and research in science and engineering. It was divided into seven departments. Einstein entered Department VI, the School for Mathematics and Science Teachers, and not the engineering section as his father had hoped.[45] Department VI was itself divided into two parts. Section VIA, where Einstein concentrated, was the Mathematics Section (mathematics, physics and astronomy), while Section VIB taught the other natural sciences. There were eleven students in Einstein's entering class. He was graduated in July 1900 as an accredited teacher of mathematics and physics at the secondary-school level.

Each student's curriculum at the Swiss Polytechnic was worked out individually. One free elective was allowed per semester. Einstein went on to take more than the required minimum, enrolling in philosophy, politics and economics courses. One might therefore assume that he flourished intellectually; he did not.

Einstein's university experience turned out to be bittersweet. The physics curriculum neglected such contemporary cutting-edge topics as electromagnetic theory. He balked at being coerced to memorize large quantities of what he judged to be inessential material. Visual thinking in scientific research was not even mentioned. Einstein returned to his autodidactic ways and began to cut classes in order to stay at his lodgings and study the masters of theoretical physics on his own—scientists such as Ludwig Boltzmann, Hermann von Helmholtz and H. A. Lorentz. As Boltzmann wrote in a book that Einstein undoubtedly studied, "Unclarities in the principles of mechanics [derive from] not starting at once with hypothetical mental pictures but trying to link up with experience at the outset."[46] Einstein, imprinted by his Aarau education, took this to heart.

Of his Swiss Polytechnic days, he recalled: "There were altogether only two examinations; aside from these, one could just about do as one pleased. This was especially the case if one had a friend, as did I, who attended the lectures regularly and who worked over their content conscientiously. This gave one freedom in the choice of pursuits until a few months before the examination."[47] Einstein's friend, Marcel Grossmann, must have done an excellent job because, in October 1898, Einstein placed first in the intermediate exam and Grossmann second. "I would rather not speculate how I might have fared without them," wrote Einstein of Grossmann's notes many years later.[48] Einstein remembered that as a student he lacked the drive to "concentrate on what is being lectured on."[49] Grossmann, a year and a half older than Einstein, was also a student in Section VIA specializing in mathematics. He eventually wrote a doctoral dissertation on a topic in non-Euclidean geometry. Although Einstein was somewhat a loner, he grew close to Grossmann, meeting him once a week at the Café Metropol for conversations that lasted many hours. Grossmann was impressed with his friend and described him to his father as "someone who will one day be great."[50]

But not yet. One of the Swiss Polytechnic's leading professors had little appreciation for Einstein's independent thinking, and his sarcasm exacerbated the situation. The professor was Heinrich Friedrich Weber, who had gone out of his way to encourage Einstein after his failure on portions of the entrance examination. Weber was then at the height of his powers as an internationally known researcher in electrotechnology. The great German industrialist Werner von Siemens had provided funds, in 1886, toward establishing an institute for science and engineering at the Swiss Polytechnic with the proviso that its director be his friend Weber. In his midforties, Weber radiated Prussian authority and was not good at dealing with independently minded students like Einstein. In fairness, Einstein praised Weber's lectures for their style and content but found them not up-to-date enough.

It certainly did not help that Einstein went out of his way to address Weber as "Herr Weber" instead of "Herr Professor Weber."[51] In 1897, a leading scientific journal reported that Weber performed experiments on himself in order to determine the effect of shocks from an alternating current source.[52] No doubt this contributed to the use of electricity as a means for execution. One can only wonder how Einstein and his friends interpreted such experiments.

On one occasion, in a state of exasperation, Weber said to Einstein, "You are a clever young fellow, Einstein, a very clever young fellow! But you have only one fault: One can't tell you anything."[53] Nevertheless Weber had the last word. Of the four students receiving a degree in Section VIA in July 1900, Einstein was the only one not asked to stay on as an assistant to one of the professors. After this insult he forever held Weber in contempt.[54] More than pride was at stake: Einstein was madly in love and had to plan for the support of his future wife.

ALBERT, MILEVA AND PHYSICS

At the Swiss Polytechnic Einstein and a small group of friends often met at a local café to discuss scholastic as well as social issues, just as do stu-

FIGURE 3.3
Einstein while a student at the Swiss Polytechnic Institute, Zurich.

dents today.[55] His deep intellectual involvement sometimes resulted in a comical absentmindedness. Einstein recalled that "when I was a young man I visited overnight at the home of friends. In the morning I left forgetting my valise. My host said to my parents: 'That young man will never amount to anything because he can't remember anything'."[56] Apparently he so often forgot his key that waking up his landlady late at night became commonplace: "It's Einstein—I've forgotten my key again."[57]

Einstein's studied indifference to dress, his easygoing manner and unruly mane of dark hair, coupled with his love of music and philosophy, gave him the air of a poet rather than a scientist.

Yet behind this bohemian look was an almost brutal severity in the pursuit of intellectual matters that would lead him further and further into the life of the mind, to the eventual exclusion of all else.

Einstein fell in love with the only girl in Section VIA, Mileva Marić. Four years older, a native of Titel (then part of the Austro-Hungarian Empire, now in Serbia), Mileva was a highly motivated woman, determined to make her way in an almost entirely male profession. A professor at the Swiss Polytechnic, the eminent mathematician Adolf Hurwitz, recalled that, "All in all Mileva is an unusual woman."[58] From October 1896 through August 1899 they were just close friends. From then until

FIGURE 3.4
Mileva Marić, about 1896.

their marriage on 6 January, 1903, their love affair had all the trappings of Romeo and Juliet, with a good bit of *La Bohème*.

Albert and Mileva were an odd match. While Albert was of medium height (5'8"), well built and outgoing, Mileva stood only up to his shoulders, limped from the effects of childhood tuberculosis, and had an air of Slavic melancholy. When the seriousness of their romance became clear, some of Einstein's friends became concerned and asked him why someone so successful with women would choose one so unsound in health. Einstein replied, "Why not? She has a sweet voice."[59] Besides her voice, Mileva was more than pleasant-looking with striking eyes and dark hair. Some remembered her as "a gloomy, laconic and distrustful character."[60] But the few who got to know her better saw an open-minded, modest person. Whatever others thought, Mileva and Albert were passionately in love.

Intensely personal beings, with similar political views and high aspirations for physics research, they passed sunny days sailing on Lake Zurich and hiking in the Alps. They exchanged passionate letters during periods of separation, which included summer vacations and when Albert stayed at his family's home in Pavia, where Mileva was unwelcome. These letters, discovered in 1986 and published as a book, *Albert Einstein: The Love Letters,* in 1992, convey some of the hothouse atmosphere of their physics discussions.[61]

At first Albert addressed Mileva as a friend, "Liebes Fräulein." Starting in August 1899 Mileva became "Liebes Doxerl [Dear Dollie]," and Albert became "Johanzel [Johnnie]." Diminutives and verbal playfulness are interspersed with pledges of lifelong love. "Without you I lack self-confidence, passion for work, and enjoyment of life—in short, without you, my life is no life," wrote Albert on 14 August, 1900.[62] Another letter includes a sketch of Johnnie's foot.[63]

Is there another scientist whose love letters merit publication? Any of us could have written the purely personal parts of these letters, and probably better. But these were by Albert Einstein and they provide insight into a part of his life and thinking previously almost uncharted. Reading through the letters one "is struck at the outset by an asymmetry in the voices of the two lovers—his self-assured, dramatic, masterful in the command of his native German; hers, often understated, self-effacing, tentative in a language foreign to her native Serbian."[64] Einstein's punning in Swabian is nowhere matched by Mileva. As the affair progressed and she fell more and more under the spell of her lover, so does her prose become that much more unsure.

From Einstein's letters one glimpses the strong influence of the German philosopher Arthur Schopenhauer, whose ideas provided Einstein with a conceptual framework as well as a vocabulary for expressing his feelings toward Mileva.[65] Schopenhauer's portrait of the lonely intellectual pitted against philistines inspired an identity for an entire generation of young Germans living in the excessively materialistic society of Wilhelmine Germany. The genius has the higher role of leading the blind masses merely by living amongst them. Schopenhauer's biting comments on the weaknesses of the masses resonated with Einstein's cynicism and sarcasm. According to Schopenhauer there were two pleasures permitted the solitary genius: a mate with whom to share a life of the mind with the barest of material needs, and music. To this short list Einstein and Mileva added physics. In this period of his life, Schopenhauer offered Einstein a more benign guide than what Nietzsche offered Picasso.

We learn from their letters that for appearances the couple kept separate flats, but spent a great deal of time in one or the other.[66] There is an interesting pattern in Einstein's letters. On 10 August, 1899, for instance, in the midst of a highly romantic or at least lustful passage—"You're such a robust girl and have so much vitality in your little body"—he abruptly switches to physics: "I returned the Helmholtz volume and am now rereading Hertz's propagation of electric force."[67] More physics follows, and then Einstein abruptly takes up a Schopenhauerian vein: "If only you could be here with me for a while! We understand one another's dark souls so well, and also drinking coffee & eating sausages etc."[68] It is of course clear what Johnnie meant by "etc."

This patchwork style appears again in a letter posted sometime in late August or early September 1900.[69] After signing off "With tender kisses, your Albert," he adds a postscript, "How's your little throat," and then, "I'm investigating the following interesting question [regarding the propagation of electrical energy in space]." His enthusiasm for his work and his love for Mileva are everywhere mingled in these letters. They were the perfect Schopenhauerian pair. Poetry appears, too: "My dear little one, Four quartets: Oh my! That Johnnie boy!/So crazy with desire,/While thinking of his Dollie,/His pillow catches fire."[70]

No Lord Byron was Einstein. The love prose he wrote to Mileva is indistinguishable from what he sent Marie or any other lover we know of. What emerges more clearly than ever before, however, is the way Einstein's personal life affected the dynamics of his scientific work. He wrote excitedly to Mileva about his new ideas, and presumably she replied in kind in letters and in person. She gladly helped and encouraged him during the early years of their relationship. Then things changed. Mileva became Einstein's Fernande.

PHILOSOPHY AND PHYSICS AT THE FIN DE SIÈCLE

In his "Autobiographical Notes," Einstein's pithy summary of the philosophical and scientific situation at the end of the nineteenth century was that "dogmatic rigidity prevailed in matters of principles."[71] Scientific research was to be carried out with total reliance on data of the laboratory. The suggested method was to include atoms as a means for calculation, but no objective reality was to be claimed for them. What you saw was what you got, and as Ernst Mach was fond of saying of atoms, "Have you seen one?"[72]

This method for doing science had a time-honored past and was not to be taken lightly.[73] In the early 1880s, Mach had formulated a variant of Auguste Comte's positivism to fight the malaise that had come over physics. By then it was clear that the promise of Newton and his followers, of a complete physics based entirely on Newton's theory of motion, had fallen short of the mark. Phenomena of optics, electromagnetism and thermodynamics resisted all attempts at explanation using models based on how pulleys and springs work. Even worse, deeper questions regarding the nature of force, velocity, mind, and so on, seemed scientifically unapproachable. People had begun to speak of "the bankruptcy of science." The field offered no solutions, or even promising approaches, to the deep and pressing questions of life.[74]

Mach attacked what he took to be the root of this malaise: certain assumptions at the foundations of Newtonian science that he considered unwarranted, in the sense that they could not be tested in the laboratory. In 1883 Mach developed these views in his book *Science of Mechanics,* which went through sixteen editions by the time of his death in 1916. Mach's hard-hitting criticism dismissed as "an idle metaphysical conception"[75] such untestable notions in Newtonian science as absolute space and absolute time. Besides being unmeasurable, they had a distinct theological aura. Newton, as a typical man of his time, had viewed science as another means of understanding the mind of God, and motions as occurring against a background of absolute space, the unmoving sensorium of God.

Mach was no less outspoken against atomic theories. Atoms were merely "things of thought"[76] since they cannot "be perceived by the senses."[77] For Mach the goal of science is to describe experimental data as economically as possible. By advocating strong dependence on experimental data to the exclusion of the unmeasurable, Mach returned the moral high ground to scientists. They would leave the unexplainable to artists and theologians, and content themselves with determining the equations governing observable phenomena. Nothing else was "real."

Variations on Mach's positive message abounded. In all of them, experimental data were the sine qua non of scientific discovery. Readings of electrical current from meters, or even X-ray photographs, could be interpreted without invoking invisible atomic sources. Yet a deeper understanding of phenomena meant seeking causes. The apparent reality of electrons, discovered in 1897, would eventually cause some prominent supporters of positivism to jump ship.[78] Nevertheless, many scientists continued to toe the positivist line as best they could, which meant an exclusive reliance on laboratory data.

In 1897, Einstein's friend Michele Besso introduced him to Mach's *Science of Mechanics*. What impressed Einstein was not so much Mach's positivism but his "incorruptible scepticism and independence [which] shook this dogmatic faith."[79] Einstein admired Mach's courageous questioning of the accepted foundations of the physical world, such as space and time.

At the end of the nineteenth century, the cutting edge of theoretical physics was electromagnetic theory, which had been unified with the theory of light in the early 1860s by the great Scottish scientist James Clerk Maxwell. Since the time of Newton there have been two ways to envision light—as particle or wave. Newton conceived of light as particles, which he assumed to behave like any other object such as planets or apples. By the mid-nineteenth century, however, experimental evidence weighed heavily against Newton's particle theory and toward a wave theory of the sort advocated by Newton's Dutch contemporary Christian Huygens.

A principal experimental result that wreaked havoc with Newton's particle theory of light is the phenomenon of interference, which we have all seen occur with water waves. Imagine dropping two stones into a still pond of water. Spherical waves spread out from where the stones struck the water and soon these waves cross each other. This thought experiment is realized in Figure 3.5a.

Imagine further that corks are floating in the pond. We observe that at points where the waves intersect, some corks bob up and down much more than others, while some do not move at all. The corks that bob up and down the furthest do so because they are situated where the two sets of waves cross at their maximum heights. The result is called constructive interference (Figure 3.5b). The ones that remain stationary do so because the two waves cross where one is at a maximum height and the other at a maximum depth. This result is destructive interference (Figure 3.5c).

In 1803 the English polymath Thomas Young did such an experiment with light. The experimental setup in Figure 3.6 is similar to the one he actually used. Light enters via a slit S_0 in screen A and then passes through slits S_1 and S_2, in screen B. Figure 3.7 contains the data from registering light intensities by passing a photographic exposure meter over the screen C.

Young, who did not have an exposure meter but obtained the same pattern, reasoned correctly that this was an interference pattern caused by the interaction of two sets of waves from two slits. Figure 3.6 can thus be redrawn as Figure 3.8, where the spreading spherical waves are light waves, instead of water waves.

We should bear in mind that no one has ever seen a light wave. When you turn on a light in a dark room you do not see spherical light waves spreading out from the bulb. Light waves are an abstraction from the phenomena of water waves—a visual image or *Anschauung*. They are also a visual representation of the mathematical formalism of optics, not coincidentally called "wave equations."

The analogy, however, goes further. How can there be water waves without water? The waves are pulses of energy moving through a stationary medium. Light energy, the reasoning went, must likewise be waves *in* some medium. The milieu that supports light waves and all other electromagnetic disturbances was called the ether, and had been proposed by Huygens in his early wave theory of light. By definition, light travels

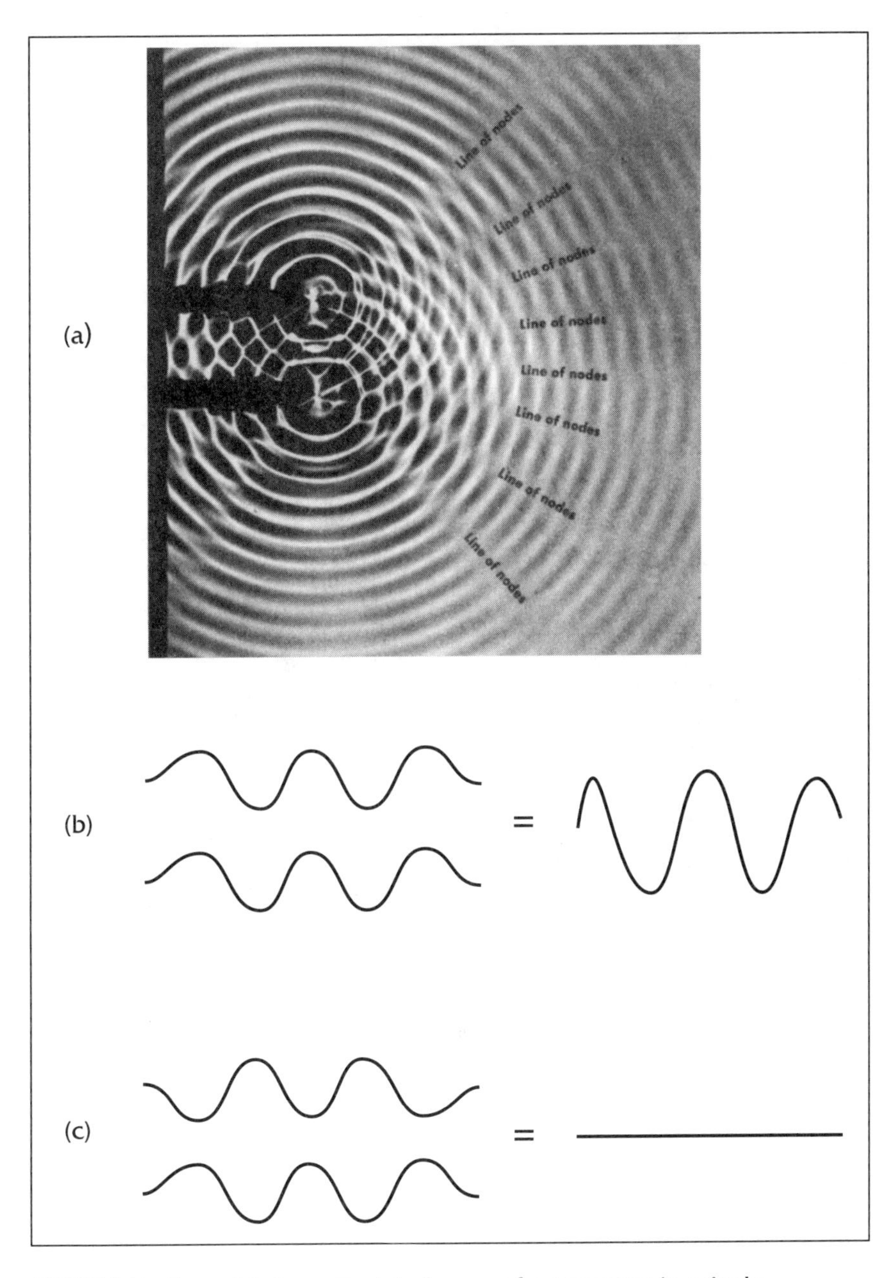

FIGURE 3.5 Figure (a) shows the interference of water waves in a ripple tank. The waves are generated by two vibrators that strike the water's surface synchronously, producing two expanding spherical waves. There is destructive interference along the lines marked "line of nodes" and constructive interference between these lines. Figure (b) shows how constructive interference occurs when two waves cross with their maximum heights in alignment. In Figure (c), destructive interference, or canceling out, takes place when two waves cross with their maximum and minimum heights aligned.

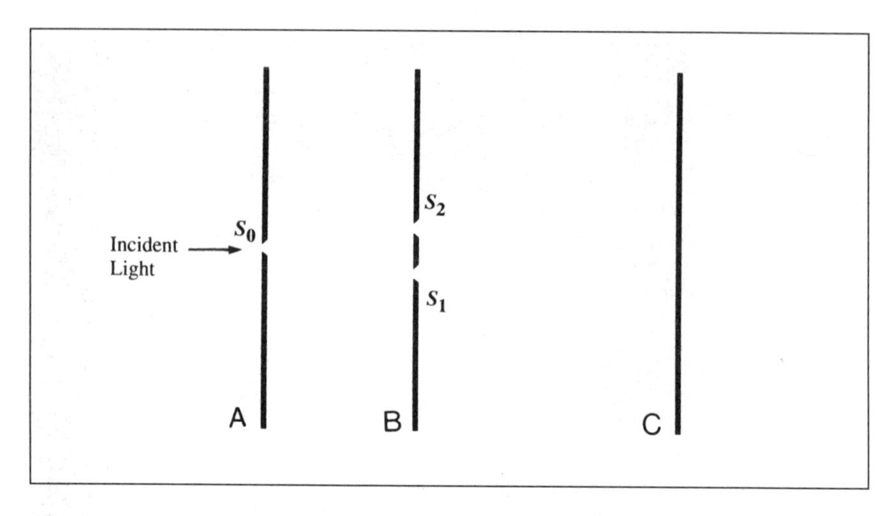

FIGURE 3.6 This shows the experimental arrangement in which light is incident on the slit S_0 on screen A. Another screen, B, with slits S_1 and S_2, is in front of screen C.

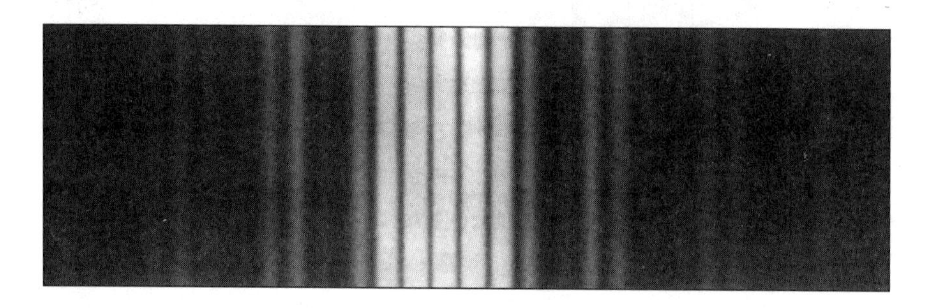

FIGURE 3.7 The data contained in the photograph of the effect on screen C from Figure 3.6 caused by light incident first on screen A and then on screen B.

through the ether at the speed of light—186,000 miles per second, a value measured first by astronomical methods.[80]

Besides the analogy with water, there was another intuitive reason for an ether. For bodies to act on one another at a distance through empty space is entirely foreign to our daily experience. As Einstein put it in 1936, the "whole of science is nothing more than a refinement of everyday thinking."[81] If in the world about us we notice that objects always affect each other by direct touch, it was felt, so should be the case in science. Consequently, electric and magnetic objects were assumed to set up disturbances in the ether called electric and magnetic fields respectively, by means of which they attract or repel one another. We can obtain a visual image of the magnetic field through the arrangement of iron filings on a

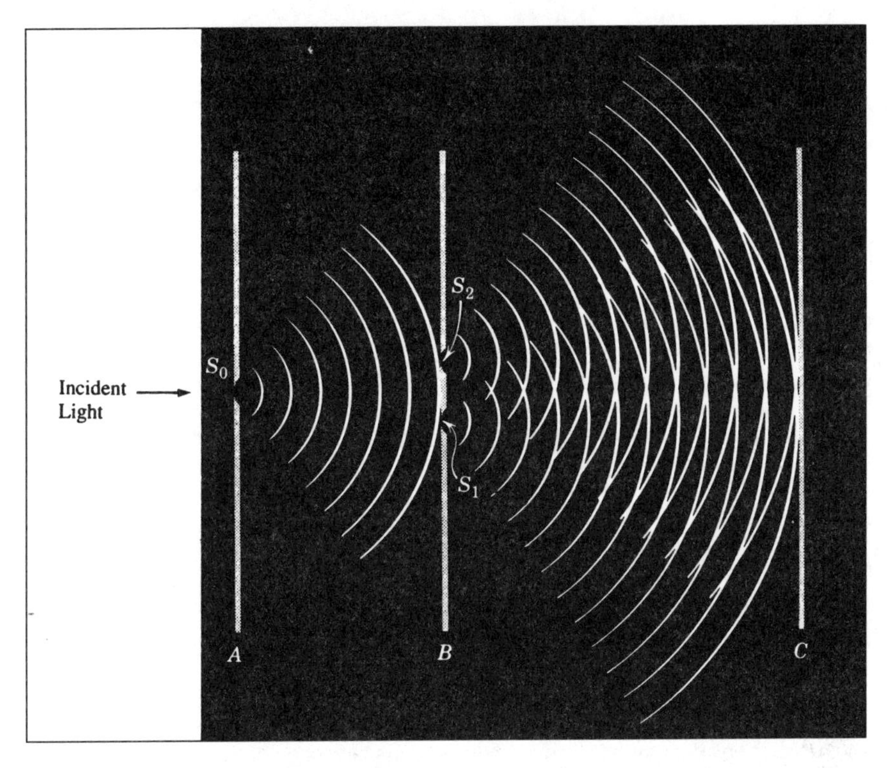

FIGURE 3.8 This figure is the result of assuming that the source is emitting spherical "light waves" in analogy with the phenomena of water waves as shown in Figure 3.5.

piece of paper placed above a bar magnet (see Figure 3.9). These dispositions can be abstracted to what are called "magnetic lines of force," which are another *Anschauung*.

Since the ether was so fundamental, there ought to be a way to detect it, or else it would be mere metaphysical baggage. The agreed-upon way was to measure the velocity of light from a laboratory on the moving Earth, and it went basically as follows: Ether theories of light and electromagnetism were formulated from the viewpoint of an observer at rest in the ether. By definition, these observers always measure the velocity of light as 186,000 miles per second. Then, by mathematical means, the problem situation is transferred to a laboratory on the moving Earth. The velocity of light, so the thinking went, should differ from 186,000 miles per second by the speed of the Earth's motion through the ether. This difference was expected to manifest itself through minute but predicable changes in such phenomena as optical interference. While everyone agreed that the ether existed, the

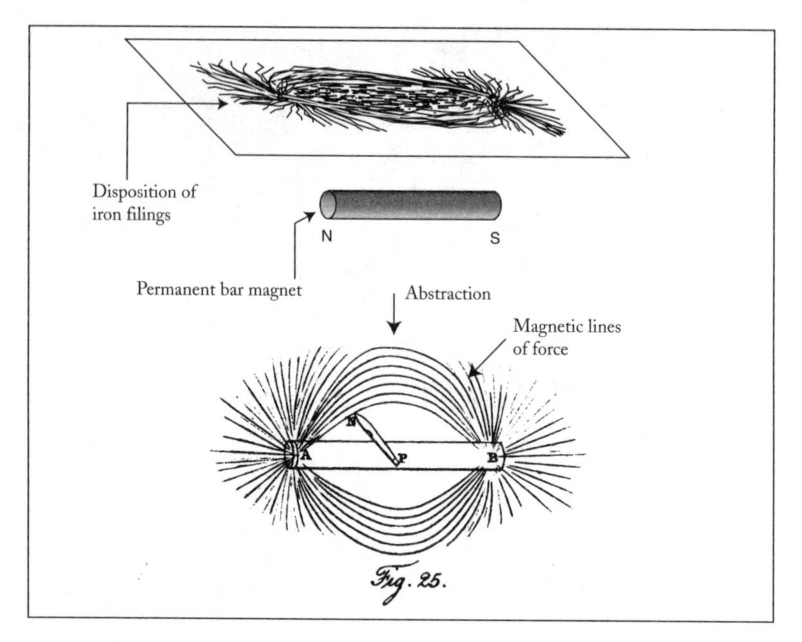

FIGURE 3.9 This illustrates the process of abstracting from phenomena that we have actually witnessed in the world of perceptions. In this case the disposition of iron filings on a piece of paper placed above a permanent bar magnet is abstracted to a visual image of magnetic lines of force that are assumed to permeate all of space. This representation is an *Anschauung*. The figure in the lower part was drawn by Michael Faraday in the course of his pioneering researches on the magnetic field in the 1830s (see Faraday 1965).

issue was whether it was at rest or some portion of it was dragged along by the moving Earth. If the ether were at rest, then the Earth's velocity relative to it is a combination of its various motions through space. These are the Earth's orbital velocity about the Sun, its daily rotation on its axis, and the solar system's velocity through the universe of which little was known at the end of the nineteenth century.

As an illustration, let us consider a variant of Einstein's thought experiment in Figure 3.2, which he may also have had in mind. Let us assume that the source of light is at rest in the ether and so, by definition, point *P* moves at 186,000 miles per second, relative to the source. It is intuitively reasonable that the thought experimenter would measure the velocity of light to be less than 186,000 miles per second because he is catching up with the point *P.* By performing optical experiments in his laboratory the thought experimenter ought to be able to detect predicted changes from phenomena as they were expected to occur if the laboratory were motionless relative to the ether.

Experiments designed to detect the motion relative to the ether were often referred to as "ether-drift experiments." For simplicity, physicists took the Earth to be at rest and the ether "streaming" by in an "ether wind" opposite in direction to the Earth's motion about the Sun. The result is that light rays are "blown about" by the ether wind and so their velocity has a value other than 186,000 miles per second. Ether-drift experiments were state-of-the-art, carrying the imprimaturs of the great theoretical and experimental physicists of the nineteenth century. Yet one by one, all of these experiments failed because the measured velocity of light turned out to be the same as if the Earth were *actually* at rest in the ether. But we know that the Earth is not at rest.

The electromagnetic theory that had the greatest success in dealing with the ether-drift experiments was a version of Maxwell's theory proposed in 1892 by H. A. Lorentz. Considered at age forty-two to be the greatest physicist in the Netherlands since Huygens, Lorentz would go on to receive, among many other awards, the 1902 Nobel Prize.[82] Lorentz was equally at ease with problems of theoretical physics or very practical ones. His work in hydraulic engineering, for instance, was instrumental for constructing a key enclosure dike on the Zuiderzee. He personally supervised the construction as well.

In 1895, Lorentz published his latest developments in a book-length treatise that Einstein read while a student at the Swiss Polytechnic.[83] The unique aspect of Lorentz's theory was that it could explain most of the available data on electromagnetic phenomena in addition to explaining the failure of a large class of ether-drift experiments. It accomplished this by proposing a species of submicroscopic charged particles, which Lorentz originally called "ions," that move about in an all-pervasive stagnant ether that has no properties other than to support light in transit.[84] In 1897 Lorentz's ion was discovered and dubbed the "electron."

All of the ether-drift experiments performed in the nineteenth century were of the same low order of accuracy* except for the one undertaken in 1887 by two American scientists, Alfred A. Michelson and Edward W.

---

*From now on what I mean by "low order of accuracy" is the ratio of two terms: the relative velocity of the Earth with respect to the ether and the velocity of light as measured by observers in the ether, which is 186,000 miles per second. For an estimate of this ratio physicists used the Earth's orbital velocity about the Sun, which is about 18.6 miles per second. The ratio is one in ten thousand. Whereas the accuracy of the Michelson-Morley experiment was expected to be two powers higher in accuracy, an astounding one in one hundred million.

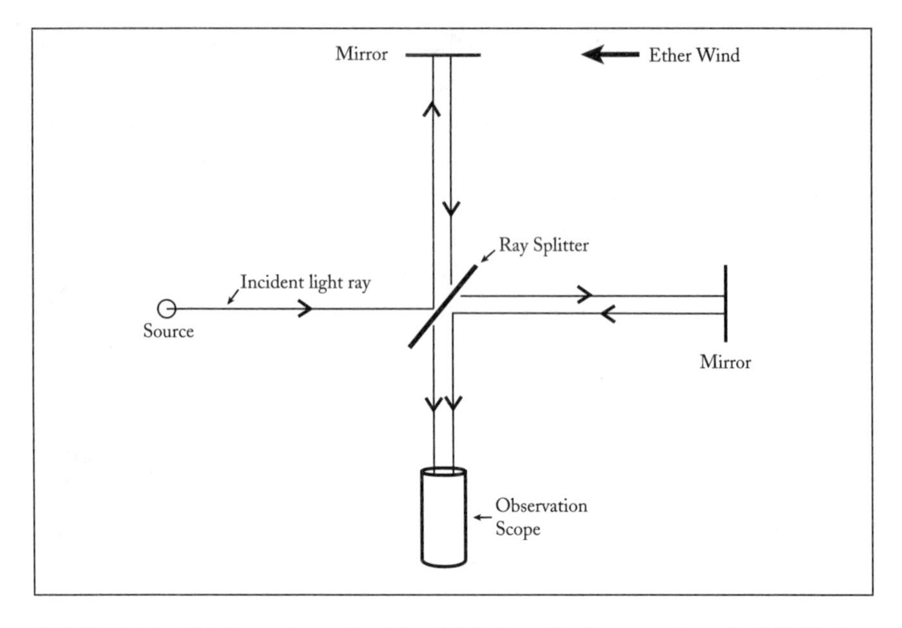

FIGURE 3.10 This is a schematic of the Michelson-Morley experiment of 1887. An incident ray is split by a half-silvered mirror into two rays that race at right angles to each other over equal distances and are then recombined to produce interference effects such as those in Figure 3.7. The Earth is assumed to be at rest, and so an ether wind streams by in a direction opposite to its motion about the Sun.

Morley, of the Case School of Applied Science, Cleveland, Ohio.[85] Michelson and Morley's apparatus split an incoming light ray into two rays (see Figure 3.10).

They travel equal distances at right angles to each other and are reflected back to the point where they were split, and are then recombined. The observed result is a series of light and dark lines called "interference fringes," much like the ones in Figure 3.7.

Since the ether wind causes an inequality in times for the light rays to make their to-and-fro journeys, the interference pattern was predicted to undergo a measurable displacement. To explain this let us make an analogy to two swimmers in a swiftly moving river. One of them swims first with and then against the current, while the other swims back and forth across the current. If they both make the same speed through the water, their round-trip times will differ: the one who swims back and forth across the current will take longer. In the Michelson-Morley experiment the swimmers are light rays, the current is the ether wind, and their velocity relative to the ether is 186,000 miles per second. The difference in times predicts a measurable movement for the fringe pattern.

But within experimental errors, Michelson and Morley measured no fringe shift.

The high accuracy of their experiment led them to propose that the ether is not absolutely at rest but that a portion of it is dragged along with the Earth. In this case there would be some relative motion between the Earth and the ether apart from the motions in which the Earth participates. Michelson and Morley could only conclude that the relative velocity between the Earth and the ether was much less than the Earth's orbital velocity about the Sun.

There were about ten experiments of the lowest accuracy genre. Lorentz could systematically explain their failure by employing, instead of the usual time, a strictly mathematical entity he called the "local time."[86] The usual time, the one that squares with our sense perceptions, is not affected by our movements. A ship's chronometer can stay synchronized with a clock in Bristol even when the chronometer, and the ship carrying it, are halfway around the world. This assumption, a cornerstone of Newtonian science, carried over into electromagnetic theory. It is the source of the prediction that the velocity of light measured on the moving Earth differs from the one measured by observers at rest in the ether.

Lorentz's "local time" is another matter altogether, because it contains two terms. One of them is the usual time, while the other depends on the clock's motion relative to the ether. This extra term is so extremely small as to be beyond our perceptions, a result that was taken as further evidence for this term's, and so, too, the local time's, lack of any real physical content.[87] Consequently, the local time was not considered to be the physical time. Yet explaining the failure of the ether-drift experiments to lowest-order accuracy required hypotheses that included the local time. In this way terms could be introduced into Lorentz's electromagnetic theory to cancel effects predicted but not measured. Einstein would keep the local time in mind. But he did not need it just yet.

Lorentz's local time, however, could not explain the Michelson-Morley experiment. Michelson, the world's premier high-precision experimenter on light, claimed an accuracy far beyond the lowest-order ether-drift experiments. According to the Michelson-Morley experiment the ether was not absolutely at rest, a result that ran counter to Lorentz's fundamental assumption about the ether.[88] To rescue his theory from this offending result, Lorentz proposed the most fantastic hypothesis yet: objects contract in the direction of the Earth's motion through the ether. This was the famous Lorentz contraction. Consequently, in Figure 3.10 the arm of the measuring instrument that runs parallel to the ether wind is supposed to contract by just the amount necessary to equalize the times for both sets of light rays to travel to-and-fro, thereby negating the predicted fringe shift. But it was a rogue hypothesis, literally inserted by hand to save the theory from the result of a single experiment. Most physicists regarded the Lorentz contraction as over-the-top.

ALBERT, MILEVA AND RELATIVITY

Let us return to Albert's letter to Mileva of 10 August, 1899, and to some of the physics that abruptly follows his compliment on the "vitality in your little body." "I'm convinced more and more," the letter continues, "that the electrodynamics of moving bodies as it is presented today doesn't correspond to reality, and that it will be possible to present it in a simpler way."[89] The phrase "electrodynamics of moving bodies" stands out because it will be the title of Einstein's 1905 relativity paper. We can thus glimpse the advanced physics that Einstein was thinking about as a mere third-year university student. He was concerned about the artificial way in which certain theories represented the ether.[90] There had to be a "simpler way," simpler even than Lorentz's theory where the ether had no mechanical properties and served only to support light in transit.

On 10 September, 1899, Einstein advised Mileva that he had thought of "investigating the way in which a body's relative motion with respect to the luminiferous ether affects the velocity of the propagation of light in moving bodies."[91] This was most likely a variant of a certain type of ether-drift experiment that he had just been reading about. Two years passed before Einstein again brought up the ether and relative motion. Then, in a letter of 27 March, 1901, he wrote to Mileva about bringing "our work on relative motion to a successful conclusion."[92] Unfortunately we know nothing about this work.

On 4 April, 1901, Einstein told Mileva about a four-hour session he had with his friend Besso in which they touched on the "fundamental separation of luminiferous ether and matter, the definition of absolute rest."[93] Seven months later Einstein informed her that he was "busily at work on an electrodynamics of moving bodies, which promises to be quite a capital piece of work. I wrote to you that I doubted the correctness of the ideas about relative motion, but my reservations were based on a simple calculational error."[94] His "capital piece of work" would appear four years later. But we have no details about the 1901 version. Presumably

Albert and Mileva discussed these topics when they were together. Mileva may have served as a sounding board for Albert's ideas, much as others would during his Bern period, when Albert and Mileva drifted apart. The merit of the letters is that we now know that Einstein was actively thinking about relative motion and the ether even as a student.

EMPLOYMENT ANGST

Part of the passing-out examination from the Swiss Polytechnic was a *Diplomarbeit* or fourth-year dissertation. Einstein and Mileva chose the topic of heat conduction, which they prepared in Weber's laboratory. Their grades were 4.5 and 4.0, respectively, out of 6 points. Additionally, Einstein's success in the intermediate examination was not repeated on the passing-out. Of the five candidates, he placed fourth with 4.91 out of a possible 6 points. Mileva failed and arranged to retake the exam in 1901, but Einstein remained optimistic about the paid position as an assistant to Weber because there were so few eligible students in his section. The combination, however, of the little respect he had shown Weber and his mediocre diploma essay and exam grades led Weber to appoint two mechanical engineering students instead. This was a blow for the young couple, who hoped to be financially independent enough to marry. Nevertheless, Einstein remained enrolled at the Swiss Polytechnic in order to work on a doctoral thesis in Weber's laboratory. By spring of 1901, however, a final falling-out between them led Einstein to abandon his project and leave.[95]

At home, things never went well for Albert and Mileva. Neither set of parents approved of the relationship. In his letters to her, Einstein referred to the situation as the "Dollie affair."[96] In a letter of 29 July, 1900, he described the outburst from his mother that followed his announced intentions to marry Mileva. Pauline steadfastly opposed her son's marrying someone who she perceived to be from a lower social station, an intellectual, and older as well: "Like you, she's a book—but you ought to have a wife. By the time you're 30 she'll be an old witch."[97] Then came the timeless warning of mothers to sons: "If she gets pregnant you'll really be in a mess."[98] To which Einstein "denied that we had been living in sin."[99] Evidently a letter from Mileva arrived to the effect that Einstein should have handled the situation more tactfully instead of losing his temper and speaking sarcastically to his mother. Einstein agreed, but the disapproval persisted. Mileva came to believe that his parents were opening her letters.[100]

Meanwhile, Einstein accompanied his father on business trips, perhaps entertaining the notion of going into the family electrical concern: "I'd like to learn a little bit about business administration so I can take Papa's place in an emergency."[101] A classmate at the Swiss Polytechnic, Jakob Ehrat, had just recommended Einstein for a position as an assistant "with the insurance office in which he is presently working." Einstein was repulsed at the "mindless drudgery" of such a job.[102] He still counted on the possibility of a scientific position.

But none was forthcoming. Einstein could find only intermittent employment during 1900–1902, when the couple lived almost hand-to-mouth. "Neither of us two has gotten a job and we support ourselves by private lessons—when we can pick up some, which is still very questionable. Is this not a journeyman's or even a gypsy's life? But I believe that we'll remain as cheerful as ever," wrote Mileva to her friend Helene Kaufler.[103] Einstein always suspected that Weber was somehow behind the rejections from academic institutions. He wrote to Grossmann, "I could have found one long ago had it not been for Weber's underhandedness. All the same, I leave no stones unturned and do not give up my sense of humour. God created the donkey and gave him a strong hide."[104]

Einstein continued to apply for a university position as assistant to a major physicist. He wrote to universities in Berlin, Leiden, Vienna and in Italy, as well. No luck—in most cases not even a reply. On 4 April, 1901, he wrote to Mileva, "I will have soon graced all the physicists from the North Sea to the southern tip of Italy with my offer!"[105]

In a particularly heartrending series of letters, on 19 March, 1901, Einstein wrote to the esteemed Wilhelm Ostwald, Professor of Physical Chemistry at the University of Leipzig, whose work Einstein read assiduously and which formed an important part of his first publication. Having received no reply, on 3 April Einstein took the desperate measure of writing again, with the excuse that "I am not sure whether I included my address" in the previous letter.[106] Still no reply.

Einstein must have mentioned this great disappointment to his father because, unknown to him, on 13 April, 1901, Hermann wrote to Ostwald, "Please forgive a father who is so bold as to turn to you, esteemed Herr Professor, in the interest of his son."[107] Hermann summarized his son's career at the Swiss Polytechnic and reiterated Albert's desire to further his education. Despite all setbacks, Albert "clings with great love to his science." No reply has been found nor was Einstein ever aware of this letter. Ironically, in 1909, Ostwald nominated Einstein for the 1910 Nobel Prize and later did it twice more. But in 1901 there was no inkling whatsoever that Einstein would recover from his mediocre performance at the Swiss Polytechnic.

On 4 June, 1901, Einstein wrote to Mileva that he was about to send off to the eminent physicist Paul Drude criticisms on Drude's recent electron theory on which "he will hardly be able to offer a reasonable refutation, as my objections are very straightforward."[108] Drude's reply greatly displeased Einstein. "It is manifest proof of the wretchedness of its author. . . . From now on I'll no longer turn to such people, and will instead attack them mercilessly in the journals, as they deserve. It's no wonder that little by little one becomes a misanthrope."[109] A recently discovered fragment from Einstein's letter to Mileva of 8 July, 1901, indicates that he also criticized Boltzmann's work. Drude's reply, in part, was meant to inform a beginner like Einstein that the great Boltzmann could not be wrong and furthermore was above criticism.[110]

His back against the wall, Einstein declared his "irrevocable decision" that he will "look for a position *immediately*, no matter how modest. My scientific goals and personal vanity will not prevent me from accepting the most subordinate role." The insurance job now did not look quite so bad: "First thing in the morning I'll write old man Besso and go to the director of the local insurance company for additional advice."

Einstein's only steady employment had been as a substitute teacher at Winterhur Technical College from 15 May to 15 July, 1901. This involved some thirty hours a week teaching. Some of the subjects were new to him, "but the valiant Swabian is not afraid."[111] After Winterhur he moved to Schaffhausen to work as a private tutor.

In December 1901 Einstein visited Alfred Kleiner, Professor of Experimental Physics and Director of the Physics Institute at the University of Zurich, who was supposed to have read a doctoral thesis that Einstein had submitted the month before.[112] Einstein wanted to discuss his new ideas with Kleiner. With a mixture of sarcasm and forced gratitude, he reported to Mileva that

> Today I spent the whole afternoon with Kleiner in Zurich and explained my ideas on the electrodynamics of moving bodies to him & otherwise talked to him about all kinds of physical problems. He is not quite as stupid as I'd thought, and, moreover, he is a good guy. He said I may refer to him anytime I need a recommendation. Isn't that nice of him? . . . He advised me to publish my ideas about the

> electromagnetic light theory of moving bodies together with the experimental method.[113]

Despite Kleiner's encouragement, Einstein did not publish on this subject for almost four years. In February of 1902 his thesis was rejected because of Einstein's sharp criticisms, again judged inappropriate, of Boltzmann's theory of gases.[114]

Living in Schaffhausen, near Zurich, Einstein wanted to use the library at the university during Christmas break. Kleiner not only delayed reading Einstein's doctoral thesis but had to be prevailed upon to leave the library open during this time. We know this from a letter of about 17 December, 1901, that Einstein wrote to Mileva from Schaffhausen. He began by describing the almost monkish life he led "in the sense of Schopenhauer's solitude."[115] Then he described the visit that he had arranged with "the interminably slow" Kleiner to talk physics and also

> to convince him to let me work during Christmas vacation. I wonder if I'll succeed. It's really terrible, all the things these old philistines put in the path of people who aren't of their ilk. They instinctively view every intelligent youth as a danger to their fragile dignity, or so it seems to me. But if he dares to reject my dissertation, then I'll publish his rejection along with my paper and make a fool of him. But if he accepts it, then we'll see what good old Herr Drude has to say. A fine bunch, all of them. If Diogenes were alive today, he'd be looking in vain for an *honest* person with his lantern.

Mileva, not surprisingly, detected an increase in her Albert's usual high level of cynicism and criticality. Sometime between late November and mid-December 1901, she wrote to her close friend Helene Savic that "it is not likely that [Albert] will soon get a secure position; you know that my sweetheart has a very spiteful mouth."[116]

Not too many years later a friend recalled similar episodes in which Einstein was treated "rather contemptuously by the professors at the Polytechnic. [He was] locked out of the library, etc."[117] Einstein never forgot these slights and humiliations. Ten years later, as a professor at the German University in Prague, on hearing of Weber's death in May 1912 he wrote to his close friend Heinrich Zangger that "Weber's death is a good thing for the Polytechnic."[118] Even later, at the very pinnacle of success as professor at the University of Berlin, he still felt bitter. When, in 1918, the Swiss Polytechnic made an extraordinary offer to lure him back, he wrote to Besso, "How happy I would have been eighteen years ago with a simple little place as an *Assistent.*"[119]

Yet another problem surfaced. Before beginning his substitute position at Winterhur, at the beginning of May, Albert took Mileva on a short vacation in the Lake Como region. On about 28 May, 1901, he wrote to her from Winterhur. The letter begins as usual with news of what he was currently reading in physics. Then there is the customary abrupt shift. His mother's worry had become a reality: "How are you darling? How's the boy?"[120] Mileva was pregnant and Einstein was delighted. On 17 December, 1901, Einstein wrote of how much he longed to be with her, "even if you do have a 'funny figure'."[121]

Many people would have been driven to the brink of despair by this deluge of disappointments, problems, and new responsibilities, but not Einstein. With persistence and tenacity, he withdrew from the realm of the "merely personal." On 13 December, 1900, he submitted his first paper to the prestigious German physics journal, *Annalen der Physik,* and it was published on 1 March, 1901.[122] He included reprints with his job-inquiry letters. On 30 April, 1902, Einstein submitted a second paper, which appeared later that year. This was also included with inquiries.

The first paper explored capillarity and the second one molecular forces. Einstein took an atomistic approach to these issues that sought connections between chemistry and physics, much as Bernstein had suggested in the popular books he had read as a boy. Despite being excited about his first two publications, by 1907 he thought of them as "my two worthless first papers."[123] Yet in the one on molecular forces a theme emerged that runs like an Ariadne's thread throughout Einstein's research: unification of seemingly disparate realms of nature. In this case Einstein assumed that the mathematical form of the intermolecular force was analogous to the gravitational force. Early in the research, when a connection seemed imminent between phenomena in the small and the large, Einstein expressed his delight to Grossmann: "It is a wonderful feeling to recognize the unity of a complex of phenomena which appear to direct sense observation as totally separate things."[124]

THE SWISS FEDERAL PATENT OFFICE

The year 1902 began well for Einstein. Overjoyed at the news of Lieserl's birth, he expected mother and child to join him in Bern where he was now supporting himself with private mathematics and physics lessons, in addition to a small allowance from his family. Einstein had moved from Schaffhausen to Bern in February in a calculated gamble that he would get a position at the Swiss Federal Patent Office. Marcel Grossmann's father had prevailed upon his good friend Friedrich Haller, the Patent Office's director, to consider his son's unemployed college pal for the next opening. Clearly Grossmann had convinced his father that Einstein really had a future, because Haller was not easily impressed. A graduate of the Swiss Polytechnic in 1872, Haller was a tough Swiss railroad engineer from the earliest days of the system, in which tracks were being laid up and down the Alps and, whenever necessary, tunnels were blasted through them. In 1888 Haller was appointed the first director of the Patent Office, where he served until retirement in 1921. On 14 April, 1901, Einstein wrote to Grossmann, "I am deeply moved by your devotion and compassion which did not let you forget your old luckless friend."[125]

Meanwhile, Einstein's mother continued to rail against Mileva, whose own parents had already accepted the situation. This time Einstein did not rebel. Mileva was left to bear the responsibility of Lieserl on her own in the town of Novi Sad (then in the Austro-Hungarian Empire, now in Serbia). Max Talmud, Einstein's boyhood mentor, happened to pass through Milan and visited the Einstein family, whom he found uncommunicative when it came to Albert. They said little more than that he was living in Bern. Concerned, Talmud went there and found Einstein living in dismal poverty and blaming his sorry situation on others who intentionally set obstacles in his path.[126] Throughout his life Einstein suffered from stomach problems caused by malnutrition during this period.[127]

In spring of 1902 Haller called Einstein for an interview and all went well. On 23 June, 1902, promptly at 8:00 A.M., Einstein reported for work at the Swiss Federal Patent Office as a Technical Expert (Provisional) III Class with a salary of 3,500 Swiss francs per year. Early on in the application process an acquaintance had cautioned Einstein about accepting a position of this sort. But as he wrote to Mileva, "I'm sure I'll like it and will be grateful to Haller for as long as I live."[128] And he was.

Einstein's duties involved assessing patent applications, which included rewriting accepted ones in order to protect the inventor against possible infringements. The job called for knowledge of patent law and the ability to read technical specifications, in addition to engineering and physics. Among the patent applications he inspected were many for electrical dynamos. In a terse refusal of one such patent application he wrote, "This patent application is incorrect, inexact and unclearly written."[129] Einstein

FIGURE 3.11
Einstein at the Patent Office, Bern, about 1905.

was not exaggerating when he recalled that the Patent Office trained him to think logically and with clarity.[130] "More severe than my father, [Haller] taught me to express myself correctly."[131]

Thus commenced Einstein's Bern period, 1902–1909, during which he "was set free from everyday worries to produce [my] best creative work."[132] In those years he published on the order of fifty papers. The four published in 1905 would set the tenor of science in the twentieth century.

LIFE IN BERN

By the fall of 1902 Einstein began to have second thoughts about Mileva's bringing Lieserl to Bern. To have a child out of wedlock in the staid society of northern Switzerland might well offend the Swiss authorities and jeopardize his position at the Patent Office. When Mileva joined Albert in Bern, toward the end of December 1902, she left Lieserl with her parents in Novi Sad. In August 1903 Mileva returned to take the child to Belgrade to be handed over to strangers. Einstein never set eyes on his daughter. In the midst of her sad journey from Budapest, on 27 August, 1903, Mileva wrote in emotional panic to Albert: "It's going quickly, but badly. I'm not feeling well at all. What are you up to Johnnie? Write me soon, okay? Your poor, Dollie."[133] Mileva was pregnant again.

By this time Albert and Mileva were married. The ceremony had taken place 6 January, 1903, at the registry office in Bern. The two witnesses were Conrad Habicht and Maurice Solovine, two close friends of whom we will learn more in a moment. After a small party the couple returned to Einstein's new flat. As usual he had forgotten his key and had to awaken the landlady.

The circumstances surrounding the marriage were not auspicious. Einstein's father gave permission on his deathbed, 10 October, 1902. What followed haunted Einstein for the rest of his life: "When the end came, Hermann asked all of them to leave the room, so he could die on his own. His son never recalls that moment without a sense of guilt."[134] Almost fifty years later, Einstein recalled his "inner resistance" to the marriage.[135]

His lack of enthusiasm was evident in a letter to Besso written two weeks after the wedding: "Well, now I am a married man and am living a very pleasant, cozy life with my wife. She takes excellent care of everything, cooks well, and is always cheerful."[136] Einstein had done the honorable thing.

The face of Mileva that Einstein saw was the one seen only by longtime friends. She had changed, and it seemed that the cause was something for which she held Albert responsible. But she insisted on saying nothing.[137] The hard years after Albert's graduation had been doubly hard for Mileva. In 1901, she flunked her second and last opportunity at the passing-out examination from the Swiss Polytechnic and would never receive a diploma. Her depression at being pregnant and alone surely contributed to this failure. Einstein's view was that he married out of a "sense of duty" and "embarked on something that simply exceeded my strength."[138] Mileva was now completely dependent on her husband, a burden both financially and emotionally. "She was gloomy and moody [and] generally she was very cool and suspicious toward anybody who, in some way or other came close to me," wrote Einstein some years later.[139]

FIGURE 3.12 The founding members of the Olympia Academy, Conrad Habicht, Maurice Solovine and Einstein, Bern, 1902–1903.

THE OLYMPIA ACADEMY

Still, Einstein felt that things were going well. He had a wife to take care of him, three close friends in Bern, a job and lots of ideas about physics. Immediately on arriving in Bern, in order to earn some money, Einstein had advertised his services in a local paper, the *Berner Stadtanzeiger*, as a private tutor in mathematics and physics, with the added bonus of "free trial lessons." Among the first to answer was Maurice Solovine, who had come to the University of Bern to study physics and philosophy. Finding the lectures rather shallow, particularly the ones in physics, he sought private lessons.

Solovine went to the address in Einstein's advertisement with the hope that "perhaps this man could introduce me to the mysteries of theoretical physics."[140] Upon ringing the doorbell to his first-floor flat, the one that Talmud recalled as small and poorly furnished, Solovine "heard a thunderous '*herein,*' and then Einstein appeared. As the door to his apartment gave on to a dark corridor I was struck by the extraordinary brilliance of his huge eyes."[141] Soon afterward the regularly scheduled paid lessons ended, and they began to meet for freewheeling discussions of philosophy.

Einstein and Solovine were soon joined by Conrad Habicht, whom Einstein had met during his brief stay at Schaffhausen. The three men were the charter members of an informal discussion group to which they gave the grandiose name "Olympia Academy." The ribald tone of their correspondence is an indicator of their close friendship. The fourth member of the group was Lucien Chavan, who took private mathematics and electrotechnical lessons from Einstein starting in 1903. Chavan worked for the Federal Postal and Telegraph Administration, which was, until 1907, housed in the same building as the Patent Office.

Besides generally horsing around, the Olympia Academy had a serious "curriculum," probably set up by Einstein because most of it concerned philosophical topics related to physics. They read John Stuart Mill's *System of Logic*, Karl Pearson's *Grammar of Science,* David Hume's *Treatise on Human Nature,* Plato's *Dialogues,* works by Leibnitz, and Henri Poincaré's *La Science et l'hypothèse* in its excellent German translation of 1904.[142] Solovine recalled that Poincaré's book "profoundly impressed us and held us spellbound for weeks on end."[143]

During the heyday of the academy, 1902–1904, they met regularly at one or another's apartment. Prelude to discussion was a frugal meal usually consisting of sausage, Gruyère cheese, fruit and tea.[144] After Einstein's marriage the meetings often took place in the Einsteins' flat. Sometimes Mileva attended the discussions, but she never intervened.[145]

Every so often the conversations took place outdoors or in cafés. Habicht and Solovine sometimes met Einstein after work at their favorite hangout, the Café Bollwerk, near the university. Other times they would walk in the hills around Bern and along its empty streets late at night. No comparison can be made, of course, between the intellectual climates of Bern and Paris. Picasso lived in the cultural capital of Europe, worked among artistic and literary friends who were rising stars, and had patrons of the intellectual quality of Leo and Gertrude Stein.

Einstein, on the other hand, was essentially isolated from the intellectual and scientific worlds. His closest friends were not research scientists, artists or literati. The Olympia Academy's only similarity with *la bande à Picasso* was that Einstein's circle were high-spirited young men who shared a genteel poverty and whose deepest bond was one of ideas. In the same spirit as *la bande à Picasso,* Solovine recalled fondly, "*What a beautiful thing joyous poverty is!*"[146] Although the Olympia Academy's bohemianism paled by comparison, it was no less earnest in its intense give-and-take of ideas. Solovine recalled that typically they "read a page, a half page, sometimes only a phrase and the discussion, when the problem was important, went on for many days. I often went at noon to seek out Einstein as he left his office and go back over the discussion of the previous evening. You said that . . . , but you do not believe that. . . . Or well: What I said last night, I would like to add this. . . ."[147]

For Einstein's birthday, Habicht and Solovine bought expensive caviar. During a monologue by Einstein on a topic in Newtonian physics, they opened the tin and pretended that they were putting slices of sausage on bread, as usual. Einstein was oblivious and just continued to eat what was on the platter without missing a beat. When informed of what he was eating, Einstein said, "It's all the same to me."[148] Good conversation was all that mattered.

No doubt with the dream of being surrounded on a daily basis with friends to discuss physics and philosophy, Einstein suggested to Habicht that

> If an opportunity arises, I will mention you to Haller; perhaps it would be possible to smuggle you in among the patent slaves, which for all that, you probably would find relatively pleasant. Would you actually be willing to come? Keep in mind that besides the eight hours of work, each day also has eight hours for mischief, and then there's Sunday. I would love to have you here.[149]

During those eight hours of "mischief," and on Sunday, Einstein pursued his ideas on radiation theory, atomic theory of gases and relativity.

Almost half a century later, Einstein recalled to Solovine that "it was indeed a wonderful time in Bern" and "our Academy was far less childish than those respectable ones that I later got to know."[150] This group revolved about its central sun, Einstein, and became his think tank. Especially so when the Academy's fifth member, Michele Besso, arrived in Bern in 1904 and Einstein smoothed the way for a position at the Patent Office. Einstein would immortalize him with an acknowledgment in the relativity paper.

FIGURE 3.13
Michele Besso and his bride, Anna Winteler, 1898.

MICHELE BESSO

Michele Besso was born 25 May, 1873, in a town near Zurich. Although his father was from Trieste, the family had moved to Zurich so that he could take up a post as director of an insurance company.[151] Besso was a precocious child and read by age five. He was intensely interested in science and particularly in mathematics. At nine he wanted to understand what is meant by a point at infinity. Besso shared with his future friend Einstein the ability to exasperate teachers. Unhappy and disappointed with how mathematics was taught at his Gymnasium, Besso and a friend circulated a strongly worded petition. For their forthrightness the boys were expelled, whereupon Besso's parents sent him to Rome to complete his secondary studies. After a year at the University of Rome, in 1891 he entered the Swiss Polytechnic to pursue engineering studies, and was graduated in 1895.

Sometime in late 1896 or early 1897, at a musical soirée, Besso met Einstein.[152] They found that in addition to a shared interest in the violin, they had in common the "same preoccupations, the same tastes, the same thirst for knowledge."[153] Einstein recalled how stimulating Besso was for him in those days when they often ran across each other at musical evenings.[154]

In 1898 Besso married Anna Winteler, one of Mama and Papa Winteler's daughters, and Einstein became a frequent guest at their home in Milan. Besso arranged for some consulting jobs for Einstein with the company he worked for, the Development of Electrical Enterprises in Italy.[155] From their voluminous correspondence, we know that Besso's knowledge of physics and philosophy was deep and wide. Yet he could never apply himself toward a scientific career, preferring instead to read widely. Besso was the paradigmatic noncompleter. He preferred to be a perpetual student, forever auditing classes at the University of Bern. During the winter semester of 1909, for example, he audited courses in political legislation, the politics of banking, the physiology of the central nervous system, English literature and celestial mechanics.

Einstein lovingly referred to him as "an awful schlemiel[156] [who is] unable to pull himself together enough to do anything in his life or in his studies, but he has an extraordinarily keen mind, the disorderly workings of which I observe with great enjoyment."[157] He described to Mileva one of Besso's vintage "schlemiel" stunts, related to him by Besso's boss, with whom Einstein played music. Besso was asked to inspect some newly installed power lines at the Casale power station southwest of Milan in the Piedmont region. "Our hero," reported Einstein, thought he could get a jump on things by leaving the evening before, but managed to miss the train. Being somewhat absentminded, the next day Besso remembered his assignment too late. Finally, on the third day he caught a train. But at the station in Milan he realized that he had forgotten his assignment. So, on the spot, Besso wrote a card requesting that "'instructions should be wired!!' I don't think this fellow is normal."

But Besso was normal enough to talk physics with Einstein. "Last night we talked shop eagerly for almost four hours. We discussed the fundamental separation of luminiferous ether and matter, the definition of absolute rest, molecular forces, surface phenomena, dissociation."[158] Everything! Einstein realized that Besso, with his critical mind, was just the person he needed to bounce ideas off. "A better sounding board I could not find in all of Europe."[159] They became inseparable, as did their families. To Einstein, "The circle of [Besso's] interests seemed truly without limit."[160] He encouraged Besso to come to Bern and work in the Patent Office, which Besso did on 15 March, 1904.[161] Exactly a year later, the Einsteins moved to a residence nearby the Bessos and the two men began their fifteen-minute walk home from work together. "Our conversations on the road home were of an incomparable charm—it was as if everyday contingencies did not exist at all."[162]

Besso left the Patent Office in 1908 and focused on self-study while holding a number of odd jobs, which included a stint as a consultant engineer. In 1916 he was appointed *Privatdozent* at the Swiss Polytechnic in patent law. He returned to the Patent Office in 1920 and retired in 1938. Einstein intervened on his behalf when, in 1926, the new director tried to dismiss him. Einstein addressed his letter of support to Zangger, who had informed him of Besso's dire situation.[163] It is a marvelous assessment of his dear friend's abilities.

The problem was Besso's inefficiency in writing up patent appraisals. Against this weakness in bringing tasks to completion, Einstein balanced Besso's extraordinary sphere of knowledge. Besso, he wrote, was "a talent of the first order" whom "everyone at the Patent Office knows" they can count on for rapid advice. Firing him would be a serious mistake because it would deprive the Patent Office of someone possessing extraordinary qualities. Einstein suggested a solution that amounted to a new job description. Besso should be the one to "objectively clarify the cases—and someone else should be assigned to write up the official dossier." The intervention succeeded.

This episode is testimony also to Haller's patience and understanding of the young men who worked for him. After all, one of them spent a significant part of his day dreaming about the nature of space and time, while another was the instant expert on everything but who had difficulty writing up anything. Besso and Einstein always spoke of Haller in tones bordering on veneration for his fairness, as well as for the methods and taste for work he had imparted to them.[164]

The Olympia Academy met until about 1906, when Solovine moved to Paris and Habicht was only intermittently in Bern. On 27 April, 1906, Einstein suggested to Solovine that if things were not going well in Paris "then there still is some possibility that you might find some work, and in due course even a permanent appointment here in the patent office."[165] Since the breakup of the Olympia Academy Einstein was feeling rather isolated: "As for my social life, I have not been meeting with anyone since you left. Even the conversations with Besso have now come to an end; I haven't heard anything whatever from Habicht."[166] Having moved that month, he no longer lived near Besso, which ended their daily walks home.

ACADEMIC ASPIRATIONS

Through a coworker, Josef Sauter, Einstein began to make contacts in the scientific world of Bern starting in 1902. Although Sauter was also a graduate of the Swiss Polytechnic, they had never met owing to Sauter's being eight years older. But they had experiences in common, including Sauter's discouragement with the Swiss Polytechnic's curriculum, which resulted in his also doing independent reading. Sauter introduced Einstein into the Natural Science Society, Bern's version of the great scientific societies of Berlin and Paris. Through conversations with members, Einstein began to think again about a university position.

In the German academic system, one started out at the bottom as a *Privatdozent*, a position that carried no remuneration from the university, but a small fee paid by the *Dozent*'s students. The prerequisites for such a position were a doctorate and a *Habilitationschrift*, or original piece of work beyond a doctoral thesis. Einstein had neither. But he found a loophole in the rules: In "exceptional cases" both the doctorate and *Habilitationschrift* could be replaced by other "outstanding achievements."[167]

This escape clause was meant for established scholars, not mere tyros, but Einstein believed he had proof of outstanding achievements in his—in retrospect—"two worthless beginner's works."[168] In January 1903 he advised Besso, then in Milan, of what he was up to: "I have recently decided to join the ranks of *Privatdozenten*, assuming, of course, that I can carry through with it. On the other hand, I will not go for a doctorate, because it would be of little help to me, and the whole comedy has become somewhat boring."[169] Needless to say, the University of Bern rejected Einstein's request. He took it badly. "The university here is a pigsty. I will not lecture there because it would be a shame to waste the time."[170] The comedy continued.

FIRST SCIENTIFIC DISCOVERY

On 15 April, 1904, Einstein wrote Habicht a few lines of friendly banter: "Do come at once, you have a bike, after all! We are expecting a baby in a few weeks. I have now found the relationship between the magnitude of the elementary quanta of matter and the wavelengths of radiation in an exceedingly simple way."[171] In his usual manner, he slides from the hope of a visit to the impending birth of Hans Albert to the abrupt announcement of his first scientific discovery. The *Annalen der Physik* received his paper on 29 March, 1904.[172] It would be Einstein's fifth publication and the third in a series of papers in which he explored the foundations of the atomic theory of gases and of thermodynamics.[173] Einstein arrived at the Swiss Polytechnic already an advocate of atomism, based on his childhood readings of Bernstein's books and then his detailed study of Boltzmann.[174] While he appreciated Boltzmann's rigorous mathematical development of the atomic theory of gases, Einstein's own style was more speculative, linking different fields much as he had first seen in Bernstein's popular series.[175]

Einstein found an exceedingly simple way to study fluctuation phenomena—the behavior of a gas before it reaches equilibrium—that could be used as well to calculate Avogadro's number, which is the number of molecules per mole of matter, and through this the dimensions of an atom.[176] There was more: Einstein demonstrated that one system in particular exhibited measurable fluctuations, namely the radiation inside a cavity dug out of a heated metal object. This sort of radiation was especially interesting because its characteristics are independent of the emitter. There is something universal about it. In 1900, the doyen of German physics, Max Planck, at the University of Berlin, deduced an equation for its brightness that agreed with available experimental data. Further investigations into this radiation law's atomic roots led Planck to the surprising result that the energy of cavity radiation was divided up into discrete packages, or energy quanta, that were not arbitrary but integral multiples of a basic unit. This violated radiation theory as it was understood through Lorentz's electromagnetic theory, because in the customary representation of light as a wave, its energy is expected to be distributed continuously.

Most physicists politely ignored Planck's interpretation of his radiation law. At best, they tried to rederive it without using energy quanta. Einstein also tried this and failed. He was devastated: "It was as if the ground had been pulled out from under one, with no firm foundation to be seen anywhere, upon which one could have built."[177] But he was alone in this conclusion: Every major physicist regarded cavity radiation as a sideshow. By consensus, in 1904, the fundamental problem was understanding the electron's structure.

Here, for the first time, emerged Einstein's unique ability to know when to stop asking for a derivation of certain statements. He decided to take the enormous step of accepting Planck's radiation law as axiomatic—that is, beyond experimental or theoretical proof—in order to ascertain "what general conclusions can be drawn from [it]."[178] It was from cavity radiation that Einstein deduced the inverse relationship he described to Habicht on 14 April, 1904. This result was exciting because it successfully unified the invisible microcosm of fluctuating radiation with its observable spectrum. Einstein planned to study the mysteries of cavity radiation further. He would drive it to a "very revolutionary" result, as he wrote to Habicht, in the magic year of 1905.[179]

A young physicist who visited the Patent Office recalled Einstein's pointing to his top desk drawer and saying that it was his department of theoretical physics.[180] We can imagine Einstein sitting at his desk in the Patent Office with sunlight streaming through the windows. Every now and then he looks around to see if Herr Haller is observing him. If not, then Einstein carefully slips open the drawer and sneaks a look at some of his own calculations. As in his years of near starvation, Einstein's mind soared beyond this earthly world, beyond the "merely personal," into one where the secret of the compass needle lay, a world structured according to the axioms of the holy geometry booklet, a world where thought experimenters chased light waves. The Patent Office was his "secular cloister," the nearest he ever came to Heaven on Earth.[181]

4

How Picasso Discovered *Les Demoiselles d'Avignon*

Les Demoiselles d'Avignon . . . was my first exorcism painting.

—Pablo Picasso

The scientific, mathematical and technological roots of Picasso's *Les Demoiselles d'Avignon* are a neglected subject in Picasso studies.[1] My interest is not so much *how* Picasso formulated the *Demoiselles* the way he did, but *why*. This *why* includes Picasso's quest to produce a work of art that would measure up to the magnificent achievements of science. To understand this we will delve into the artist's mind using available sources.

While this is no easy task with any artist, it is especially difficult with Picasso. Not only did he leave no reflective correspondence during the two periods, or campaigns, in which he began to conceive of the project that led to *Les Demoiselles d'Avignon*—the end of 1906—and of his actual work on it—March through July 1907,[2] we have to contend with his recollections, which often contradict history as well as each other.

Because many art historians have resisted considering any scientific dimensions of Picasso's thinking, it is necessary to discuss the sources I will use. The first question, of course, is why Picasso himself denied any scientific roots to the *Demoiselles*.

SOURCES

Picasso's contradictory statements about the *Demoiselles'* origin and that of cubism generally are well known. For example, he claimed to have been unaware of *art nègre*, or African art, before completing the *Demoiselles*, whereas we know this was not true.[3] The reason for this denial is clear. He was angered by published comments to the effect that so complex a painting could be simplistically interpreted as emerging from a single source.[4] This "led to remarks on his part downplaying (and later denying) the role of 'art nègre'—to the point that from World War II onward, Picasso consistently claimed not only to have seen no tribal art prior to painting the *Demoiselles,* but that his celebrated visit to the Trocadéro had followed rather than preceded the painting's execution."[5] In fact, as we will see, African art gave Picasso the inspiration to continue along the conceptual path he had already embarked on after his May 1906 discovery of Iberian statuettes.

Picasso's answers to Alfred H. Barr in October 1945 must be seen in the same light.[6] Barr writes: "Picasso when asked if he had discussed mathematics or the fourth dimension with [Maurice] Princet replied that he had not (questionnaire, October 1945). Concerning the somewhat mysterious Princet, Picasso said that he was an actuary."[7] The "mysterious" Princet was actually a frequent visitor to the Bateau Lavoir. Just as with African art, however, Picasso did not want to grant primacy to any single origin for the *Demoiselles,* and Princet's name had been used in two articles that intentionally distorted the history of cubism.

The first was by the art critic Louis Vauxcelles. In a sarcastic backward look at the history of cubism, Vauxcelles claimed for Princet the honor of having been the "inventor of Cubism."[8] The other, more malicious article was by Picasso's friend from the Bateau Lavoir days, Vlaminck. This piece, which appeared in *Comoedia* on 6 June, 1942, appears to have been a Nazi-motivated attack.[9] Vlaminck accused Picasso of "having dragged French painting into a most mortal impasse, into an indescribable confusion, impotence, death. . . . Picasso was the crowd pleaser of [cubism], Guillaume Apollinaire the mid-wife, Princet the godfather."[10]

Even when he was not fending off specific attacks, Picasso could be a difficult interview. His responses were most disappointing when interviewers asked him to define cubism or inquired into how his work is going. When a journalist asked him about cubism in 1911, for instance,

Picasso replied: "Il n'y a pas de Cubisme [There is no cubism]," and then left to feed his monkey El Mono.[11] He said to Florent Fels, in an interview in 1920 regarding African art, "Negro art? Never heard of it."[12] This particular revisionist statement reflects Picasso's disgust at the incredible rise in price of African art after his supposed reliance on it became a widely circulated story.[13]

Regarding the question, How is your work going?, Fernande Olivier remarked that Picasso gave the one reply that can be expected of any creative person: "Awful."[14] And that would be that.

We do not know under what circumstances the interviewer for *The Arts* in 1923 asked Picasso about the effect of science on his work. Picasso answered that "mathematics, trigonometry, chemistry, psychoanalysis, music and whatnot have been related to Cubism to give it an easier interpretation. All this has been pure literature, not to say nonsense, which brought bad results, blinding people with theories."[15] I suspect that the question was put to Picasso too directly. In his younger days he would have pulled out Jarry's pistol and shot the interviewer.

In 1926, Picasso sent an unsolicited letter to the Soviet paper *Ogoniok* in which he singled out the mathematician Maurice "Princet [who] used to be present at our discussions on aesthetics."[16] He later denied taking any part in this signed letter.[17] Yet as we shall see, Princet played an important role in communicating science to Picasso and his circle.

Examples multiply of Picasso's inconsistencies and attempts to denigrate his search for a new means of representation, despite the documentable highly intellectual conversations among *la bande à Picasso*.[18] Many of Picasso's pronouncements are in the style of Alfred Jarry, who relished every opportunity to mislead critics and interviewers: any seriousness was to be countered "with non sequitur, absurdity, *blague*."[19]

An extremely useful interview is one with Christian Zervos in 1935, and published as "Pablo Picasso, Conversation, 1935."[20] Picasso met Zervos sometime in the second decade of the twentieth century when Zervos was a student in Paris. They remained friends until Zervos's death in 1970. An indicator of their close friendship is Zervos's lifelong labor of love, the first catalogue raisonné of Picasso's works, which occupies a monumental thirty-three volumes. Zervos claimed to have written the interview notes immediately after their conversation at Picasso's home in Boisgeloup in 1935. Picasso read the notes and approved them informally.[21] What emerges is Picasso's thoughtful commentary on photography, the nature of art and his own creativity. Sitting with a close friend, under no pressure to reply to journalistic-type questions, Picasso felt no need to make outrageous Jarryesque comments.[22]

What about the testimony of others in *la bande à Picasso*? Art historians have divided up the firsthand witnesses into two camps: Picasso's first principal dealer, Daniel-Henry Kahnweiler, and Apollinaire/Salmon.[23] During the First World War, from exile in Switzerland, Kahnweiler wrote a Germanic treatise entitled *The Rise of Cubism* in which he clothed cubism in a heavy coat of homespun Kantian philosophy.[24] Kahnweiler's scholarly technique and his proximity to Picasso during the early days of cubism have led art historians to prefer his testimony to that of Apollinaire/Salmon, with their flights of metaphoric fancy.[25] But Kahnweiler was never part of *la bande à Picasso*, being too conservative in attitude, repartee and lifestyle: He was neither an artist nor a writer and certainly not a Bohemian.

Art historians tend to distrust the writings of Apollinaire and Salmon for their errors of fact and lack of proper emphasis on Braque's contribution. For example, in his 1912 essay "L'Histoire anecdotique du Cubisme [An Anecdotal History of Cubism]," Salmon writes of six *Demoiselles*, when there are five, and mentions a vacation taken by Picasso during work on the *Demoiselles*, whereas Picasso insisted to Daix that there was none.[26] Picasso did, however, leave Paris during the summer of 1908 for La Rue-des-Bois. So, Daix argues, if we assume that Salmon's article is about Picasso's *Trois femmes* [*Three women*][27] rather than the *Demoiselles*, then the account is correct.[28] But Daix then turns around and writes that Picasso may have continued to retouch the *Demoiselles* into 1908. If so, then maybe Salmon was speaking about the *Demoiselles* after all. Regarding Braque, in a letter to Braque of 31 October, 1912, Picasso wrote that[29] "Salmon's description of the genesis of cubism is revoltingly unjust as far as you are concerned."[30] But he did not mention any errors as far as *he* was concerned.

What these quibbles overlook is that Apollinaire and especially Salmon were present in Picasso's atelier during his immense struggle with *Les Demoiselles d'Avignon*. Evidence is in Picasso's caricatures of Salmon during June and July 1907.[31] Salmon writes in 1912: "Let those who were inclined to consider the cubists merely as audacious farceurs or some canny merchants, deign to take a complete look at the real drama that occurred at the birth of this art."[32] That is the point of his 1912 article.

Apollinaire was a source of inspiration to Picasso and was sensitive to changes in his moods. Despite Picasso's annoyance at Apollinaire's florid language and shabby treatment of Braque, among the poets he was the artist's closest friend. Early on, Apollinaire shared a "corner of the napkin" during his many dinners at Picasso's atelier.[33] Regarding questions of creativity I will therefore take seriously Apollinaire's *Les Peintres Cubistes,* published in 1913,[34] as well as Salmon's "Anecdotal History of Cubism."

This is, in short, how I plan to deal with Picasso's responses and which of his friends I will focus on. That Picasso never made any direct comments on the role of science, mathematics and technology in his struggles with the *Demoiselles* can be attributed to the fact that no interviewer or friend, such as Zervos, asked him about it in a manner likely to elicit a thoughtful response. Picasso's recorded comments regarding Princet, as well as his disagreement with the manner in which science was used to interpret cubism, must be seen within the context and circumstances in which he was queried. In his many comments on his career, Picasso never emphasized any one source, or even any core of circumstances, as critical to his conception of the *Demoiselles* as the painting stood in mid-1907.

LES DEMOISELLES D'AVIGNON

As it hangs in the Museum of Modern Art today, the painting represents five prostitutes in a bordello. Although in close proximity, they do not interact with each other, only with the viewer—the client.

As we look across the expansive canvas—244 centimeters in height by 233.7 centimeters in width—we see a partially clothed demoiselle on the far left with an Egyptian-Gauguinesque face, whose seemingly disembodied arm is pulling open a curtain; then there are two more attractive demoiselles of Iberian-Oceanic likeness, the second one on the left in an impossible standing position (note the location of her left foot, directly below her right knee—she should topple over). The standing demoiselle on the far right is also parting a curtain, while the squatting demoiselle is in a grotesquely impossible posture, with her back facing the picture plane and her head turned 180 degrees as if on a swivel, eyes distinctly different and off line, nose almost like a *quart de Brie* (wedge of Brie) and a face that is shockingly hideous in comparison to the others. This painting contains no narrative of the traditional sort, rather the representation is iconic. The head of the squatting whore, the most advanced in geometrization and experimental representation, underwent the most extensive metamorphosis

FIGURE 4.1 Pablo Picasso, *Les Demoiselles d'Avignon*, 1907. Paris, summer 1907.

in Picasso's working drawings.[35] It is the key to Picasso's discovery of the geometrization that became the hallmark of cubism.

In the "Anecdotal History of Cubism," Salmon tells us that "Picasso's new painting was spontaneously baptised 'the philosophical brothel' by a friend of the artist." One is supposed to infer that the friend was Salmon himself.[36] We know that in 1910 it was still untitled because it appears in an article by Gelette Burgess with the caption "Study by Picasso."[37] The unsigned and untitled painting finally received its permanent title in 1916, when it was exhibited at Salon d'Antin, part of a dress shop converted into a gallery, for an exhibit entitled *L'Art Moderne en France*. The exhibition was organized by Salmon and held during 16–31 July. Salmon dubbed it *Les Demoiselles d'Avignon*.[38]

SOME RECEIVED INTERPRETATIONS OF THE *DEMOISELLES*

Most art historians agree that Picasso's choice of a bordello was his response to treatments of this theme by Cézanne, Lautrec, El Greco and Ingres. Picasso's earliest attempt at the theme was the oil painting done in 1906 while at Gósol, *The Harem*.[39] Of course, if he wanted to shock the public in a Jarryesque vein, a bordello scene was the way to do it.

Whereas brothel photographs from any period are meant to titillate, the *Demoiselles* do not. Picasso tried to depict the most jaded view of sex that he possibly could, presenting it merely as commerce. These are hard-core whores, detached and slightly bored. They are strictly business and even somewhat threatening. One threat was syphilis, which was fatal in Picasso's day and thought of somewhat as we think of AIDS today. Newspapers reported on its horrible consequences. Picasso had, in fact, firsthand knowledge of the ravages of venereal disease from his visits to Saint Lazare prison sometime during 1901–1902.[40] Saint Lazare housed prostitutes, many of them with venereal disease, some pregnant with the distinct possibility of giving birth to children with congenital syphilis, which brought horrible facial disfigurations.[41] Venereal disease has to be taken into account as one tile in the mosaic of Picasso's thinking toward the *Demoiselles.* Yet until Leo Steinberg's 1972 article "The Philosophical Brothel," the sexual theme of Picasso's *Les Demoiselles d'Avignon* went unexplored. This is rather extraordinary given that Picasso saw himself as a creative dynamo fueled by an insatiable sex drive.[42]

The art historian William Rubin interprets *Les Demoiselles d'Avignon* as an exercise in Freudian self-analysis in which Picasso exorcises his own psychological demons. These include tensions from his deteriorating relationship with Fernande in conjunction with fear of death and venereal disease. Rubin assumes that Picasso's ravenous desire for women is in conflict with a loathing of their bodies. The African masks supposedly provided Picasso with inspiration for an iconic representation that spans the polarities between Eros and Thanatos—beauty and ugliness, sex and death.[43]

Interesting as they are, Freudian conjectures are highly adventurous inferences drawn from Picasso's dramatic life. They can be generalized to explain too much. Other Montmartre artists had emotional problems akin to Picasso's, but they never produced works of his calibre.

To some degree African art entered Picasso's life through a theme emphasized by Richardson, Picasso's competitiveness.[44] At the Salon des Indépendants, which opened 20 March, 1907, Picasso was astonished at Matisse's *Blue Nude (Memories of Biskra)* and Derain's *Bathers.* All of the notoriety gained by Matisse and Derain, in addition to the obvious quality and inventiveness of their work, meant to Picasso that they had moved ahead of him. While their paintings at the Cage aux Fauves at the 1905 Salon des Indépendants had already caused a scandal, the ones of 1906 provoked an uproar on account of their primitiveness and barbarity. In the intervening year, both men had found in primitivism a means to move from fauvism.[45] Matisse and Derain, who had their own collections of African objets d'art, alerted Picasso to African art in Spring 1906. A year older than Picasso, Derain felt more camaraderie with him than with the older and austere Matisse. The two men met through Apollinaire, and by the end of 1906, Derain left Matisse's circle and began to focus on Picasso's more conceptual methods. Derain's expertise spanned philosophy, science, mathematics, music and art history. We can imagine him filling in gaps in Picasso's "education" left by Apollinaire and Jacob.[46]

Since 1901, through his visits to the Louvre and elsewhere, Picasso had already been exposed to both Egyptian and primitive art. All of this prepared him for his Louvre visit in May 1906 to see the exhibition of Iberian reliefs from Osuna. By March 1907 Picasso possessed his own Iberian sculpture in the form of two stone heads purchased from Géry Pieret, a friend of Apollinaire.[47]

In early June of 1907 Derain suggested that Picasso visit the Musée Ethnologie du Trocadéro to see the African exhibit. Picasso returned several times and realized that something profoundly disturbing had happened to him: "I understood why I was a painter."[48] In later years he used terms such as "shock" and "revelation" for what he saw.[49]

When Picasso visited Trocadéro he was at an impasse and had ceased work on the *Demoiselles.* The African masks offered a way out. Unlike his earlier viewings of them in friends' ateliers, where they were displayed as ornaments with no apparent relation to any of his own work in progress, they now meant something. Still, although the sharp change in the right-hand demoiselles occurred after Picasso's visit to Trocadéro, we will see that the situation was more complex than a mere response to African art.[50] It turns out that African art supported his conceptual approach and convinced him of the deep meaning of geometry as the language of the new art.

The art historian Ron Johnson has gone beyond the effect of various artists on Picasso to explore his cultural milieu. He has looked at the influence of Nietzsche's view of women as slaves or tyrants incapable of friendship; and of Nietzsche's Will to Power, which states life's goal as creating art in an ambience of sexual fulfillment. Johnson has also probed Jarry's influence: his notions of stark juxtaposition of opposites, and the use of masks in theatrical settings, in which the *Demoiselles* can be seen as puppetlike figures that are progressively dehumanized.[51] Having opened the field to cultural concerns, Johnson's work led to examinations of the political and social dimension of the avant-garde and its effects on Picasso.[52]

Yet none of the various interpretations of this great turning point in twentieth-century art includes anything about the role played by science, mathematics or technology. Can this dimension permit us a further glimpse into the painting's overall composition and trend toward geometrization? To address this position we must explore the conditions in which Picasso thought and worked.

ALONENESS AND ANXIETY

Apollinaire knew Picasso since the Blue Period, and in their almost daily meetings, he witnessed and encouraged his transformation of style into the Rose Period. But during 1906–1907 a much greater change occurred, about which Apollinaire's comments offer invaluable insight.

There are, he wrote, two kinds of artists and poets. Although the first sort produce prodigiously, they do so like "poetic or artistic instruments"[53] because they never struggle very much. In stark contrast stand those who "must draw everything from within themselves, for no demon, no muse inspires them. They live in solitude. . . . Picasso was the first type of artist. There has never been a spectacle as fantastic as the metamorphosis that he underwent while becoming an artist like the latter."[54]

Unfortunately, Apollinaire wrote nothing more about Picasso's "metamorphosis." His only eyewitness comment from the period of Picasso's struggles with the *Demoiselles* is a diary entry from 27 February, 1907, when he had dinner at the Bateau Lavoir: "Evening, had dinner with Picasso, saw his new painting: even colours, flesh pinks, flowers, etc. . . . Women's heads, all the same and simple, and men's heads too. A wonderful language that no literature can express because our words are defined beforehand. Alas."[55] Words failed a master of language, even for the preparatory sketches.

Picasso was in the habit of inviting friends over to view his works in progress.[56] In this case their comments were not encouraging. For example, Vollard and the important art critic Félix Fénéon "left without understanding a thing," recalled the dealer Wilhelm Uhde in 1938.[57] The intense ridicule would come later, after the painting's completion. We have the scenario of Picasso working in incredible loneliness: very close friends offer support; others are speechless; and some are downright puzzled. The consensus is not promising. To add to this he is living with a woman who has no understanding of what he is doing or what he is up against. The Steins were sensitive enough to the space and emotional problems Picasso faced that sometime toward the end of 1906 or beginning of 1907 they rented another studio for him under the main one.[58] In this subsidiary studio Picasso could lock himself away from everyone, *la bande à Picasso* and Fernande included. This superprivate studio was also a venue for affairs.[59]

The "solitude" and independence that Apollinaire commented on in 1913 were also noted by Kahnweiler, who wrote—probably more with the reception of the finished painting in mind—"I shall always admire Picasso's solitary enterprise. He did it alone, terribly alone. He really had to be the admirable genius that he is to stick to what he was doing. And indeed, he stuck to it."[60] Derain, Kahnweiler reports, spoke in awe of the "terrifying spiritual solitude" that Picasso experienced while working on the *Demoiselles*.[61] It was at this time that Picasso realized, as he said forcefully some years later: "The important thing is to create. Nothing else matters; creation is all."[62]

For the first time in his life Picasso not only experienced aloneness but also "knew anxiety [*l'inquiétude*]"[63] because he was on the threshold of something new, something that he desired with all his heart, something absolutely groundbreaking. But how to proceed? He decided to focus entirely on searching for a new mode of representation. He "turned his paintings to the wall and threw down his brushes."[64] On the financial level Picasso passed up lucrative sales by ceasing to do any more Rose Period studies. He put just about everything on the line.

Picasso's anxiety is akin to that experienced by research scientists on the cutting edge of their field, where the problem they have posed for themselves may have no solution at all. One proceeds by stepping into the void with one's very career perhaps at stake. Research at this level carries a high likelihood of no payoff whatsoever.

Another possible rendering of the French *l'inquiétude* is "restless striving"; from time to time I will use this meaning when the term "anxiety" is not appropriate.

CÉZANNE

Before Picasso's decision in the summer of 1906 to focus on studying new forms, he had already seen examples of the master of Aix's art and could not have failed to be impressed by Cézanne's concept of *passage*.[65]

Cézanne's influence is clear in the robust females that inhabit early sketches for the *Demoiselles.* By the end of the first campaign—March to about the end of May 1907—they have been transformed to an Iberian likeness. It is almost as if Picasso had read Cézanne's letter of 1905 to the artist-critic Emile Bernard: "The Louvre is the book in which we learn to read. We must not, however, be satisfied with retaining the beautiful formulas of our illustrious predecessors."[66] Whereas Picasso began the *Demoiselles* with inspiration from Cézanne, El Greco and Ingres, among others, their modes of representation eventually disappeared.

But Cézanne himself remained an inspiration for Picasso in his most trying creative period and then through the *Demoiselles'* less than enthusiastic reception. Cézanne's solitary nature, the long periods he spent studying a scene and then painting it, shifting his viewpoint a little bit each day, had become legendary among young artists.[67] In his 1908 catalog for Braque's exposition at Kahnweiler's gallery, Apollinaire wrote of "the solitary and determined toil of Picasso."[68] Picasso lamented the "unbelievable solitude" required of someone trying to produce a new style.[69] "You think you aren't alone," he continued, "and really you're more alone than you were before." Picasso attempted to clarify what he meant by anxiety to Zervos in 1935: "It's not what an artist does that counts, but what he is. Cézanne's restless striving [*l'inquiétude*] is what interests us. That is his lesson."[70] This is a source of Picasso's use of the term "research" for the lengthy series of preparatory sketches that usually preceded his paintings. "Paintings are but research and experiment. I never do a painting as a work of art. All of them are researches. I search constantly and there is a logical sequence in all this research."[71] Picasso spoke in this vein to André Warnod in 1945: "The painter's studio should be a laboratory. There, one does not make art in the manner of a monkey, one invents. Painting is a play of the mind [*l'esprit*]."[72]

Fernande, who both admired and feared Picasso's deep creative moods, wrote: "[Picasso] has always needed to make discoveries through painting. Like Max Jacob or Guillaume Apollinaire, he always felt an overpowering need to work, a need to awaken his mind and to learn to use it as he wants to."[73]

The technical dimension of Cézanne's painting was also important. Here was an artist capable of producing a still life with four different perspective points depending on how you view it, or none at all at other viewing points.[74] Cézanne was a master of "technique and originality."[75] The scientific dimension of his art did not escape Picasso.

At the 1906 Salon des Indépendants Picasso was struck by how Matisse and Derain, in their own search for new forms, had moved away from the fauve style. In addition to primitivism, these men reacted directly to the influence of Cézanne and produced a style that was conceptual and sculptural. The exhibit shocked Picasso, piquing his competitiveness, leading him to concentrate even more on experimentation with new forms and adding to his *l'inquiétude.* Cézanne's influence on Picasso thus ran the gamut from the *how* to the *why.*

WORK HABITS

Picasso drew inspiration from the silence and calmness of late night. For night work he lit his atelier at the Bateau Lavoir with a large petrol lamp, and used a handheld candle to study details. When he went out to dinner, the circus or movies he often returned by ten o'clock and worked until about five or six in the morning. He then slept until late afternoon.

He kept this schedule, that is, until he began working on *Les Demoiselles d'Avignon.* From mid-1906 through 1907, Fernande tells us, Picasso started to go out less and "his friends got used to coming regularly to see him at the times he wasn't working." Otherwise, no matter how insistent they were, the door remained shut.[76]

The couple tried to maintain their weekly social calendar. Mondays they went to Jacob's for a "meeting of conspirators . . . against the established order in all the arts."[77] Tuesdays they walked across Paris to the literary salon at the Closerie des Lilas in Montparnasse. Wednesdays they visited Apollinaire's apartment. On Fridays, *la bande à Picasso* usually went to the cinema, and Saturday evening was generally reserved for dinner at the Steins.

His manner of painting was intensely physical. As Sabartès recalled in 1901:

> I generally found him in the middle of his studio, not far from the stove, seated on a dilapidated chair, perhaps lower than the ordinary chair, because discomfort does not bother him, and he seems even to prefer it as if he delighted in self-mortification and enjoyed subjecting his spirit to tortures so long as they spur him on. The canvas was placed on the lowest part of the easel, and this compelled him to paint in an almost leaning position. . . . He surrenders body and soul to the activity which is his *raison d'être*.[78]

Picasso's studio was an unbelievable mess. But this was of no consequence: He focused only on the work at hand.

In 1906–1907 Fernande was in the midst of an affair, a response to Picasso's increasing coolness toward her. She believed that to keep a lover you have to keep him jealous.[79] Picasso, as always, was seeing other women, often in his private studio on the floor below. Fernande was angrily certain that somewhere in the *Demoiselles* was her image as a whore, a feeling that Picasso's "jokingly" telling friends that Fernande "was one of the girls in his brothel" did not help dispel.[80] Then there was the episode with the little girl Raymonde, whom they adopted in April 1907.

Fernande had had a miscarriage in 1901 that left her unable to bear children; Picasso was guilty over having forced his previous mistress, Madeline, to abort. Frustrated at Picasso's single-minded absorption in his work, Fernande insisted they adopt a child. They convinced themselves that somehow Raymonde would ease their various guilts while restoring stability to their relationship. Raymonde was adopted as one would acquire a pet. Although members of *la bande à Picasso* doted over her, she began to wear on Picasso. Having always been sexually excited by young girls and perhaps seeing in Raymonde something of his sister Conchita, who had died of diphtheria in 1894 at age eight, he found her presence difficult to cope with. Fernande could not fail to miss the sexual part of this problem, which came to a head with her discovery of several of Picasso's explicit drawings of Raymonde. Raymonde became *de trop*. To avoid a traumatic disaster, Fernande returned her to the orphanage. She said nothing at all about her in her memoirs.

On 24 August, 1907, Fernande wrote to Gertrude Stein of "being tired and often depressed."[81] By that time the couple had decided to split up.

Picasso was waiting to receive money from Vollard who, in February of 1907, bought out Picasso's studio for 2,500 francs. Vollard finally appeared on 14 September with the first payment of 1,400 francs. Picasso gave Fernande half and she promptly moved into her own flat on rue Caulaincourt.[82] They continued to attend the Steins' dinner parties together. In late November Fernande returned to the Bateau Lavoir and they coexisted in some semblance of peace until 1912, when Picasso left her for Eva Gouel, "Ma jolie."

While Picasso was at work on the *Demoiselles*, the antics of *la bande à Picasso* continued, if not with the customary regularity then with the usual intensity. Although work, sex and tobacco remained Picasso's principal interests, drug taking was another, and one that he long concealed. In a conversation on 14 March, 1953, with Jean Cocteau,[83] Picasso revealed that from the summer of 1904, when he moved into the Bateau Lavoir, until 1908, he and his friends smoked opium two or three times a week with occasional sessions using ether, morphine and hashish.[84] In this way they spent many dreamy evenings "in an atmosphere of heightened intelligence and subtlety," recalls Fernande with a touch of cynicism.[85] Although the effect of drugs on Picasso's painting cannot be discounted, the high intellectuality of the *Demoiselles* precludes overemphasis.[86] Picasso's addictions remained work, sex and tobacco.

CINEMA, LITERATURE, MUSIC AND THEATER

In 1906 Picasso's taste in classical theater was not at all sophisticated, due to his inadequate understanding of spoken French. To please Jacob, he permitted himself to be dragged to classical French theater, sometimes with disastrous consequences rooted in Picasso's horseplay. They were once expelled for eating sausages during a performance.[87] Classical music moved him not at all: He was insensitive to musical subtlety and preferred Spanish gypsy songs accompanied by guitar and castanets.[88]

Picasso fancied the circus and the cinema. As early as 1904 he had been attending movies, particularly with Jacob, at the cinema on rue de Douai where they saw the latest in cowboy films and films by Georges Méliès.[89] Within a year this became a weekly activity: Each Friday, *la bande à Picasso* eagerly trooped over to rue de Douai to see the latest motion picture.[90] Especially at the time of his painting of *Les Demoiselles d'Avignon,* the "cinema was Picasso's other passion."[91] With his flair for self-dramatization, Picasso developed the menacing look of silent screen actors that can be seen in contemporary self-photographic portraits. Fernande began to look like a screen vamp.

With an entry price ranging from around one franc, down to thirty centimes, the cinema was accessible to just about everyone. Besides Picasso's interest in anything new visually, something else attracted *la bande à Picasso*. Cinemas were almost exclusively located in Montmartre and were considered marginal environments: When one went to the movies one literally went slumming. This snubbing of bourgeois society definitely appealed to Picasso and his friends.[92]

What did Picasso read?[93] We know that he read newspapers, an indication that, whatever his difficulties speaking or writing French, he had more than a little facility with reading it. Picasso's small library at the Bateau Lavoir ran the gamut from dime-store westerns about Buffalo Bill and Nick Carter detective stories to books by Verlaine, Rimbaud and Mallarmé. Raynal has written of Picasso's attraction to eighteenth-century French literature.[94] In a letter of 22 February, 1905, to an old Spanish friend from Barcelona, Jacint Reventós, Picasso prides himself in his knowledge of French literature: "Tell me if you know of Rabelais (*Gargantua*), *Gargantua*, you know of it perhaps in Spanish, but what's the difference. *La Bruyère*, and all those that we call here classics. One of these days I will send you a book by Pascal that you perhaps do not know about."[95] Kahnweiler wrote of Picasso's feel for French poetry, even "when he spoke little French,"[96] and he was under the tutelage first of Jacob and then Apollinaire. The implication is that by around 1906 Picasso could have begun to read classics in French.[97]

Whether Picasso could actually read Bergson, Poincaré, Jarry, Rimbaud or Mallarmé is beside the point. He was surrounded by a circle of poets, writers and literary fantasists who kept him informed of the latest in the avant-garde. Poets of the magnitude of Jacob and Apollinaire wrote him letters. Would they have done so if Picasso could not read them? By 1906 Jacob felt comfortable using nontrivial vocabulary.[98]

Cubism was an intense intellectual effort for all concerned. Picasso, seeking to free himself from previous modes of thought, looked deeply into what the avant-garde offered. Any hint at all from science, mathematics and technology could only have been of value. Like Einstein, he was an intellectual opportunist.

Earlier we discussed X rays and the way they were described in newspapers and literary journals as a means to "see" beyond the visible. Another not unconnected aspect of cutting-edge science was non-Euclidean geometry and the fourth dimension.

POINCARÉ AND NON-EUCLIDEAN GEOMETRY

The philosophical impact of non-Euclidean geometries was enormous because they lent support to the relativity of knowledge and therefore constituted an attack on positivism.[99] Time-honored Euclidean geometry had now to share the stage with a host of other geometries. French literary journals became the locus of a vigorous debate on the foundations of geometry. An issue of concern was why Euclidean geometry is preferred when intellectually so many geometries can exist. The positivists dismissed this issue because experiment always came out in favor of Euclidean geometry. But philosophers more prone to speculation were unsatisfied with this resolution. Among them was Henri Poincaré.

Poincaré spun a philosophical system that could explain how we can intellectually discover any sort of geometry in any number of dimensions, and yet three-dimensional Euclidean geometry turns out to be the most "convenient" for our daily activities.[100] He discussed this view, known as "conventionalism," in his immensely successful book *La Science et l'hypothèse.* The sheer sweep of topics is breathtaking, transporting even the educated lay person in crystal-clear prose to the very frontier of mathematics, science and philosophy. Poincaré emphasized, as well, the free constructive element in scientific thinking. Once the proper set of experimental data has been agreed on, of all the possible paths of generalization from these data, the scientist is guided by "simplicity,"[101] a value difficult to quantify or reduce to sense perceptions. Poincaré's underlying message is that the positivist view of science is a mere caricature.

Picasso had heard about X rays and the occult from friends who read of these topics in sources such as *Mercure de France.* In France, however, there were no popular science magazines where one could learn in any detail about something like non-Euclidean geometries or the fourth dimension.[102] Only someone au courant with details could deliver the real thing. For Picasso, this person was Maurice Princet, who became known as "*le mathématicien du Cubisme.*"[103]

LE MATHÉMATICIEN DU CUBISME

Six years older than Picasso, in 1906 Princet passed his examination as an insurance actuary, the line of work that would be his career.[104] He also read widely about the new geometries, including Poincaré's masterful exposition of this topic in *La Science et l'hypothèse,* on which Princet based the informal lectures he delivered to *la bande à Picasso.* While Jarry supplied inspiration and Bergson offered flowing and vague prose, it was Poincaré, via Princet, who provided the details on what the new non-Euclidean geometries meant and a hint as to what the fourth dimension is all about.

Another source for someone like Princet was Esprit Jouffret's 1903 *Traité élémentaire de géométrie à quatre dimensions* [Elementary Treatise on Four-Dimensional Geometry]. Jouffret reviewed the literature on the fourth dimension, including discussions by Poincaré, and supplied impressive geometrical illustrations with a high degree of faceting such as would later appear in Picasso and Braque's paintings.[105] Jean Metzinger, to whose circle of cubists Princet gravitated by 1909, recalled that Princet "conceived of mathematics like an artist and evoked continua of *n* dimensions as an aesthetician. He liked to interest painters in the new views of space . . . and he succeeded in doing this."[106]

How did someone like Princet, who was neither artist nor writer, enter the Bateau Lavoir circle? Among the women with whom Picasso was involved in spring 1905 was the twenty-one-year-old Alice Géry, who since adolescence had been Princet's mistress. Alice was notoriously unfaithful, however, preferring hot-blooded Spanish artists. Since 1901 she had been a fixture in the Spanish community that lived in the Bateau Lavoir.[107] Alice introduced Princet to Picasso in 1905, and he soon became a member *en marge*[108] of *la bande à Picasso.* Leo Stein recalled "a friend of the Montmartre crowd, interested in mathematics, who talked about infinities and fourth dimensions."[109] This was almost certainly Princet, whom Picasso evidently brought along to some of the Steins' Saturday night soirées.

Princet also took part in conversations at Le Lapin Agile, where he is remembered as having professorial airs, intelligence and charm, yet with a turn of mind that could turn caustic and diabolical—just the sort of person who was welcome in the Bateau Lavoir circle. Francis Carco recalled of Princet, "We respected him because he very much earned his own living and, always well dressed, made at Frédé's table the sort of figure of a washed out gentleman, mocking and melancholic."[110] Princet usually sat at the corner of a bistro table and, with notebook in hand, held forth on "some elementary principles of geometry in space."[111] He claimed to be happy only with artists and poets, who treated him "as much 'companion of painter's bistros' as 'sagacious mathematician'."[112]

Princet's regular presence in the Bateau Lavoir is recorded in Carnet 8, dated May-June 1907, where the artist scribbled Princet's name three times over a multiplication table on one of the carnet's covers.[113] There are several instances of Princet's joining the group for opium sessions, among others of Picasso's "closest friends," writes Picasso's biographer John Richardson.[114] Fernande recalled an opium session with Apollinaire, Jacob and Picasso that commenced with hashish pills taken with dinner at the bistro Azon. A small group adjourned to Princet's nearby flat where "Princet simply wept about his wife, who had just left him."[115]

This dates the session as just after September 1907. Much to Picasso's disgust, the previous March Princet had finally convinced Alice to marry him. Although Picasso served as a witness, he is reported as saying, "Why should they marry simply in order to divorce."[116] Picasso prophesied well. The Princets moved to the outskirts of Paris, but Alice missed Montmartre life. Soon after she returned to a flat near the Bateau Lavoir, Picasso invited her to lunch with the intent of fixing her up with Derain. It was love at first sight, and by September Alice had dumped Princet. Although furious with Picasso, Princet remained in *la bande à Picasso,* at least for the time being.

Salmon devoted one of his columns in the *Paris Journal* to Princet:

> We are promised for this publication season, a curious book on aesthetics.
>
> The author, Maurice Princet, is a mathematician for whom the efforts of modern painters have inspired curious reflections, of which it is too early to say whether they are judicious or condemnable.
>
> M. Maurice Princet preoccupies himself especially with painters who disdain ancient perspective. He praises them for no longer trusting the illusionary optics of not long ago and takes them for great geometers. . . . In brief, for the next winter, long discussions of perspective (it is the case to call it that) offered to painters and to art lovers.[117]

Salmon's column on Princet for a major newspaper demonstrates the high regard in which he was held by a member of the inner circle of *la bande à Picasso*.[118] Alas, it has proven impossible to locate Princet's book. For all I know, he never completed it.

Salmon again mentions Princet in laudatory terms in his 1919 essay "Les Origins et Intentions du Cubisme [The Origins and Intentions of Cubism]" where, among prominent frequenters of Picasso's atelier, he names "a mathematician M. Maurice Princet who alone knows his true part in what we call improperly the invention of Cubism."[119] What did Salmon mean by "Princet who alone knows his true part"? Why did not Picasso and others say more about him? I have speculated on this point, but maybe there is more to the story. What is clear is that Princet did play a substantial role in Picasso's work.

Salmon continues in this essay by giving an even larger role to Princet. He mentions the forceful effect on Picasso of his visit to the Trocadéro and the ensuing "passionate examinations and discussions in the old Montmartre atelier from which emerged Cubism."[120] This give-and-take involved artists such as Derain and Vlaminck, as well as the poets who "provided nothing more than a *plastique* [or abstract turn] . . . to the language that is necessary in order to comprehend in the face of these novelties, and the mysterious mathematician furnished his friends with reasoned preciseness."[121] So, according to Salmon, Princet made contributions at the critical juncture in June 1907. In the end, of course, "there was only Picasso."[122] In his memoirs published in 1955, Salmon recalled that the grand old days at the Bateau Lavoir included the company of "Maurice Princet the actuary who would become the legendary *mathématicien du Cubisme*."[123]

Taken together, these various descriptions tell a story in which Princet was just what Picasso needed during the second campaign in June through July 1907, when the demoiselles on the right became geometrized. As Salmon recalled in 1912, Picasso "meditated on geometry."[124] Salmon continued in this scientific vein: the demoiselles "are naked problems, white numbers on the blackboard. It is the principle posed of painting = equation. . . . Painting, henceforth, became a science and not one of the less austere."[125]

In summary, Princet brought to the Bateau Lavoir, in addition to Henri Poincaré's writings, a rigor of thought about non-Euclidean geometries that was clearly welcomed. I mention here *cum grano salis* the testimony of another of Picasso's close Catalan buddies, the painter and sculptor Manuel Manolo, who recalled in 1910 that "Picasso used to talk a lot then about the fourth dimension and he carried around the mathematics books of Henri Poincaré."[126] While I can well believe the first half of this statement, the second half seems incredible.

THE FOURTH DIMENSION AND HENRI POINCARÉ

Whereas mathematicians hotly debated the philosophical ramifications of non-Euclidean geometry and the violation of Euclid's fifth postulate, higher-dimensional geometries offered no such intellectual attraction. At the turn of the century these two geometries were generally kept separate. In popular culture, the geometry of four dimensions was all the rage owing to its connection with transcendentalism or spiritualism. So it was reasonable, for example, that the fourth dimension could not be seen by denizens of the three-dimensional world. Some well-known scientists agreed.[127]

Another major attraction to speculations on the fourth dimension was the theosophical notion of "astral sight" on the "astral plane." Astral vision meant seeing the true and absolute representation of an object as it occupies the astral plane, which extends out to infinity. Once situated on the astral plane, one sees all sides of an object at once, so that it will seem almost unrecognizable. Coincidentally, there is no perspective point as there is with the inferior physical viewing.[128] Discussions such as these appeared in the French popular literature by 1903.[129]

The fourth dimension was also the stuff of early science fiction. In his 1895 tale *The Time Machine,* H. G. Wells referred to a fourth dimension as not one of space but of time. Wells's book was serialized in *Mercure de France* in December 1898 and January 1899. Jarry responded immediately to Wells's story by describing nothing less than "How to Construct a Time Machine," published in *Mercure de France*'s February 1899 issue.[130]

For Jarry this was a serious article—relative, that is, to the adventures of Dr. Faustrall, which he had just completed. Jarry's time machine is a gyroscopic device operated by the time traveler while sitting on a bicycle frame. Gyroscopes are required for the time machine to remain motionless in a complex mechanical ether while the rider observes time passing by.[131] This would be much like an observer of one of Picasso's paintings standing in one place while watching many different representations of an object unfold in time.[132] No doubt through Apollinaire and Salmon, Picasso heard about Jarry's story, as well as the relation between the fourth dimension and the theosophist's astral plane, a topic close to Jacob's heart. But it is reasonable to say that Picasso was most intrigued by Poincaré's analysis of the four-dimensional world. Princet explained these topics in detail to Picasso.

Poincaré proposed to actually view the fourth dimension, raising his discussion to the realm of possibility and so separating him from positivists whose distrust of speculative thinking demanded tests of every proposed hypothesis, and occultists who deem its viewing to be either via X rays or by a medium. Poincaré suggests how "we may picture a world of four dimensions" as follows:

> The images of external objects are painted on the retina, which is a plane of two dimensions; these are *perspectives*. But as eye and objects are movable, we see in succession different perspectives of the same body taken from different points of view. . . . Well, in the same way that we draw the perspective of a three-dimensional figure on a canvas of three (or two) dimensions, so we can draw that of a four-dimensional figure from several different points of view. This is only a game for the geometer. Imagine that the different perspectives of one and the same object succeed one another.[133]

Just as a scene on a two-dimensional plane can be a projection from three dimensions, so an image on a three-dimensional surface can be interpreted as a projection from the fourth dimension. Poincaré's fourth dimension is a spatial dimension, and he suggests portraying it as different perspectives in succession on a canvas. This was a mistake. Picasso, with his visual genius, saw that the different perspectives should be shown in spatial simultaneity. Thus emerged the *Demoiselles*.

Picasso's idea of spatial simultaneity goes beyond Bergson's, which is typified in the art of Cézanne. Cézanne sets down on canvas all at once (simultaneously) the total of the views of a scene that have been stored in his unconscious over a long period. Picasso transcends, as well, the concept of time in impressionist art, such as that in Claude Monet's series of paintings of haystacks or of Rouen Cathedral, which are series of static representations. Picasso's spatial simultaneity is more radical because it is the simultaneous representation of *entirely different* viewpoints, the sum total of which constitute the object. The squatting demoiselle, represented simultaneously in full frontal and profile perspectives, is considered by Picasso as a projection from the fourth dimension. It is as if Picasso imagined himself sitting on the "astral plane."

Picasso's epiphany at Trocadéro provided the insight into *why* he had to proceed as he did with the two rightmost demoiselles and *why* the abstract language of geometry was necessary. At this point Princet's lectures fell into place. If Picasso had not appreciated the importance of the African statuettes he had seen in the studios of Derain and Matisse, now he was in need of such an idea. The multiple planes and edges of the masks were the perfect way to geometrize the multiplicity of views. Having gleaned the crucial insight from Poincaré's instructions on how to view the fourth dimension, Picasso transformed and elaborated his vision in dramatic fashion.

WHAT PICASSO'S CARNETS TELL US

The literally hundreds of preparatory drawings and paintings associated with the *Demoiselles* are "unique not only in Picasso's career, but without parallel, for a single picture, in the entire history of art."[134] There are sixteen carnets from the period autumn 1906 through spring 1908. Imposing order on them was no easy task.[135] In a way they are a diary of Picasso's life during this time. His dog Frika appears, as do Raymonde and, in related sketches, Salmon. There is also a shopping list in Carnet 8 that supports Fernande's recollection of Picasso's forbidding her to leave the house.[136] The sketches resemble a scientist's notebooks with false starts and dead ends. Such archival material provides important insight into the creative thinking of an artist or scientist. These carnets are research on the frontiers of knowledge.

The squatting demoiselle is the most challenging motif to interpret because it underwent the most far-reaching transformations. Candidates in past paintings for models of Picasso's crouching nude appear in Cézanne's *Three Bathers* (1879–1882) and *The Temptation of Saint Anthony* (1869–1870). They are the "standard" female representations of their day. The question is why Picasso transformed the squatting demoiselle into someone comprised almost exclusively of triangles, squares and circles, si-

FIGURE 4.2 *Composition study with Seven Figures for Les Demoiselles d'Avignon.* Carnet 2, p. 32R, winter 1906–1907.

multaneously in full face and profile. His carnets enable us to substantiate, elaborate and add to the conjectures offered so far.

The squatter appears in the very first of Picasso's sketches for the *Demoiselles,* in Carnet 2, which dates from the winter of 1906–1907.[137] She is a Cézannesque sort of woman with her back to the viewer and depicted as making a direct attempt to attract the sailor—seated in the middle of the sketch—with an outright show of her sex.[138] Her head is turned to the left in a reasonably comfortable manner so as to profile her face. She wears her hair in a ponytail. In a sketch in Carnet 3, March 1907, Picasso turned her face completely away from the viewer (figure 4.3). In this carnet she makes the transition from a bulky Cézannesque woman to an Iberian form resembling the Fernande-type women in *The Harem.*

Sometime during April or May 1907, Picasso, in Carnet 5, embarked on his most extreme geometrical experiments.[139] Page 1V, for example, is a remarkably innovative study of proportions in which the female is faceted into interlocking diamonds with kneecaps of circles (figure 4.4). Surely

FIGURE 4.3 *Composition study with Seven Figures for Les Demoiselles d'Avignon.* Carnet 3, p. 8V, March 1907.

Salmon, in his "Histoire anecdotique du Cubisme," had such drawings in mind when he wrote that Picasso had "meditated on geometry."[140]

We can imagine Salmon's amazement at Picasso's boldness. The only known source from whom Picasso could have learned geometry was Princet. Might Picasso have taken the idea of faceting from Jouffret's projections of four-dimensional solids onto a plane?

Jouffret's figures, with their extreme faceting, are generated by an analytical method for "rotating" complex polyhedra so as to obtain different views, or perspectives, of their four-dimensional structure seen as projections onto a plane (figures 4.5 and 4.6).[141] This sort of technique is commonplace with today's computer software.

Another surprise is on page 3V of Carnet 5. Here we find a female head with a nose that is a *quart de Brie* (wedge of Brie) and eyes off level (figure 4.7). There is no hatching on the nose and so the face appears rather benign. Picasso's experiment in geometrization is connected with the one

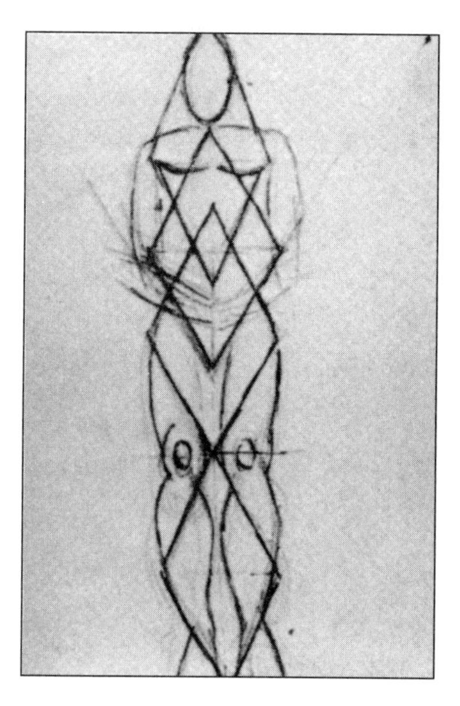

FIGURE 4.4
Standing Nude with Joined Hands (Study of Proportions). Carnet 5, p. 1V, April–May 1907.

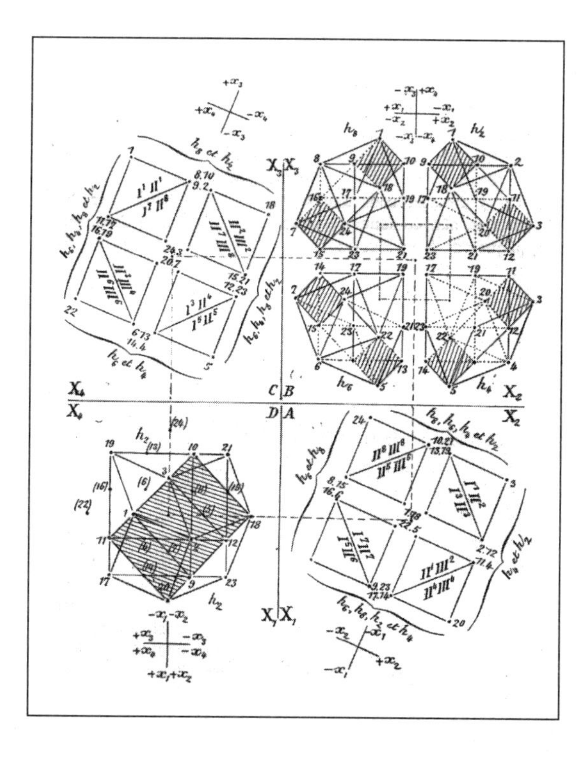

FIGURE 4.5
Projections on a plane of the sixteen fundamental octahedrons that constitute the four-dimensional ikosatetrahedroid, which is made up of twenty-four octahedrons (Jouffret, 1903, p. 152).

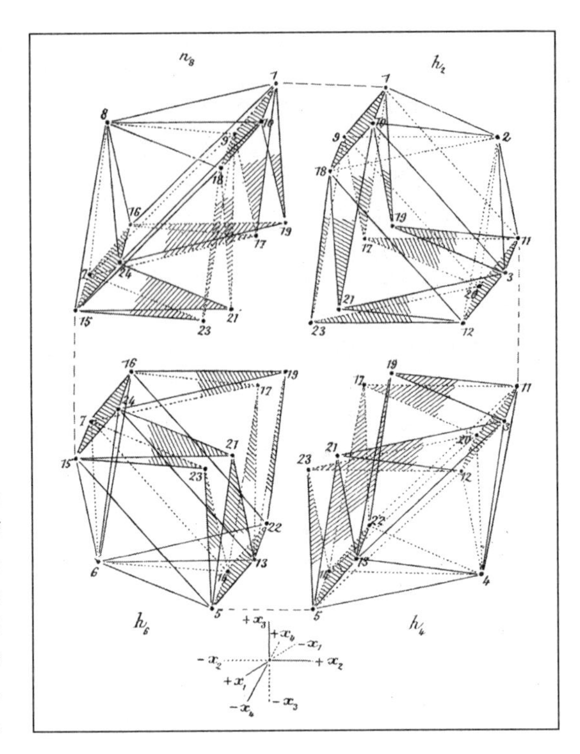

FIGURE 4.6
Projections on a plane of views of the sixteen fundamental octahedrons as the four-dimensional ikosatetrahedroid is rotated. From Jouffret, (1903), p. 153.

on page 1V (see Figure 4.4), but less extreme. This situation will change after he visits the museum at Trocadéro.

Salmon recounts how "during long days, and so many nights, [Picasso] drew, concretising the abstract and reducing the concrete to essentials. Never labour was less paid of joy, and it is without the juvenile enthusiasm of not long ago that Picasso began a large canvas which became the first application of his researches."[142] *La bande à Picasso* were no longer meeting regularly because their leader was entirely absorbed with his new work. "Juvenile enthusiasm" yielded to deadly serious solitude in which Picasso had to reach deep within himself for inspiration. He was bent on creating something entirely new on the "large canvas."

From the winter of 1906–1907 through May of 1907, Picasso transformed the bordello scene into one with five demoiselles. This is the state of the *Demoiselles* at the end of May when it was in its Iberian phase. It was at this point that Picasso "turned [it] against the wall."[143] This must have been a difficult time. Not only was he stuck, he had to stand by while Matisse and Derain were acclaimed the leaders of the avant-garde.[144] He had to do something. Trocadéro provided the clue.

It is in Carnet 8 (May–June 1907) that Picasso moves toward a final version of the *Demoiselles.* This carnet signals the beginning of the "sec-

FIGURE 4.7
Study for Woman with Joined Hands: Head of woman. Carnet 5, p. 3V, April-May 1907.

ond campaign," and Picasso wrote on its cover, "Je suis le cahier. [I am the notebook.]" Under this he writes: "belonging to Monsieur Picasso, painter. 13 rue Ravignan, Paris XVIIIème."

Picasso intercalated into Carnet 8 informational pages from the *Crédit Minier et Industriel* on which he drew sketches that tend toward a highly geometrized primitivism.[145] From approximately halfway in there is a distinct succession of geometrized figures that have clear African inspiration. This carnet demonstrates, as well, a bifurcation in Picasso's work. On the one hand there is a trend toward the more sophisticated, less violent African formalism that would result in *Nu à la draperie*.[146] On the other hand there is a savage and violent trend whose hatchings would be included in the *Demoiselles*.

Before Salmon's eyes, Picasso was trying to formulate a new way of representing reality. Following the clue from Iberian art, it was conceptual rather than perceptual representation. But the real shock had been Picasso's epiphany at the Trocadéro. "Monsieur Picasso, painter" discovered how to push the large canvas to completion. Before the Trocadéro visit Picasso had begun to experiment with geometry, among other avenues. After Trocadéro it became clear to him that geometry was the language in which he would express the conceptual message of primitivism while simultaneously realizing a new mode of artistic representation that could

FIGURE 4.8 *Head of Josep Fontdevila.* Carnet 13, p. 9R, end of June–beginning of July 1907.

FIGURE 4.9 *Study for the Squatting Demoiselle from Back to Right.* Carnet 13, p. 11R, end of June–beginning of July 1907.

take its place with the great achievements of avant-garde science and technology. The formerly informal language of art became formalized in Picasso's hands. We must not forget here, as well, Jarry's advice to abandon the formal painting styles typified by "Bougrereau" and "make geometry" instead.[147] All of this, as Salmon wrote in 1912, was part of "the real drama that occurred at the birth of this art."[148]

The carnets after number 8 focus almost exclusively on geometric studies. An added ingredient from the Trocadéro visit is a violent turn to the geometrizations through the hatchings that appear on faces and bodies. Picasso returns to the face of the old peasant from Gósol, Josep Fontdevila, increasingly geometrizing it. Fontdevila had become Picasso's "benchmark of primitive transformations that has not ceased to accompany him since that first transformation at Gósol of his face into a mask."[149] His face asks to be geometrized. The transformations in Carnet 13, along with further studies in charcoal and oil, will result in the four-dimensional representation of the crouching whore's face. We notice the gradual geometrization from Carnet 13, where Fontdevila's nose verges on becoming a *quart de Brie.*

In Figure 4.9 Picasso flips Fontdevila's nose in Figure 4.8 by 180 degrees over its vertical axis, simultaneously changing from an Iberian

FIGURE 4.10 *Head of the Squatting Demoiselle.* Spring 1907.

FIGURE 4.11 Pablo Picasso, *Squatting Demoiselle (Study for Les Demoiselles d'Avignon)*. Paris, spring 1907.

structure to a flattened one. Coincidentally Fontdevila's Iberian-shaped eyes pancake.

In Figure 4.10 a rudimentary form of hatching appears on the nose. Then, in Figure 4.11, Picasso flips the *quart de Brie* nose again by 180 degrees, while taking the eyes off level; the figure has also become asexual.

The face in Figure 4.11 is almost feminized and ready to be transferred to the painting itself, where it will undergo yet further changes, more than on any of the other demoiselles.[150] In the end it will be a projection onto the plane of the canvas from the fourth dimension, as Picasso understood this term in 1907 through Poincaré's writings. The faces of the demoiselles, especially the squatter, are almost feminine, tending toward asexual, in order for Picasso to accentuate asymmetries and movement (see Figure 4.1).[151]

Salmon describes Picasso's geometrical method in his second campaign:

> He created for it by a dynamic decomposition of light values; an effort leaving far behind attempts of neo-Impressionism and Divisionism. The geometric signs—of a geometry at the same time infinitesimal and kinetic—appeared as the principal element of a painting of which nothing, from then on, could any more stop its development.[152]

It is unclear what Salmon meant by "infinitesimal and kinetic," but "geometric signs" referred to Picasso's desire to represent realism transformed into geometry. The transformation, of course, would not approximate completion until the analytic cubism of Picasso and Braque.

An example of what Salmon meant by "geometric signs" present in the *Demoiselles* are the "noses . . . situated on the face in the form of isosceles triangles."[153] Salmon means the noses of the two demoiselles on the right, particularly the squatter's, which being rounded is not exactly an isosceles triangle but is certainly the most geometrized nose of the quintet (see Figure 4.1). The squatter's nose *en quart de Brie* that appears in Figure 4.10 did not make it into the *Demoiselles* because Picasso found it too diverse in style from the reduced noses of the Iberian demoiselles in the center.[154] The nose on the curtain parter is in a transitional stage from an Iberian one. Yet this demoiselle's breasts are represented by well-delineated four-sided figures, not unlike the faceting in Figure 1V of Carnet 5 (see Figure 4.4).

Geometrical signs are everywhere in the painting: the breasts of the other three demoiselles are "constructed" from triangles and half circles. The most geometrically delineated bodies belong to the demoiselles on the right-hand side, in keeping with the major theme of gradually reducing nature to geometry. Like her snoutlike mouth, the eyes of the squatter are without any human connection, being reduced to two almost oval shapes with dots for retinas.

The increased tempo of geometrization starting from Carnet 8 resulted from Picasso's finally realizing the depths in Princet's lectures on geometry, which probably intensified at the time. This may be why Picasso inscribed Princet's name on one of Carnet 8's covers.[155]

To give us some inkling as to what Princet was talking about, here is a recollection of the painter and art critic André Lhote, who was present at bistro gatherings among Picasso's circle. Although Lhote's recollection is most likely from 1909 or 1910, nevertheless the problem Princet posed was one Picasso was wrestling with in June 1907.

> You represent by means of a trapezoid a table, just as you see it, distorted by perspective, but what would happen if you decided to express the table as a type [*la table type*]? You would have to straighten it up onto the picture plane, and from the trapezoid return to a true rectangle. If that table is covered with objects equally distorted by perspective, the same straightening up process would have to take place with each of them. Thus the oval of a glass would become a

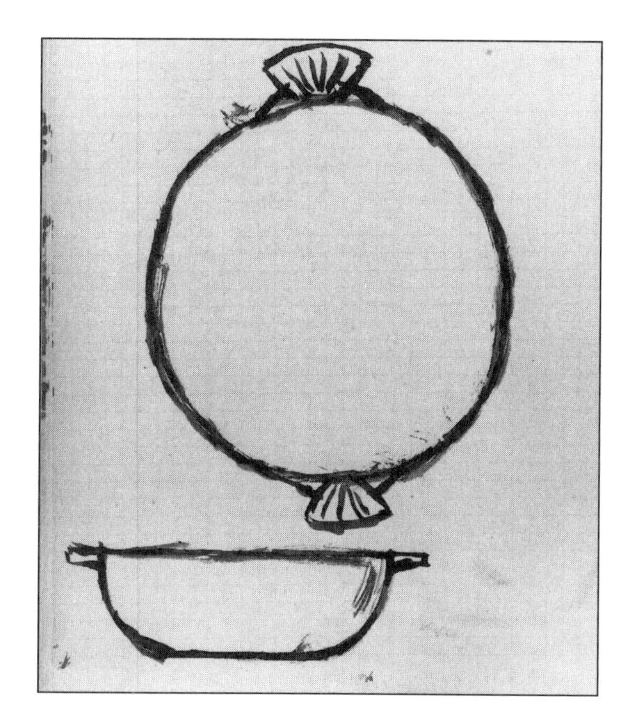

FIGURE 4.12
Pablo Picasso, *Plate and Bowl. Black ink.* Carnet 8, p. 30V.

> perfect circle. But this is not all: this glass and this table seen from another angle are nothing more than, the table a horizontal bar a few centimetres thick, the glass a profile whose base and rim are horizontal, whence the necessity of another displacement.[156]

Granted that perspective distorts, how can the artist represent an object from differing viewpoints all at once, giving equal validity to each one?

As if in response to conversations with Princet on this issue, on page 30V of Carnet 8 Picasso sketched a plate and bowl in such a way that the plate is an overhead view of the bowl seen sideways (Figure 4.12).[157] How to put them together? In Paris's highly experimental visual culture of this era, others took on this sort of problem with a direct approach, as shown in an illustration from the popular journal *Le Rire*, a play on multiple-exposure photography at which Picasso was already adept, as we will see in Chapter 5 (Figure 4.13).

Princet's conversations on Poincaré's notion of the fourth dimension could well have jogged Picasso's memory of cartoons such as the one in Figure 4.13. The next step was the move toward simultaneity of images rather than succession.

FIGURE 4.13
A caricature of a double exposure from *Le Rire,* 1901.

Salmon wrote of Picasso's strategy in the second campaign:

> Science, is it not the only director of these researchers, anxious to make us endure all the edges of the prism at once, merging touch and sight which are the factors of delights so different? To that no one has yet been able to respond in an authoritative manner. [I am] only proposing to prove that some artists, unjustly burdened, followed the dictates of inescapable laws for which anonymous genius carries responsibility.[158]

Here Salmon is struggling to understand the key role that science played in Picasso's thought, a point to which "no one has yet been able to respond in an authoritative manner." It seemed as if Picasso had tapped into "inescapable laws" of nature. It would not be an injustice to replace Salmon's "inescapable laws" with "invariant" or "unchanging laws of nature," as scientists began to put it soon after Hermann Minkowski's four-dimensional representation of Einstein's special relativity in 1907.[159]

FIGURE 4.14 Edmond Fortier, *Types of Women West Africa, 1906.*

Two last points must be taken into account. Both are spin-offs from the vibrant visual culture of Paris, which was deeply involved with such technological developments as photography and the cinema.

THE PARISIAN VISUAL CULTURE

Anne Baldassari of the Musée Picasso has recently brought to light Picasso's use of picture postcards of African women taken by Edmond Fortier, which were particularly popular at that time.[160] There are forty of them in Picasso's archives and they all date from 1906. Baldassari hypothesizes that "each of the figures of *Les Demoiselles d'Avignon* incorporates—via the many preparatory studies—formal elements derived from one or several of Fortier's postcards."[161] The resemblance, shown in Figures 4.14 and 4.15, between one of Fortier's postcards and Picasso's six-figure sketch of May 1907 is too close to be coincidental.[162]

Another passage of Salmon's "Anecdotal History" gives us a clue to further input from the Parisian visual culture. Salmon writes of Picasso's attempts to go even beyond the "barbarian imagers" and their already far-reaching conceptual representation. Picasso had to find the means to represent the three-dimensional conceptual quality of African art "on a

FIGURE 4.15 Pablo Picasso, *Study for Les Demoiselles d'Avignon.* May 1907.

surface." In order to accomplish this Picasso found that he "must create, in his turn, by situating these equilibrated figures which are beyond the laws of the academy and of the anatomical system, in a space conforming rigorously to the unexpected liberty of movements."[163]

The disembodied left arm of the curtain raiser at the left, and the floating demoiselle next to her, reflect Picasso's attempt to create a space "beyond the laws of the academy and of the anatomical system," giving rise to "an unexpected liberty of movements."[164] We are reminded here of the filmmaker Georges Méliès, whose most famous special effect was the fragmentation and reassembling of human bodies in sometimes weird and hilarious ways, and whose movies Picasso saw at the cinema on rue de Douai.

On another experimental track, there were the explorations into motion by Étienne-Jules Marey and Eadward Muybridge. Marey studied a sequence of events on a single frame, while Muybridge created closely spaced sequences of photographs (Figure 4.16).[165]

Marey's multiple exposures recalled X rays in their interpenetration of forms, and went beyond them in their breathtaking views of continuity of motion. The illustration from *Le Rire* is a play on experiments with multiple exposures. Such Marey photographs as that in Figure 4.17 must have influenced Picasso's realization of cubist simultaneity and the interpenetration of forms as these notions appeared for the first time in the squatter.

Muybridge offered Picasso something else: the idea of a "motion picture sequence" of five women with the "plot" of increased geometrization.

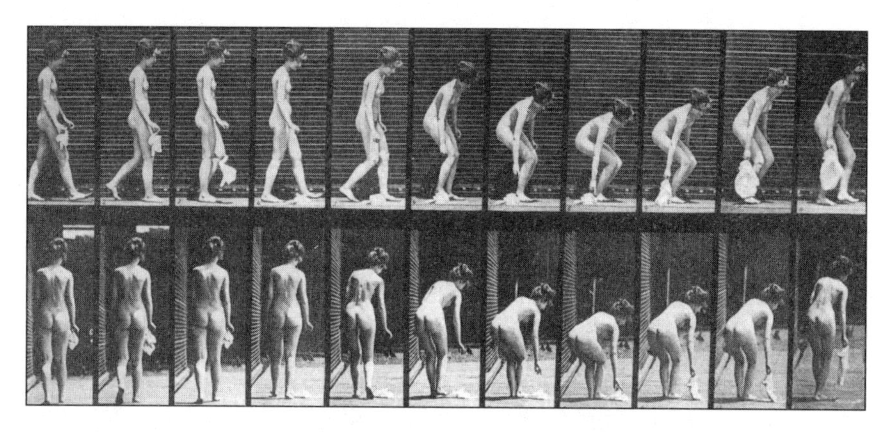

FIGURE 4.16 Eadward Muybridge, *Dropping and lifting a handkerchief,* 1885.

Picasso's interest in the new visual arts of cinematography and photography broadens our view of the roots of the *Demoiselles*. He was not much interested in exploiting the dispassionate series of poses in the motion studies of Marey and Muybridge, but in "posing" a sequence of frames in a new manner. Here the ideas suggested by Méliès's films were a great help. I believe the project of Picasso's second campaign for the *Demoiselles* was precisely to create such a sequence: to set into action a gradual geometrization of the human form as the "action" progresses from left to right. The end is a four-dimensional view of the squatting whore, the embodiment of Picasso's realization that spatial simultaneity was the essence here and not, as Poincaré had written in *La Science et l'hypothèse,* the succession of perspectives.

In the spring of 1908 Picasso took a remarkable photograph in the Bateau Lavoir (Figure 4.18). He created this specific photographic setup for two reasons: to assemble works that had led to the breakthrough in the *Demoiselles,* and as a matrix for the *Composition à la tête de mort* [Composition with Skull].[166] The *Demoiselles,* partially seen on the right, is cut off by Picasso's two wooden *iberique* sculptures. Hung side by side on the wall are two drawings of the theme that emerged from the old peasant at Gósol. Beneath the Iberian sculptures is an asexual head-and-shoulder sketch that is referred to as *Head with One Eye Blank.*[167] In the photograph's optical center is a study for *Standing Nude,* which dates the photograph to sometime in early 1908.[168] The pyramidal setup in the photograph, with the pile of books, palette and brushes, is an indicator of the foreground of *Composition à la tête de mort.* In the center of that painting is a figure of a female nude similar to the one in

FIGURE 4.17
Etienne-Jules Marey, Chronograph of a fencer, 1880s.

Standing Nude.[169] Perhaps by partially obscuring the *Demoiselles* in this photograph, Picasso was saying that it was as yet unfinished and that he meant to keep it in view.

FILLING IN THE MOSAIC

Any great work of art or science necessarily draws on many different, apparently unconnected areas. Such highly creative thinking may be likened to a mosaic of many tiles. In Picasso's case we have identified these:

- *Cinematography* provided a succession of forms side by side (or, as in Méliès's movies, hilariously rearranged) from which Picasso took the idea of revealing a transformation of forms.
- *Geometry* gave Picasso the basis—the *why*—for a new conceptual language for the emerging art, one capable of exploring the deep structure of forms. Picasso's dramatic move was in the direction of formalizing the previously informal language of art. This includes Picasso's discovery about the projection onto a plane of an image with spatial simultaneity and auras of the fourth dimension. Picasso learned about all this from Princet's discourses on Poincaré.
- *X-rays* carried the antipositivistic message of the relativity of knowledge. In conjunction with other rays flying everywhere through space—alpha rays, beta rays, gamma rays, cathode rays—they laid open bodies, giving Picasso the license to rearrange body parts.

FIGURE 4.18
Pablo Picasso, *Study for* Standing Nude *in the Bateau-Lavoir Studio*. Paris, spring 1908.

- *Photography* was used genially in a suggestive manner for painting. For the *Demoiselles,* Picasso utilized photographs to refine his composition. But additionally, Picasso was inspired by writings extolling the virtues of photography as art.
- *African art* supported Picasso's conceptual approach and convinced him of the deep meaning of geometry as the language of the new art.
- Picasso's *oedipal complex* and attempts at total emasculation of his father. He sought to accomplish this by achieving great fame in a style of art that differed dramatically from anything practiced by any other artist and even opposed social norms.
- *Competitiveness* was a means for focusing on advancement. Picasso's trip to the Salon des Indépendants in 1907, where he saw the advancements made by Matisse and Derain, fired him with the desire to do them one better.
- *Sex* is an Ariadne's thread that runs throughout Picasso's life and work. It was the fuel for a mighty creative engine.

- *Material independence* to follow a course of action that offered no hope of immediate sales, permitted Picasso to work unhindered on the *Demoiselles* and then on other simultaneous works. The important steps were Vollard's buying out Picasso's studio in February of 1907 and then Picasso's meeting with Kahnweiler in July 1907.
- *Philosophy and literature* provided the intellectual stimulus to shatter current norms and move toward a method of expression that was radically new.
- Not least, of course, there was *painting*—the usual sources including Cézanne, El Greco and Ingres, and the paintings by Derain and Matisse that piqued Picasso's competitiveness.

I would like to add evidence for another tile in the mosaic. This one comes from the philosophical culture and its presence is in the faceted breast on the right-hand standing demoiselle. This "folding card" structure will be central to Picasso's subsequent experiments toward analytic cubism. Whereas faceting may well have been suggested by Jouffret's diagrams, the very particular shading may give a clue to another intellectual stimulus on Picasso: William James, via Gertrude Stein.[170] While an undergraduate student at Radcliffe from 1893 to 1897, Gertrude admired the renowned philosopher-psychologist enough for James to admit her to graduate courses and seminars. She steeped herself in such James classics as *The Principles of Psychology,* prominently displayed his books at rue de Fleurus, remained in contact with James after graduation, and made it a point to see him during his Paris trip of September 1908. In her *Autobiography of Alice B. Toklas* she writes of James coming to see her art collection, which included Picasso's *Nu à la draperie.* "He looked and gasped, I told you, he said, I always told you that you should keep your mind open."[171]

Somehow, between Gertrude's poorly accented Radcliffe French and Picasso's less than facile command of the spoken language, during the ninety-odd sittings for her portrait in spring 1906 Gertrude communicated her views of art theory and her former Harvard professor. The question is how much of James's visual experiments she actually discussed, and whether Picasso actually incorporated any of them into his art. James's aversion to perspective was an immediate point of intellectual contact between the two men.[172]

The particular faceting and shading of the right-hand standing demoiselle's breast provides some evidence that Picasso learned something from Gertrude about James. Just as Princet showed Picasso Jouffret's diagrams,

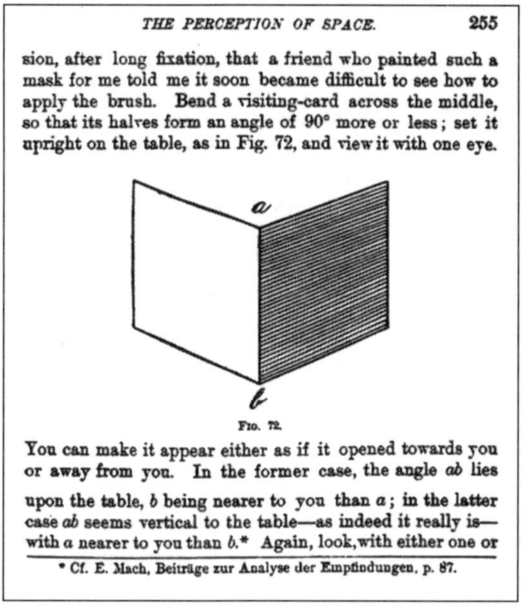

THE PERCEPTION OF SPACE. 255

sion, after long fixation, that a friend who painted such a mask for me told me it soon became difficult to see how to apply the brush. Bend a visiting-card across the middle, so that its halves form an angle of 90° more or less; set it upright on the table, as in Fig. 72, and view it with one eye.

a

b

FIG. 72.

You can make it appear either as if it opened towards you or away from you. In the former case, the angle *ab* lies upon the table, *b* being nearer to you than *a*; in the latter case *ab* seems vertical to the table—as indeed it really is—with *a* nearer to you than *b*.* Again, look, with either one or

* Cf. E. Mach, Beiträge zur Analyse der Empfindungen, p. 87.

FIGURE 4.19
William James's "folding visiting card" experiment (James, 1890, vol. 1, p. 255).

Gertrude could not have passed up the opportunity to show him James's *Principles of Psychology.* Being interested in ambiguous visual representations, Picasso was undoubtedly struck by James's "folded visual card" experiment (Figure 4.19).

In this experiment the card seems to move back and forth, yet it remains in position with unchanging light and shade. The flip-flop phenomenon would be made famous by the Gestalt school of psychology.[173] James's "flip-flop" experiment may well have found its place in the standing demoiselle on the right because it fits into the overall painting in which there are no discernible light sources for the various hatchings and shadings on the two right-hand demoiselles. Picasso's particular faceting as a sort of "folded visual card" imparts a distinctive rhythm to the *Demoiselles.*

* * *

Picasso's great breakthrough was to realize a connection between science, mathematics, technology and art. In rejecting the accepted rules and turning to a radically new intellectual framework, he turned to science as a model and mathematics as a guide, as would physicists some twenty years later. In this way he found the courage to create his own version of visual representation in *Les Demoiselles d'Avignon* and move toward the geometric language of the emergent cubism.[174]

Widening our view of the origins and genesis of the *Demoiselles* to include science, mathematics and technology enables us to gain further insights into this monumental work without falling back on such old saws as "genius," "intuition" or Freudian analysis.[175] Thus we are better equipped to tackle a wider issue, the power of the avant-garde at the beginning of the twentieth century, an era of genius unparalleled since the Renaissance. The works produced in the first decade of the twentieth century will forever be stepping stones that define the course of civilization. They are the responses of geographically and culturally diverse individuals to changes that swept across Europe like a tidal wave.

THE RECEPTION OF *LES DEMOISELLES D'AVIGNON*

We have noted already the lukewarm responses to the *Demoiselles* during the first period of Picasso's work. Worse was to come. "It is the hideousness of the faces which froze and appalled the semi-converted,"[176] wrote Salmon, who witnessed these reactions. Kahnweiler reports a comment by Derain on seeing the *Demoiselles:* "Painting of this sort was an impasse at the end of which lay only suicide; that one fine morning we would find Picasso hanged behind his large canvas."[177] Apollinaire never wrote anything on the *Demoiselles* nor made any comments that have been recorded, even in gossip, except for his response to an early version in February 1907.

Matisse was outraged. Fernande writes that after seeing the painting he "talked of getting even with Picasso, of making him beg for mercy."[178] This reaction must have particularly struck Picasso because the two men had become somewhat friendly despite their differing styles. On many Fridays Picasso would visit Matisse's studio, and they saw each other on Saturday evenings at the Steins'. They also exchanged paintings by 1907. Matisse's rage may well have been rooted in the realization that Picasso had overtaken him. Apparently Matisse and the art critic Félix Fénéon came over to the Bateau Lavoir to see the *Demoiselles* and exploded in laughter. The most that Matisse would offer Picasso was the backhanded compliment, "a little boldness discovered in a friend's work was shared by all."[179]

Since the beginning of the second campaign, Salmon recalled, Picasso felt "somewhat abandoned." He "found himself in a really tragic situation" and thought "some of his painter friends were avoiding him."[180] Even Fernande was gone, although they remained in contact. In a letter of 19 September, 1907, to Gertrude Stein, Fernande reported that Picasso seemed "very fed up."[181] After over half a year of work in almost total isolation, no one was offering any encouragement. Kahnweiler recalled that "a short period of exhaustion followed" after Picasso put the *Demoiselles* aside, and that he had to revive his "battered spirit."[182]

The first time it was seen by a wider public was in a 1910 article by the enterprising American journalist Gelette Burgess, where it appears as "Study by Picasso."[183] Commenting on the "terrible picture [that looms] through the chaos [of the Bateau Lavoir] and the monstrous, monolithic women," Burgess says he inquired whether Picasso used models. "'Where would I get them?' grins Picasso."[184] Burgess was not alone in his shock.

Although Picasso and Braque first met in about March 1907, Braque did not visit the Bateau Lavoir until November. The *Demoiselles* shocked him, too, although he shortly realized the painting's power and innovativeness.[185] By the end of 1907 things began looking up for Picasso. His meeting with Braque and growing friendship with Derain made his atelier more of a *rendez-vous des peintres.* Fernande had returned as well.

That the *Demoiselles* would remain unrecognized as a seminal work is something Picasso had to learn to live with. Its creation had been all-consuming, yet at the end of it he received little praise and much derision. The *Demoiselles* also turned out to be unsalable. The painting would not be exhibited until 1916 and not be recognized for what it is until the 1920s. Only Picasso knew its worth.[186]

5

Braque and Picasso Explore Space

At that time our work was a kind of laboratory research from which every pretension of individual vanity was excluded. You have to understand that state of mind.

—Pablo Picasso

After *Les Demoiselles d'Avignon* opened the floodgates of Picasso's creativity, there followed seven years of productivity unequalled in the history of art. In this work Picasso joined forces with Georges Braque, in a partnership without precedent.

Seven months younger than Picasso, Braque was originally trained as a decorator and housepainter. In about 1902 he decided on a career as an artist and moved to Paris, where he settled in Montmartre on rue d'Orsel, a few hundred meters from the Bateau Lavoir. After two years of formal art training, Braque began his career in the fauve tradition and achieved some success. Becoming disillusioned with fauvism, he sought another means of representation. The Cézanne retrospective that commenced at the Salon d'Automne on 1 October, 1907, gave Braque the clue as to how to proceed.

Once they became friends, the interactions between Braque and Picasso intensified until by 1910 they were seeing each other almost daily. Braque likened their closeness to a *cordée en montagne,* the rope connecting two mountaineers as they scale an unconquered peak.[1] At the time of the Bateau Lavoir meeting, Picasso was already, unlike Braque, a somewhat accomplished and moderately well-known artist. They differed in other ways, too. Whereas Braque radiated clarity and sensibility, Picasso was often moody and quick to act on artistic impulses. For Picasso, painting was like the air he breathed; it was something he was born to do. Braque claimed he never set out to be a painter. Painting was something he enjoyed doing, but he preferred to have no set goals because, quoting Nietzsche, "A goal is a servitude."[2]

BRAQUE'S RESPONSE TO THE *DEMOISELLES*

Braque had known Picasso for about seven months when he first visited the Bateau Lavoir in November 1907. What immediately impressed him about the *Demoiselles* was its obliteration of all previous styles of Western painting. What horrified him was the new style put in their place. First there was Picasso's primitivism, a manner of representation whose only attraction to Braque was that it "went against the Fauve tradition."[3] Even worse was Picasso's use of space. Braque, who obsessed over this concept, had recently already begun deviating from fauvism with a wholly original variation on Cézanne. Braque moved toward more somber colors in conjunction with a pronounced reductivism toward geometrical forms. His advancement on Cézanne emerges from minimization of detail in order to accentuate geometrization of foliage and houses.[4] Braque abstracted Cézanne's *passage* until traditional perspective completely vanished. Whereas Cézanne's paintings are the culmination of many viewings, Braque painted in a more intellectual style, without nature as a model. This method did not come easily, but it yielded greater creative freedom.[5]

Braque asked himself whether there was a way to combine Cézannesque technique with what Picasso had accomplished in the *Demoiselles*. First he tried to come to terms with the *Demoiselles* by combining Picasso's radical experiments on space, geometry and ambiguity with Cézannian structure. The result is his *Large Nude* on which he worked almost exclusively for six months. The subject is a version of the second demoiselle from the left, placed on a Cézannian still-life background. Yet like Picasso, Braque failed to achieve a balance in the woman's awkward pose. The real importance of *Large Nude* is that "it is the first move in the game of co-operation and one-upmanship that is the subtext of his and Picasso's Cubism."[6]

Simultaneously with *Large Nude*, Braque worked on a sketch entitled *La Femme*, from which evolved a painting of that name exhibited at the 1908 Salon des Indépendants along with Derain's *Bathers III*.[7] Picasso was flattered that his friends Braque and Derain were advertising their switch from Matisse and the fauves to him, and was happy to have Braque take the abusive criticism that resulted.[8] Braque had taken a step toward "raising a corner of the veil on the secrets of the Bateau Lavoir,"[9] while Picasso remained in the background as the *chef d'école*.

Since the Salon d'Automne exposition, Cézanne had become all the rage. Braque was trying to reconcile Cézanne with Picasso, while Derain pursued his breakaway style based on Cézanne and primitivism, in a manner more sculptural and less colorful than fauvist. Picasso worked on variations of Cézanne's bathers, which led to a series of paintings and sketches culminating in such works as *Three Women*. Along the way he produced sketches that began to lose figuration and stray toward abstraction, at which point he drew back. In the end the sketches displayed more experimentation than the finished painting, in which Picasso reduced the three women to Iberian statuettes. *Three Women* is nevertheless a further move toward a geometrization even more severe than in the *Demoiselles*.

In the summer of 1908, Picasso and Braque went their separate ways. Ill and nervous from the strain of creating *Les Demoiselles* and his worsening relations with Fernande, in addition to being upset over Wiegels's recent drug-related suicide, under doctor's advice Picasso vacationed at a tranquil little village, La Rue-des-Bois, forty miles north of Paris, near Creil, on the edge of the Forêt d'Halatte.[10] In need of further study of Cézanne, Braque decided to return to one of the Master's haunts, L'Estaque, a small town outside of Marseilles.

Picasso focused on landscape painting in the idyllic forest setting, which offered just the right ambience to recover from a *crise de nerfs*. Needing a break from Cézanne, he turned for inspiration to Henri Rousseau, whose dreamlike *Snake Charmer* he had seen at the 1907 Salon d'Automne. Picasso returned to Paris carrying landscapes devoid of any experimentation with *passage*. To his chagrin, they paled before Braque's work from L'Estaque.

At L'Estaque Braque immersed himself in Cézanne's *passage*, with the painter himself for inspiration. "Everything about [Cézanne] was sympathetic to me," he recalled later, "the man, his character, everything."[11]

Braque was a man who had taken the measure of himself. Unlike Picasso, he had not been a wunderkind art student. Unlike Picasso, he worked slowly and thoughtfully. Braque approached painting in an analytical way, as he imagined Cézanne must have done, and was well aware of his limitations. "Progress in art," he wrote in one of his notebooks, "consists of not extending one's limits, but in knowing them better."[12] Recalling his own awkwardness he wrote, "Cézanne is as great for his clumsiness as for his genius."[13] All of these characteristics would be essential to the *cordée,* during which Picasso was forced to think long and hard before each step he took, something he never again did with such intensity.

Cézanne's *passage* was the springboard for Braque's first contribution to cubism, which was a reassessment of space itself.

BRAQUE AND SPACE

Braque believed that the nature of space was not represented adequately in art with a single perspective point: "Traditional perspective did not satisfy me. Mechanical as it is, this perspective never gave the full possession of things. It started from a point of view and did not leave it. But the point of view is one totally small thing. It is as if someone who all his life would draw profiles would come to believe that man has only one eye."[14] More directly, to John Richardson, Braque said:

> The whole Renaissance tradition is antipathetic to me. The hard-and-fast rules of perspective which it succeeded in imposing on art were a ghastly mistake which it has taken four centuries to redress: Cézanne and after him Picasso and myself can take a lot of credit for this. Scientific perspective is nothing but eye-fooling illusionism; it is simply a trick—a bad trick—which makes it impossible for an artist to convey a full experience of space, since it forces the objects in a picture to disappear away from the beholder instead of bringing them within his reach, as painting should.[15]

Braque then describes how he went about correcting the situation:

> What attracted me—and was the principal direction of Cubism—was the materialisation of that new space that I sensed. Then I commenced to focus on still lives, because in nature there is a tactile space, I would say almost manual. I have written about it thus: "When a still life is no more accessible to the hand, it ceases to be a

> still life." That answered for me the desire that I have always had to touch the thing and not only to see it. This is the space which especially attracted me, because it was what early cubist painting was about, research into space.[16]

Braque's term "research into space" stands out because this is precisely what he set out to do in L'Estaque in reinterpreting and extending Cézannian *passage.* All space became tactile: He "always had to touch the thing and not only see it." This quest was not yet completed.

Braque was especially interested in the space between and behind objects on a canvas. Many painters, he complained, "totally ignore that what is *between* the apple and the plate can be painted too. . . . This in between space [*entre deux*] seems to me just as important as the objects themselves."[17] We can imagine Braque moving through a canvas and trying to touch these spaces. What bothered him about Picasso's *Demoiselles* was that, although it had no perspective point, its structure was incomplete. There was space between and behind the demoiselles that was not tactile and so not on a par with other objects. Braque's research into space aimed at the "materialisation of that new space that I sensed," a transformation of the space between and behind objects into a "tactile space."

How did he come to the phrase "tactile space"? Although terms like "tactile values" and notions such as the painter's converting a sense of sight into a sense of touch were often used by art critics at the turn of the century, to the best of my knowledge the concept of "tactile space" was the domain of philosophers.[18] It is highly unlikely that Braque and Picasso would either have read articles such as the one by the philosopher Charles Dunan or heard them discussed in cafés on the Butte. Braque's usage of "tactile space," however, is so close to Poincaré's that I would wager he was influenced mainly by Princet's discourses about *La Science et l'hypothèse.*[19]

According to Poincaré, there are two main sorts of spaces, geometrical and representative. Geometrical space is an abstract entity that is infinite in extent, its properties are the same everywhere, and it has three dimensions. It is the space of axiomatic Euclidean geometry, in which there are perfectly formed triangles and spheres. Representative space, on the other hand, has none of the above properties. It is the space in which we live and is comprised of three spaces, visual, tactile and motor.[20] One of the principal problems explored in *La Science et l'hypothèse* is the difference between geometrical and representative space.

Whereas geometrical space is a pristine mathematical ideal, representative space is constructed from our perceptions, and so "our representations are only the reproduction of our sensations."[21] Poincaré continues: "It is also impossible to represent to ourselves external bodies in geometrical space, as it is impossible for a painter to paint on a flat canvas objects with their three dimensions."[22] But we can project three dimensions onto two and thus "we can *reason* about these objects as if they were situated in geometrical space."[23]

Reasoning in this manner requires that we clarify the three components of representative space. Since the retina's physical properties vary from point to point, the properties of visual space are not the same everywhere. Since we have two eyes and each retina is of a two-dimensional surface, we ought to be able to see in four dimensions. Neural connections between the two eyes result in accommodation that reduces the number of dimensions to three.[24]

When he comes to discuss tactile space, Poincaré simply instructs the reader to refer back to the section on visual space and substitute the word "tactile" for "visual."[25] Thus the space of touch also has three dimensions. In conjunction with data of sight and touch, our muscular sensations constitute the third kind of space, motor space. Here too, Poincaré finds no reason to believe that the dimensionality of this space necessarily need be three. Rather "*motor space would have as many dimensions as we have muscles.*"[26] Its apparent three-dimensionality comes from the "simple association of ideas" that we form by experimenting on the perceptions that are constantly bombarding us.[27] In other words, we choose the proper geometry by examining the relationships between objects as we actually move them around.[28]

The only apparent source for Braque's idea of seeking "materialisation of that new space that I sensed" was Poincaré's suggestion to link the properties of visual and tactile space. Princet is the only person likely to have explained such matters to Braque. The two men encountered each other in the Bateau Lavoir as well as socially. Princet was sometimes seen explaining to "Picasso and Braque some elementary principles of geometry in space with the aid of a notebook on a corner of a bistro table."[29]

During the summer of 1908 at L'Estaque, Braque pushed Cézannian *passage* to dizzying heights, discovering his new pictorial space. He realized that the key was to slice up receding distances into two-dimensional planes or layers and then reassemble them to comprise the picture itself, as in a software drawing program. The reassembly is done from back-

FIGURE 5.1
Georges Braque,
Maisons à L'Estaque,
1908.

ground to foreground. Braque's version of Cézannian *passage* contains no perspective points anywhere, and all space becomes tactile, with interstitial spaces squeezed out and an almost complete geometrization of objects in the picture plane. Picasso was struck by Braque's version of *passage*, particularly in his *Maisons à l'Estaque*, of which Matisse remarked that it was built up from *petites cubes* (Figure 5.1).[30]

Braque's summer at L'Estaque was a momentous leap forward. These paintings clash with the flat pictures Picasso produced at La Rue-des-Bois, with their minimal geometrization and a Rousseau-like primitiveness. Of particular importance was Braque's technique of composing a scene from background to foreground by means of layers of planes, with variations in lighting so that they seemed to vibrate, in conjunction with simplification and geometrization of objects. With these innovations Braque had made the first steps toward a cubist "materialisation of space," in which he was forced to incorporate the geometrization of forms that he had seen in Picasso's *Demoiselles.* Further extremes of geometrization and more severe sculpturing would follow. This was territory for Picasso, for whom cubism "is an art that deals primarily with forms."[31]

FIGURE 5.2
Pablo Picasso, *Woman in an Armchair*. Paris, spring 1909.

Picasso's major advance on Braque's *passage* was to soften the geometrization of forms by carving them up into facets. His works during the autumn of 1908 and spring of 1909 culminate in *Woman in an Armchair* (Figure 5.2).[32] He would develop this technique further at Horta de Ebro during the summer of 1909.

Picasso's oeuvres during the autumn of 1908 through the spring of 1909 exhibit a distinct movement to increased faceting. Might Picasso have moved in this direction instinctively, or by some of sort of intuition? Perhaps. Such an explanation, however, dependent on a vague notion of "intuition," is not a tool for exploring Picasso's thought processes but rather a barrier. Intuitions result from a great deal of preparation. The moment of inspiration or illumination is when everything comes together. What, in the case of Braque and Picasso, is the "everything" that came together?

BRAQUE, PICASSO, GEOMETRY AND PRINCET

Braque's experiments with space could not have failed to interest Princet. I would venture that at one of their bistro sessions Princet showed up with

Jouffret's *Traité élémentaire de géométrie à quatre dimensions*, and passed around Jouffret's illustrations of projections of four-dimensional figures on a two-dimensional plane (see Figures 4.5 and 4.6 in Chapter 4). Picasso had already benefited from Jouffret's figures in the *Demoiselles,* where rudimentary faceting appears on the right-hand standing demoiselle.

The American artist and writer Max Weber, a former student of Matisse, reports that during his stay in Paris from 1905 through 1908, there was a great deal of talk about the fourth dimension.[33] He frequently attended the Steins' Saturday evening soirées, which were a haunt for young American painters in Paris, and visited the Bateau Lavoir at least once. It is reasonable to conjecture that he met Princet either at the Steins' or at the Bateau Lavoir and learned something about the fourth dimension from him. Although, as we will see, Weber was more receptive to Apollinaire's metaphorical and romantic version.

In Chapter 4 we introduced the problem that Princet, according to André Lhote, discussed with the cubists around 1909 or 1910. This problem indicated that Princet was well aware of what the painters were up against. We can imagine him giving his geometry discourse while sketching diagrams on the tablecloth of a smoky bistro, late in the evening, after a dinner with plenty of wine—possibly at Azon, then a favorite hangout of *la bande à Picasso.* Apollinaire, Braque, Jacob, and probably Juan Gris are gathered around. They listen intently because they are all trying to represent objects by simultaneously exhibiting many different viewpoints. Assuming that Lhote remembered at least the gist of Princet's words, Princet is describing what one must do in order to represent the "table as a type."

"Type" is a technical term from philosophy that had just begun to appear in the work of the British philosopher Bertrand Russell. Princet analyzes the ideal construct of a table, which, unlike the trapezoidal table he starts out with, is undistorted by perspective. The ideal construct of a table is flat against the picture plane so that it becomes a "true rectangle." Princet uses type as an ideal construct abstracted from empirical tests. "Straightening up" would be necessary also for objects on the table such as a glass whose oval rim becomes a "perfect circle."

Princet may now have looked up and everyone nodded. He then, so to speak, turns the table on his listeners. He goes on to ask, what if the notion of a "table as a type" were actually a side view? Then it becomes a "horizontal bar a few centimetres thick, the glass a profile whose base and rim are horizontal. Hence the need for another displacement." The message is clear: Perspective deforms or distorts from the object as a type.

At which point Braque and Picasso raise their filled glasses of wine and drink until the oval rims are circular against the plane of their faces. Almost in unison they thank Princet for confirming that what they are trying to do in cubism has, indeed, a scientific basis. Princet has reassured them that they were on the right track in their attempts at representing objects all at once (simultaneously) from different views.

One wonders if Princet, as a coda to his geometry lecture, went into Russell's reason for formulating a theory of types or hierarchies. It goes as follows: In spring of 1901 Russell stumbled on a logical paradox that came eventually to bear his name. He found that certain theories of logic produced statements that are true if and only if they are not true.[34] One can only wonder what Picasso and his friends thought about a situation like this and how it bore on their still-developing art form, which sought to avoid ambiguity while representing objects in a manner closer to their deep structure. Perhaps Picasso had logical paradoxes in mind when he said, "Art is a lie that makes us realise the truth, at least the truth that is given us to understand."[35]

One last point: Where might Princet have learned about mathematical logic and particularly the cutting-edge material concerning paradoxes? The most likely place was Henri Poincaré's latest best-seller, *La Science et méthode*, published in 1908, where Poincaré discusses Russell's recent researches as well as the current situation in logic.[36] Although he does not explicitly define the concept of a type here, someone like Princet could have read further in any number of sources. For example, there is Poincaré's 1909 paper "The Logic of Infinity."[37] Princet folded into Russell's hierarchy of types the antipositivist argument that because there are different ways of viewing an object, there is no single reality. Good food and good company can produce exhilarating effects.

By the summer of 1909, Braque and Picasso were actively continuing Picasso's original search for a way to represent nature that absolutely broke from impressionism and its positivistic basis. They wanted to produce paintings that were, in their favorite phrase of Cézanne's, *bien couillarde*. Perhaps the best translation of this phrase is "ballsy."[38]

PHOTOGRAPHY AND PICASSO'S CREATIVITY

In May 1909, ten years after his spiritual and intellectual transformation as a teenager, Picasso returned to Horta at another turning point in his career. Instead of his boyhood friend Pallarès, this time he arrived with his mistress. Certain locals were not happy about this arrangement and one evening a particularly militant group gathered angrily below the couple's hotel window. Picasso appeared with his pistol and dispersed the crowd with a few shots.

Picasso worked in Horta until September of 1909. The stay was not happy for Fernande, who suffered from a kidney infection and was often bedridden with little to do and less comfort from Picasso, for whom she had become a nuisance, a "suffering machine,"[39] as he recalled some years later.

As to the extremely faceted works that emerged from his stay at Horta, most art historians explain them as the honing of a technique whose basis he had set in progress the previous spring. Until recently, that is, when another tool in Picasso's artistic development at Horta came to light.

The recent discovery by Anne Baldassari of over one hundred negatives and prints by Picasso, all made before 1920, is highly significant for Picasso studies. Baldassari demonstrated "a solid link between many of these photographs and Picasso's creative process."[40] Our discussion of the *Demoiselles* has already noted Picasso's nonstandard employment of picture postcards. Rather than copy from them, he used them as a source of ideas for figure groupings and other compositional elements.

Picasso's earliest recorded comment about photography, according to Fernande, emerged during a hashish session at Princet's flat in about September 1907: "In a state of nervous hysteria [Picasso] shouted that he had discovered photography, that he wanted to kill himself, he had nothing left to learn."[41] He was undoubtedly excited over the resources that photography afforded his painting, and perhaps even more over its still-unknown potentialities.[42]

Picasso was never averse to new technological developments, and from the beginning of his contact with photography, he developed a definite view of what he wanted. In a conversation with Christian Zervos in 1935, Picasso elaborated on what he had been doing since 1901 with photography:

> It would be very interesting to fix photographically, not only the stages of a painting, but its metamorphoses. Possibly one might catch a glimpse of the road by which the mind moves toward the concretisation of its dream. But what is truly very curious, it is to observe that the painting does not basically change, that the initial vision remains

> almost intact despite appearances. . . . I notice that, when the work is photographed, that what I have introduced in order to correct my first vision disappeared and that, when all is said and done, the image given by the photograph corresponds to my first vision, before the transformations brought about by me.[43]

For Picasso, photography was not just a means for exact vision or a positivist apprehension of the world, but a creative device. He considered a photograph far more than a mere registration of a scene that would record the "stages of a painting" in chronological order. Instead he was interested in how a painting "metamorphoses": the conceptual transformations in the work itself and in its relation to other works. In its attempt "to catch a glimpse of the road by which the mind moves toward the concretisation of its dream," photography reveals the heart of the artist's creativity. Picasso went even further: despite changes on canvas, photographs retain the artist's "first vision."

Picasso's photographic oeuvre shows that he was using photography to explore the nature of space. Both the staging of works in progress and his manipulation of negatives and prints helped Picasso test prototypes of new visual approaches.

Since about 1901, Picasso had been photographing works as well as experimenting with photographic setups.[44] An early and dramatic example is his 1901 *Self-Portrait in the Studio* (Figure 5.3).

On the right are set out completed works, with some emerging from behind others. On the bottom right is *Portrait de Gustave Coquiot*,[45] above it is *La Buveuse d'Absinthe*,[46] from which emerges *Scène de Café*,[47] above it and cut off is self-portrait, *Yo Picasso*,[48] and on the upper left is *Groupe de catalans à Montmartre*[49] of Picasso and his Catalan buddies, including Casagemas.

On the left is the ghostly image of a man in a top hat. Baldassari has substantiated that it is Picasso himself. She goes on to suggest that Picasso's image is not in that position by accident, rather he deliberately left this space empty. Then, either by superposition of two negatives or by double exposure, he inserted himself into the empty space.[50] The artist's emaciated face carries the hallucinatory look seen in several subsequent self-portraits.[51] Picasso depicts himself as the bohemian artist living amidst the decadence of alcohol, drugs and sex, all of which inspire in him a youthful and electric sense of his destiny. No less than a Dürer or

FIGURE 5.3
Pablo Picasso,
Self-Portrait in the Studio. Paris, 1901.

Michelangelo, he is set aside from all others by the magical powers of his divine genius. The artist emerges from the wall and gazes beyond earthly pleasures into a future of withdrawal in order to accomplish great works of art. Picasso's faith in his destiny is clear from what he wrote on the photograph's back: "This photograph could be entitled—'The strongest walls open on my passage. *Regarde!*'"

Close inspection of Figure 5.3 leads me to conclude that Picasso's ghostly image is the result of a double exposure—two images on a single negative, rather than one print made from two negatives. What separates it from a commonplace exercise for budding amateurs is the faintness of Picasso's image compared with the hung paintings. Picasso achieved this by placing, and then removing, something opaque in front of the light impinging on the frame holding the glass negative and photosensitive paper, in order to retard exposure of this portion of the print. If he had not done this, then his entire image would have been as dark as the portion on his right side, which coincides exactly with his body shape. This remarkable print shows Picasso's command of photographic techniques—especially given what must have been very primitive equipment—and passionate involvement with this genre of the visual culture.[52]

That this is a black-and-white image of colored paintings is also highly relevant.[53] Baldassari proposes the intriguing hypothesis that Picasso was practicing "'judgement by mirror' as advocated in Renaissance treatises and, more immediately, by Édouard Manet's practice of assessing his work by looking at it in a 'small dark mirror'."[54] Manet thus freed himself from the effects of light and color, which he believed could distort the artist's judgment. Similarly for Picasso, whose "photographic observations of works in progress probably helped him free his painting of pointlessly 'added' colour,"[55] and so prevented his straying too far from his "first vision." As Picasso told Zervos, "I often see light and shadow."[56]

An indication of the complexity of the link between photography and art in Picasso's work is his desire to maintain the integrity of the photograph from which he is working while "demonstrating—and demonstrating to himself—that painting could borrow all of photography's effects and wield them much more impressively."[57] The 1903 painting *The Soler Family*,[58] commissioned by a Barcelona tailor, is a good example. The anonymous black-and-white photograph that served as a model for this painting is a family grouping consisting of three small children flanked by the mother and father, with a dog in the foreground. The harsh light from a glass roof and white backdrop in the photographer's studio results in a flat scene in which the subjects' faces are without shadow and devoid of emotion. The photographer has added rustic props to give the impression of a family picnic. With Manet's *Le Déjeuner sur l'herbe* in mind, Picasso rearranged the group from top to bottom instead of from front to back. Other changes further accentuate the scene's flatness while maintaining the basics of the photograph. The painting's visual ambiguity results from combining a photograph with the freedom offered by painting, as if in a conscious attempt to "wield [photography's effects] much more impressively."

The ambiguity proved too much for Soler, who obtained Picasso's permission to have an artist add a wooded background such as in Manet's painting. In 1913 Kahnweiler bought the Soler painting and Picasso set about retouching it. First he attempted to work along cubist lines but became dissatisfied. Instead he took out the added woods and restored the plain background as it had appeared in the original photograph, which contained Picasso's "first vision."[59]

FIGURE 5.4
Pablo Picasso,
Photograph of Ricardo Canals. Paris, 1904.

PICASSO AS PHOTOGRAPHER

Picasso's 1904 photographic portrait of his Catalan friend Ricardo Canals exhibits a high degree of virtuosity. Picasso presighted the camera on lap level and "signed" it with his own image in the mirror (Figure 5.4).

On the fireplace's mantel is a photograph of Canals's wife, Benedetta. If not at first, then certainly soon afterward, Picasso could not have resisted closing the circle between the artist, the model and the reproduction.[60] The following year he painted Benedetta from the photograph in the photograph. Perhaps to make this point abundantly clear, some years later Picasso wrote in French on the back of a print of Canals's photo: "portrait of the painter Ricardo Canals of Barcelona/in the glass *moi PICASSO*/on the chimney a photo of Benedetta wife of Ricardo Canals/photo taken in Paris in Canals's atelier in 1904."[61]

FIGURE 5.5
The handheld camera used by Picasso and referred to as a "detective." The self-timer is the device on a cable. (Frisch 1899)

The cameras Picasso used were known as "hand cameras," to distinguish them from the large unwieldy apparatus used by professionals. Since, at first, no one recognized these devices without a tripod as a camera, they were also popularly called "detectives." They appeared in the 1880s after the invention of gelatin-bromide plates, which replaced cumbersome wet-plate photography. From the famous hashish session at Princet's, we can date Picasso's purchase of his own first camera to about the end of September 1907. Until then he borrowed cameras from friends such as Canals.

The detective Picasso used had a magazine clipped onto its back that held several 12 x 9 cm. glass plates. It has a lens of about 57 mm. focal length, an aperture of f-9 permitting good depth of field, a shutter speed of one-twenty-fifth second, and an apparatus for self-portrait: The photographer could set the shutter on a ten-second delay in order to become his own subject. The image to be photographed was normally viewed through a ground glass with the camera held chest-high (Figure 5.5). In Figure 2.2 of Chapter 2 a detective is placed on the table as if "imposing itself into the scene as a fourth person."[62] Its lens is directed toward the point where the three models' eyes converge. It is included in the family being photographed while simultaneously forming the "image in process of being photographed."[63] This was very likely Picasso's camera. The presence of a detective in photographs of and by Picasso substantiate that it was his camera of choice.[64]

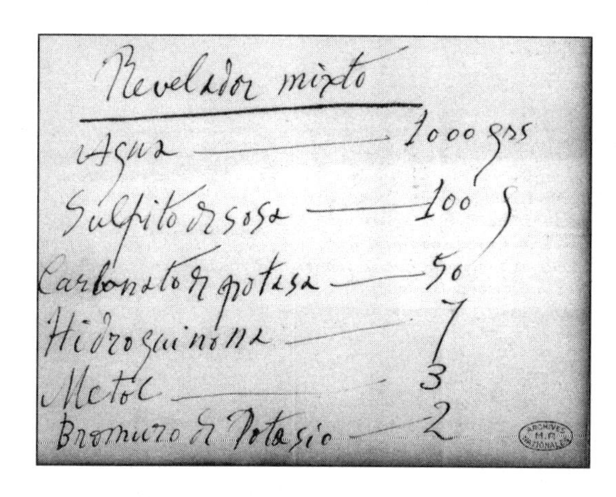
Revelador mixto
Agua — 1000 grs
Sulfito d sosa — 100
Carbonato d potasa — 50
Hidroquinina — 7
Metol — 3
Bromuro d Potasio — 2

FIGURE 5.6
Developer mixture in Picasso's handwriting (1908–1909).

Until 1914 Picasso used glass plates, which meant that his only controls over the photographic process were in the development procedure and the type of photosensitive paper he used—usually chloride-gelatin of silver. The printing process consisted of a frame in which the glass plate and photographic paper are placed in direct contact and then exposed to the sun. The paper was then developed, fixed and washed. From photographs in the Picasso archives we know that he made numerous prints from the same negative and experimented with different developer formulas (Figure 5.6).

Picasso's joy in taking pictures is illustrated by his inscription on one of the envelopes in which he kept his photographs (Figure 5.7). The "PH" stands for photography. The Spanish *Ojo* means eyes and functions as a quick self-portrait.

Picasso became highly adept with technical and practical aspects of photography. He edited prints by cutting them up or sometimes retouching directly on the print and negative. An example of the former method is his photograph taken at his studio in the Bateau Lavoir in spring 1909 (Figure 5.8). The print is pared down from its original, which was 12 x 9 cm., to 11 x 5.1 cm. It is cut so as to be parallel to the arm of *Seated Nude* and to include most of her face; there is an indentation almost parallel with her nose. By carefully positioning the canvas of *Woman with a Book* against the easel, and the two of them against the background of *Female Nude with Guitar*, Picasso produces the illusion of a real white band running vertically. The light and dark striations of the white stool and the wall supports, together with the skewness of the painting *Seated Nude,* and the apparent blending of the right lower portion of *Female Nude with Guitar* and *Seated Nude,* combine to give an illusion of the space of natural per-

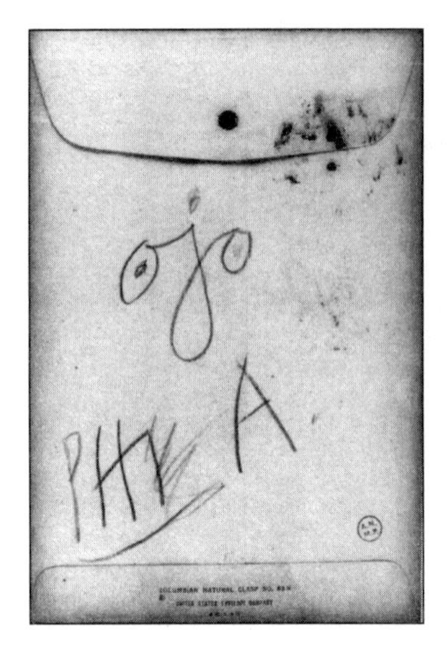

FIGURE 5.7
Picasso's handwriting on the back of an envelope.

FIGURE 5.8
Pablo Picasso, *Woman with a Book.* Photograph taken in Bateau Lavoir, Paris, spring 1909.

ception being literally contradicted by a system of visual counterweights.[65]

An example that combines Picasso's editing techniques with his exploration of pictorial space is a studio setup from 1913.[66] Picasso photographed this staging in his atelier for *Construction with Guitar Player* (Figure 5.9). We note the complex confrontation between the real world of the guitar, the bottle of anisette and the table, with the imaginary space of the *papier collé* entitled *Au bon Marché*,[67] the cut-out newspaper hands holding the guitar, the drawing of a typical Anis del Mono bottle. Picasso searches for structural possibilities as follows: He crops the photograph in Figure 5.9, sketches a sort of grid on it in ink, and obtains Figure 5.10. Then he masks the negative of Figure 5.9 and produces Figures 5.11 and 5.12, which are complementary.

By this process Picasso could consider that he had solved the following problem: How can visually and materially heterogeneous elements be forged into a visual unity that is sufficiently convincing to the eye? His further productions involved making half-tone copper plates of the photographs in Figure 5.11 and 5.12 and then eventually an engraving, and finally to such cubist constructions as *Guitar and Bottle of Bass.* As Baldassari has noted, "The transfer of one medium to another . . . thus confirms photography's active intervention as a tool of semantic transformation at several key moments in Picasso's pictorial development."[68]

A NOTE ON PHOTOGRAPHY AS ART

Retouching and manipulating of photographs were widespread and controversial. For example, the entry rules for a photography contest sponsored by *Le Monde Illustré* on 3 November, 1906, stated, "This is a photo contest and contestants are forbidden to submit what has been retouched."

Some photographers disagreed. The 11 January, 1905, issue of *Paris Journal* contains an anonymous article entitled "L'Art Photographique" which pleads for a reexamination of the possibilities offered by retouching, and of the difference between photography and painting generally: "No one denies that the camera exaggerates perspective in a manner so ridiculous that when it reproduces a man who is flexing his arms, his hands seem larger than his head." "Happily," the author continues, "the 'modern style' photograph—often very amateurish—is a

FIGURE 5.9
Pablo Picasso,
Photographic Composition with "Construction with Guitar Player." Paris, spring–summer 1913.

FIGURE 5.10
Pablo Picasso,
Photographic Composition with "Construction with Guitar Player." Paris, spring–summer 1913.

FIGURE 5.11
Pablo Picasso,
Photographic Composition with "Construction with Guitar Player." Paris, spring–summer 1913.

FIGURE 5.12
Pablo Picasso,
Photographic Composition with "Construction with Guitar Player." Paris, spring–summer 1913.

thousand times more revolutionary than works by the 'art-nouveau' writer or painter."

The article continues by extolling the virtues of retouching and playing with perspective, remarking that "proprietors of furnitures and gardens are enchanted that, thanks to some artifices, the facades seem higher and flowers more abundant." Photographers can "play with light outside of their usual set ups . . . and impart to their images an imprecision a thousand times more expressive than that past mathematical clarity of portraits."

The view of photography as a creative, expressive medium was part of a movement, called pictorialism, that dominated photography from about 1890 into the early twentieth century.[69] It insisted that photography as a visual art was not beholden to strict realism; only the quality of the final image mattered. Touching up photographs was perfectly satisfactory. Pictorialism was in tune with the antipositivist leaning of the avant-garde.[70]

Even from its very beginning in 1839, photography did not render exact reality. Yet scientists touted it as doing just that, declaring it their goal to use photography to see everything in nature.[71] The camera would be the scientist's retina. While scientific discourse attempted to toe the positivist line, experimentation with photographic processes was rendering the ideology suspect. X-ray photography, the work of Marey and Muybridge, and the astonishing photographs of hysterical women surrounded by ghostly and hypnotic auras made the camera very different from the retina.[72] Photography turned out to extend the scientist's vision in unexpected ways, and right from the start it raised the possibility of a scientific relativism. What you see is definitely not what you get.

The author of the 1905 *Paris Journal* article was enthusiastic over the future of photography, "of its evocativeness and suggestiveness, and its beauty. [Photography] is an art." We can imagine that such articles inspired Picasso to explore photography, which he did throughout his life. More than 5,000 photographic documents in his archives testify eloquently to its importance. Picasso made it a point to be photographed by such eminent photographers as Brassaï, Henri Cartier-Bresson, Man Ray, Robert Doisneau and David Douglas Duncan. In the critical years at the beginning of the twentieth century, Picasso explored for himself the practical and technical aspects of photography. Despite his rudimentary camera, the old adage holds: It is the hand behind it that counts.

PHOTOGRAPHY AT HORTA AND THE GEOMETRIZATION OF NATURE: SUMMER OF 1909

In the magic year of 1909, Picasso began to apply Cézanne's advice on painting—make it *bien couillarde*—to photography, too. As early as February 1909, Picasso was eagerly making plans to journey to Horta to pursue Cézannesque themes. On the back of a postcard sent to him that month by Leo Stein, he outlined essential working materials to take along, giving photographic equipment equal priority with art supplies. Fernande and Picasso arrived in Barcelona on 11 May and left for Horta on 5 June. In the interim, besides seeing friends and family, and taking care of the ailing Fernande, from their hotel window Picasso accomplished a series of ink drawings that showed he was prepared to dramatically develop Braque's advances on *passage* (Figure 5.13). He embarked on an extreme geometrization of houses such that everything was reduced to a single plane of geometrical forms. This was an enormous advance in cubist notation, which Picasso would continue to explore over the next three months.[73]

Besides the large amount of art supplies that Picasso hauled by donkey up the winding mountain trails to Horta, he took an ample number of glass photographic plates, processing apparatus and development chemicals. One of his first acts was to take photos of the village inhabitants, who turned out happily for the occasion. Fernande wrote to Gertrude Stein's companion Alice B. Toklas that "people here think we are photographers, and everyone has been delighted to have themselves 'done in portrait,' as their saying goes. [I]t is all due to the camera that Pablo unjustly clutches, since he barely knows how to use it."[74] In all their years together Fernande never realized Picasso's talent as a photographer.

As Baldassari points out, for a long time it was not clear who took the vast majority of the photos of Picasso himself. Through detailed analysis she showed that they are self-portraits.

In Picasso's self-portrait at Horta (Figure 5.14), he stares into the camera as if lost in a trance, perhaps thinking of the paintings that will eventually fill the vacant wall behind him. The camera is placed low and the artist sits slumped in a chair so as to take advantage of the camera's distortion of objects close to it, in this instance to exaggerate legs and crotch. This distortion effect was well known[75] and Picasso could easily have read about it in places like the *Paris Journal*. At Horta he showed extraordinary

FIGURE 5.13 Four drawings by Picasso of *View of Barcelona Courtyard*, May 1909.

photographic skills. As he wrote to the Steins, "I intend to do photography here."[76]

Picasso was after more serious game with his camera than a touristic photographic romp in the Spanish countryside. In his photos Picasso captured examples in nature of the geometry in the Barcelona ink drawings. Cézannian *passage* emerges dramatically. An excellent example from Horta is in Figure 5.15, in which the background dramatically tips into the picture frame while the rooftops and mountainside merge with the sky, resulting in a blending of planes.[77]

FIGURE 5.14
Pablo Picasso,
Self-Portrait. Horta
de Ebro, 1909.

FIGURE 5.15 Pablo Picasso, *Landscape, Horta de Ebro (The Reservoir).* Horta de Ebro, 1909.

FIGURE 5.16 Pablo Picasso, Photograph from Figure 5.15, and painting by Pablo Picasso, *Houses on the Hill, Horta de Ebro.* Summer 1909.

In this photograph Picasso shows his understanding that lines need not merge into a single perspective point. Scenes in nature can exhibit divergent lines to great effect. Like Braque, Picasso had realized that scientific perspective is a "ghastly mistake nothing but eye-fooling illusionism."[78]

For all the wrong reasons Gertrude Stein placed Picasso's photo of rooftops in Figure 5.15 of Horta side-by-side with his painting *Houses on the Hill, Horta de Ebro*[79] in order to show that the painting was a "too photographic view of nature."[80]

Close inspection of the painting reveals instead the extremes of *passage* and geometrization unified in a literal folding together of depicted space. This new result of cubism is even more dramatic in Picasso's *Reservoir, Horta de Ebro.*[81] In this painting, he succeeded in folding together the views of at least three photos. This early tour de force of cubist volumetric analysis depicts different views of an object on a single canvas in an instant of spatial simultaneity.[82]

Picasso kept Gertrude up-to-date with photos of his atelier, showing works in progress. The photographs reveal preliminary drawings for paintings, paintings in various developmental phases—metamorphosing as Picasso puts them up and takes them down; sometimes totally visible, sometimes not, being partially obscured on purpose, and then emerging

FIGURE 5.17 Pablo Picasso, *Atelier at Horta de Ebro (Nude Woman).* Horta de Ebro, 1909.

fully when completed. Figure 5.17 is typical of his photographic studies of ateliers he occupied during 1904–1914. It shows on lower left the *Head and Shoulders of a Woman (Fernande)*,[83] in the center *Female Nude*,[84] on the right *Head of Fernande*,[85] and just above *Head and Shoulders of a Woman (Fernande)* is *Head of a Woman in Mantilla (Fernande)*,[86] partially obscured and still incomplete. *Female Nude* is also incomplete and so it, too, is partially obscured. We note that not only the woman but the entire space as well is now faceted.

In Figure 5.18, *Female Nude* has been turned around and we see Picasso's *Reservoir, Horta de Ebro*[87] and *Factory at Horta de Ebro.*[88] Both of these paintings encompass all of the photographic views of Horta, but in a highly geometrized manner with extreme *passage.* Incidentally, there were no palm trees in Horta.

In Figure 5.19 *Head of a Woman in Mantilla (Fernande)* emerges almost completed. To the upper right of *Head of a Woman in Mantilla (Fernande)* is *Bottle of Anis del Mono*,[89] a much advanced version of Braque's play with Cézannian *passage* and so with geometrization of space.

Figure 5.20 is a studio setup in which *Head of a Woman in Mantilla (Fernande)* emerges completed. In this version her head is as if it were carved out of Santa Barbara mountain. There is pretty much complete unification of planes. The centerpiece is *Seated Woman*,[90] with the most

FIGURE 5.18 Pablo Picasso, *Atelier at Horta de Ebro (Reservoir, Horta de Ebro* and *Factory at Horta de Ebro).* Horta de Ebro, 1909.

FIGURE 5.19 Pablo Picasso, *Atelier at Horta de Ebro (Bottle of Anis del Mono).* Horta de Ebro, 1909.

FIGURE 5.20
Pablo Picasso, *Atelier at Horta de Ebro (Seated Woman).* Horta de Ebro, 1909.

angular of the faceting that Picasso experimented with at Horta. It is interesting that some of the most dramatic faceting, as in the forehead of *Woman in Mantilla*, is of the William James "folding card" genre that we discussed in Chapter 4. Notable as well are the three heads of Fernande, drawn as if her head were rotating about a vertical axis.[91] This can be construed as three frames of a moving picture, much as in Muybridge's cinematography.

Picasso also experimented with perception and symmetry. Figure 5.21, for example, shows his interest in a variation on binocular vision, visual dissonance. Asymmetry can also be pleasing. On the left is *Nude in an Armchair*[92] and on the right *Woman with Pears (Fernande).*[93] *Nude in an Armchair* exhibits a high degree of faceting, with greater freedom than in most of this summer's works, and anticipates advances by Picasso during the next year.[94] While the paintings are somewhat similar in angular attitude, the photograph plays off their similarities and dissimilarities. Picasso invites the viewer to look quickly back and forth, altering the images by passage of shadows or change of expression. It is interesting that Picasso was curious about stereoscopy at the moment of the conceptual definition of cubism.[95]

Figure 5.22 displays another of Picasso's photographic experiments, printing a superposition of two negatives. Baldassari has analyzed this in

FIGURE 5.21 Pablo Picasso, *Atelier at Horta de Ebro (Woman with Pears* and *Nude in Armchair).* Horta de Ebro, 1909.

FIGURE 5.22 Pablo Picasso, *Atelier at Horta de Ebro.* Horta de Ebro, 1909.

exquisite detail.[96] She ascertained that one of the negatives was the one that produced Figure 5.21. The other has since been lost but can be reconstructed to have contained on the right *Nude Woman*, at the center *Woman with Vase of Flowers*,[97] higher in the space between the paintings and the wall the *Head and Shoulders of a Woman (Fernande)*, and very likely on the left, *Standing Nude with Raised Arms*.[98] Picasso made two prints from this photographic setup. The tantalizing question is whether he also experimented with other negatives.

Here Picasso is exploring how the fusion of two sets of cubist images can give rise to a second-order cubist reality. The artist has succeeded in generating a rhythm by setting into motion about the painting's center an oscillating shockwave, a visual vibration rooted in the almost stereographical layering of negatives. The resulting structure is a stunningly complex and ambiguous transparency of planes, with an oscillating pattern of facets. Images like this would be invaluable to Picasso in his move toward a completed analytical cubism. Additionally, we can imagine that he was continuing the experiments with the concept of spatial simultaneity that commenced with the *Demoiselles.*

Picasso's ateliers, at Horta de Ebro during the summer of 1909 as well as in Paris, had become research laboratories where he experimented boldly on the nature of cubist space. He was using one of the great tools of positivist science, the camera, to show that what can be passively collected on the film or retina is not all there is to see. What you saw was not what you got.

LA CORDÉE, 1910–1912

At Horta, Picasso took the lead from Braque in terms of formal invention. By late 1909 they were meeting more often as the *cordée* commenced in earnest. Their works quickly grew increasingly complex and abstract. In Picasso's *Portrait of Vollard*[99] and *Portrait of Uhde* (Figure 5.23), objects and background are fused into what is almost a continuum of facets. At the same time, the closed form of Uhde is fragmented. These portraits are a "culmination of the explorations begun at Horta."[100] Braque, Picasso, subsequent cubists such as Metzinger, and commentators such as Apollinaire always considered cubism to be an art of realism.[101] Cubist works start with a subject that is then represented in ways that attempt to illuminate its deeper structure. In any cubist painting one can always identify objects, however faceted and fractured they may be. But starting in the latter part of 1909 the paintings became so complex that the artists

FIGURE 5.23
Pablo Picasso,
Portrait of Uhde,
1910.

sometimes had to insert prompts to pull them back from complete abstraction. In paintings, space was fused with objects that were so shattered with faceting that the whole canvas became a dazzling unity that left the viewer somewhat disoriented.

An example from Braque's style of *passage* is his 1910 painting *Violin and Palette,* which he built up by overlaying planes from rear to front (Figure 5.24). In conjunction with facetation this process materializes all space in the picture plane. The resulting image of interlocking of planes gives an impression of unprecedented complexity. Braque provides signs to identify a violin. But just to be sure to bring the viewer back to reality, he nailed the palette to the wall with a trompe l'oeil nail, complete with shadow. As he recalled in 1954: "When the fragmentation of objects appeared in my painting around 1909, it was as a technique for getting closer to the object in the limit in which the painting permitted it to me."[102] In his *Portrait of Uhde*, Picasso added a key to the drawer on Uhde's left.

The mountaineers had not yet reached the summit. Both men wanted to get even closer to objects and were bent on pushing their abstraction to greater heights. Many years later Braque explained how he was coming to

FIGURE 5.24
Georges Braque, *Violin and Palette,* 1910.

terms with the nature of pictorial space: "The fragmentation [of objects] helped me to establish space and movement within space, and I could only introduce the object after having created the space."[103] Through his interest in representation of forms, Picasso was the one on the right track.

In spring 1910, Picasso painted *Standing Female Nude*, which set the stage for the complete opening of form that summer (Figure 5.25). Like many of Picasso's pre-Cadaqué nudes it "exhibits the prominent pelvic hip bones apparent in full-body X-rays."[104] We know that by 1907 Picasso was aware of X rays and their properties. He doubtlessly saw Fernande's X rays from her illnesses and so had seen at first hand how they open up forms and reveal on photographic film the image of objects that appear abstract but are not. X rays, with their blurring of the distinction between two and three dimensions and their rendering opaque objects as transparent, were powerful weapons in the battle against positivism. It is reasonable to assume that they provided a powerful clue to Picasso's breaking the closed form of bodies and objects.

FIGURE 5.25
Pablo Picasso,
Standing Female Nude,
spring 1910.

Another hint was Braque's use of the grid. This device appears in his painting *The Rio Tinto Factories at l'Estaque* (Figure 5.26) in which the subject matter is recognizable and the painting is set out "in a loose grid or framework suggested by the outlines of buildings."[105] The simplified geometrical forms open up to reveal an intricate network of interconnecting planes. Picasso then pushed Braque's concept of grids toward an even more abstract representation in his *Harbour at Cadaqués*, in which an anchor at bottom right offers a key.[106] In breaking the closed form, Picasso here moves to the very brink of abstraction. At times he could not even remember the ostensible subject in many of the sketches and preliminary paintings he brought back to Paris in September 1910.

With these clues Picasso cracked the problem during the summer of 1910 at Cadaqué, a small town on the east coast of Spain near the French border. One product from this summer is his magnificent *Female Nude* (Figure 5.27), which reaches what for him at that time was the height of abstract geometrization.

Picasso's masterpiece of autumn 1910, *Portrait of Kahnweiler,* takes geometrization yet further (Figure 5.28). Whereas, in *Female Nude,* form

FIGURE 5.26
Georges Braque, *The Rio Tinto Factories at L'Estaque*, 1910.

FIGURE 5.27
Pablo Picasso, *Female Nude,* autumn 1910.

FIGURE 5.28
Pablo Picasso, *Portrait of Daniel-Henry Kahnweiler*, autumn 1910.

is reduced to geometrical structure with its own abstract rhythm, in *Portrait of Kahnweiler* closed form is not just broken but shattered: There are geometrized bits of Kahnweiler everywhere.

Paintings like this are difficult to read. At first we are unaware of images that materialize gradually from intricate and complex interacting parts only to be reabsorbed into the painting's dominating spatial rhythm. Picasso pulls *Portrait of Kahnweiler* back from complete abstraction by providing signs such as lips, hands and eyes—disjointed, of course, and not unlike Méliès's movies in which body parts are flying around helter-skelter. Through this network of signals we can read the picture.

The faceting of objects and space, with their concomitant fusion and increased abstraction, accelerated between 1909 and 1912. By 1911 Braque had to resort to considerably more drastic measures than nails to snap the canvases back to figuration. In *Le Portugais* (1911), he introduced stenciled letters: "They could not be deformed because, being flat, the letters were beyond space and their presence in the painting, by contrast, permitted one to distinguish objects situated in space from those which were outside of space."[107]

In early 1912 Picasso countered by inventing *collage*, as in *Still Life with Chair-Caning*.[108] This is a turning point in twentieth-century art in which a "readymade" is created by applying to canvas objects other than paint. Braque's stenciled letters, in their flatness, emphasized the flatness of the canvas itself. Picasso then sought to liberate art from two-dimensionality by adding objects that are themselves raised from the canvas. Later in 1912, Braque did Picasso one better with the invention of *papier collé* in the painting *Compotier et Verre*, where strips of wallpaper are used to simulate wooden graining. Picasso, in turn, took Braque's *papier collé* to daring heights by including brightly colored and elaborate shapes of paper. It quickly became "impossible to distinguish completely *papier collé* from *collage*."[109]

Kahnweiler recommended that "cubist pictures should always be provided with descriptive titles" because "titling will . . . prevent sensory illusions."[110] Analytical cubism, he continued, was able to reduce the representation of objects to their primary qualities, which he defined as an object's "form and its position in space," while leaving the viewer to envision such secondary qualities as color and texture.[111] This observation underscores Picasso's scientific bent and even echoes a turning point in the history of scientific ideas, when Galileo realized that in order to formulate a consistent science of motion he must consider only qualities that are amenable to objective or mathematical representation. In the physical sciences they include velocity, mass and force. An object's secondary qualities, such as its color, smell and other "essences," are excluded because they cannot be rendered mathematically.[112]

Having discovered a similar reduction in the *Demoiselles*, Picasso elaborated it further between 1908 and 1912 with a great deal of help from Braque as well as from developments in mathematics (the fourth dimension), science (X rays) and technology (cinematography and photography). In the new manner of representation that emerged, forms are reduced to basic geometrical components that are themselves shattered so as to blend into their surrounding space, creating a simultaneous geometrization of space and bodies.

APOLLINAIRE'S ELABORATIONS ON THE FOURTH DIMENSION

Around 1911, various articles appeared that attempted to formalize and popularize cubism. Neither Picasso nor Braque took a hand in this. The principal contributors were writers and critics such as Salmon, Apollinaire, Raynal, Pierre Reverdy, Olivier-Hourcade and others, as well as theoretically inclined cubists such as Alfred Gleizes and Jean Metzinger.

Avant-garde artists and writers in Picasso's circle took account of the fourth dimension's connotations of the occult, which they interpreted as a manifestation of spatial simultaneity. As part of the opening of the *Section d'Or* that ran in conjunction with the October 1912 Salon d'Automne, Apollinaire lectured on the fourth dimension and modern painting. Princet was listed on the organizing committee along with Apollinaire and the group of cubists who took, in their own view, a formal and intellectual view of the new art form based in mathematics as Princet had explained to them. Picasso and Braque did not exhibit. Painters like Robert Delaunay, whose work tended to transcend the abstraction cubism offered, did.

For Apollinaire the fourth dimension was not scientific but metaphorical, and it contained the seeds of a new aesthetic. This is abundantly clear in Part 3 of Apollinaire's 1913 book *Les Peintres Cubistes*, an expanded version of his 1912 article "La Peinture nouvelle."[113] He writes of cubism as an art of the fourth dimension: It transgresses the naturalistic representation of a three-dimensional reality and its linear perspective, which was by then thoroughly stigmatized. The term "fourth dimension" is a metaphor for the immensity, indeed infinitude, of space that "gives objects the proportions they merit in works of [the new] art."[114] Whereas Greek art "took man as the centre of the universe"—and a finite one at that—"the art of the new painters takes the infinite universe as its ideal."[115] Like Kahnweiler, Apollinaire saw cubism's parallels with the scientific revolution. It replaced a finite universe, with man at its center, with an infinite universe.

In *Les Peintres Cubistes*, Apollinaire took care to mention the importance of geometry in the new art: "Some have violently reproached the new artist-painters for their geometrical preoccupations. However, geometrical figures are the essential of drawing. Geometry is to the plastic arts what grammar is to the art of the writer."[116] The fourth dimension, he concludes, is a kind of shorthand for the various methods that Picasso brought to bear on three-dimensional figurative space in order to open it up. Because an infinite universe has no center, neither should a painting have a single perspective point such as occurs when using Euclidean geometry.

This material occurs in the first paragraph of Apollinaire's Part 3, which he separated with a small space from the metaphorical discussion that followed.[117] In the early drafts, he mentioned Princet as one of the defenders of "Scientific Cubism," along with Picasso and Braque.[118] But Princet was deleted from the published version. For the most part, Apollinaire appropriated the metaphorical interpretation of the fourth dimension from Max Weber's article.[119] Weber's and Apollinaire's ideas about the fourth dimension were rooted in discussions at the Steins' that may have owed a debt, either directly or indirectly, to Princet.

Despite his oblique prose and lack of in-depth knowledge of art, Apollinaire's assessment of Picasso's creativity and genuine interest in spreading the message of cubism were welcomed by Braque and Picasso.[120] Apollinaire reflected Picasso's attitude perfectly when he wrote in *Les Peintres Cubistes,* "*Scientific Cubism* is one of the pure tendencies. It is the art of painting new ensembles with elements borrowed, not from the reality of vision, but from the reality of knowledge [*la connaissance*]."[121] The received translation of this passage turns on rendering *la connaissance* as "insight."[122] Besides being plain wrong, this reading dilutes Apollinaire's profound understanding of Picasso's original goal of seeking a representation based in conception rather than perception. This could only be achieved through perception *aided* by understanding from science. We understand the world about us by perception *and* cognition. The understanding we obtain from processing perceptions can be dramatically transformed by advances in science. Apollinaire used *la connaissance* in direct opposition to the highly sense-dependent and positivistic representations offered by impressionism. This rendering squares with Picasso's assessment of cubism as an art of realism.

Juan Gris said of Apollinaire in 1924 that he "would often ask us what we thought of this or that and, at times, he reproduced our words verbatim."[123] Apollinaire's high regard for Picasso and his intense desire to please him are reflected in the fact that Picasso's "name reappears in almost every chapter, whatever the apparent subject, and his spirit is diffused in the entire book."[124] Certain passages in *Les Peintres Cubistes* were "more or less dictated by Picasso, which, beneath their poetic allusions were quite precise."[125] Apollinaire's note on "scientific Cubism" may be one of these passages.[126]

SOME VARIATIONS IN AND ON CUBISM

Despite Picasso's notoriety, until late in 1912 his and Braque's advances were not familiar to artists or the public generally. Picasso did not exhibit publicly at the big salons, and what information was available came from Vauxcelles's comments about Braque's exhibition at Kahnweiler's in November 1908 and a Picasso show held in May 1910 at Uhde's Notre-Dame-des-Champs gallery. Both men's works could be viewed by invitation only at Kahnweiler's gallery in the rue Vignon or at the Steins' flat.[127] In 1909 Vauxcelles once again provided some reportage in his critical review of the Salon des Indépendants in which he described two paintings of Braque's as *bizarreries cubiques.*[128] Also showing were two future cubists, Jean Metzinger and Henri Le Fauconnier.

Metzinger has for decades been shortchanged by the received history of cubism, spun by Kahnweiler while in exile in Switzerland as an enemy alien during World War I. Beginning with Kahnweiler, art historians often referred to the period after about 1913 as the "synthetic" phase of cubism, while the previous period was the "analytic" phase.[129] In the analytic phase artists broke down objects into geometrical forms and vice versa for the synthetic phase. For Braque and Picasso, however, there never was a pure analytic phase. Having broken with traditional perspective, they were nevertheless "combining or 'synthesising' various views of an object into a single image."[130] Since in Kahnweiler's history of cubism no artist other than Braque, Picasso and Gris evolved through analytic and synthetic phases, all others are deemed unimportant. Not coincidentally, these three artists were also under contract to Kahnweiler's gallery.[131]

Yet Metzinger was the first person to bring cubism to the public's eye. Of the group known as the Salon Cubists, because they exhibited in the major salons, he was the one who knew Picasso first and best, in addition to having been inspired by him.

Arriving in Paris from Nantes in 1902, Metzinger began showing as a neoimpressionist and then commenced an almost annual switch of styles.[132] Through his friendship with Max Jacob, in 1907 he met Apollinaire and Picasso, who was two years older. Metzinger was unique among the Montmartre artists in that he was also a writer, having published poetry in small Parisian reviews. In 1910 he tried his hand at cubism and exhibited two paintings at that year's Salon d'Automne, *Landscape* and *Nude*, the latter an interesting, complex study in the style of Picasso's *Portrait of Kahnweiler.*

In fall 1910 Metzinger published the first of several highly intellectual analyses of Picasso's style. In "Note on Painting," he wrote glowingly that "Picasso does not deny the object, he illuminates it with his intelligence and feeling. With visual perceptions he combines tactile perceptions. He tests, understands, organises."[133] The terms "visual" and "tactile" stand out as intended: Metzinger was well versed in Poincaré's *La Science et l'hy-*

pothèse. Metzinger continues with a passage that sets him apart among analysts of Picasso's art: "[Picasso] lays out a free, mobile perspective, from which that ingenious mathematician Maurice Princet has deduced a whole new geometry."[134]

This essay is an important document in the history of cubism because it is the first accurate account of what Picasso was after, which had nothing to do with formal analysis or analytic concepts. Rather, Picasso was interested in removing all vestiges of Renaissance perspective and replacing it with a "mobile perspective," or multiaspect view of a scene. In 1911 Metzinger elaborated, in a way that has become synonymous with the popular conception of cubism, "[Cubists] have allowed themselves to move round the object, in order to give . . . a concrete representation of it, made up of several successive aspects."[135]

Moreover, Metzinger's 1910 essay connects him with Princet. The attraction is understandable in light of Metzinger's passion for mathematics when a student at the Lycée in Nantes. In 1952 Metzinger recalled that, along with Gris, he "undertook a study of geometry under the direction of Princet in order to explore" possible connections between non-Euclidean geometries and the fourth dimension with the concept of space in cubist art.[136] His essay is additionally significant because he was the only cubist painter who spoke with Braque and Picasso at this time.[137]

Purely by chance, at the 1910 salon Metzinger's works were hung beside those of Albert Gleizes and Henri Le Fauconnier. The relations that critics saw between their styles were amplified in conversations among writers in Left Bank cafés such as Closerie des Lilas. With help from literary friends such as Apollinaire and Salmon, Metzinger and his circle persuaded the general committee for the 1911 Salon des Indépendants to permit them to exhibit together. This occurred in Salle 41 which created something of a sensation.

Toeing the line of Braque's and Picasso's earlier cubism, Metzinger meant his contribution, *Le Goûter* [Tea Time] (Figure 5.29), as a representation of the fourth dimension. A naked faceted woman is shown simultaneously in full and profile views while drinking from a cup of tea painted from side and top views. This is almost a verbatim rendering of the problem that Princet had posed over a bistro table some two years earlier. It is a straightforward multiple viewing, as if the artist were moving around his subject. In this way *Le Goûter* is a step back from Metzinger's earlier *Nude* of 1911 where he emulated Picasso's new style in *Portrait of Kahnweiler,* in which space is exploded and so the subject matter is increasingly difficult to read. Nevertheless, *Le*

FIGURE 5.29
Jean Metzinger,
Le Goûter, 1911.

Goûter "was hailed as a breakthrough . . . and opened the eyes of Juan Gris to the possibilities of mathematics."[138] The absence of Picasso and Braque at these exhibitions established Metzinger as the foremost spokesman of cubism. Primarily on the merits of *Le Goûter,* he was considered as cubism's *chef d'école.*

By the end of 1911, Gleizes, Gris, Le Fauconnier, Metzinger and Princet were joined by Francis Picabia, Roger de La Fresnaye and the brothers Jacques Villon, Raymond Duchamp-Villon and Marcel Duchamp in a discussion group that also included literati and met on Sundays at Jacques Villon's studio in Puteaux. They became known as the "Puteaux circle." Princet played a major role in these discussions, although Duchamp recalled that his pupil Metzinger sometimes outshone him.[139] This highbrow study group had two important outcomes.

The first was the high point of the cubist movement, the *Section d'Or* exhibition at the Galerie de la Boétie in October 1912, where more than thirty painters showed cubist works. Apollinaire acted as impresario, and the list of organizers included Jacob, Raynal, Salmon and Princet. Once again, Braque and Picasso remained aloof.[140] Their absence gave the public a very different and far less conceptual view of cubism than it would otherwise have had.

Fernand Léger tried to reduce scenarios to industrial-like geometric scenes, which Vauxcelles dubbed "tubism."[141] Robert Delaunay went his own way, with vibrant colors.[142] Only Metzinger remained close to the founders of cubism, but his paintings were essentially academic studies of earlier works by Braque and Picasso that lacked any real appreciation of what they were presently up against.[143]

The second important result of theoretical deliberations among the Puteaux artists was the crystallization of their thoughts in Gleizes and Metzinger's 1912 book *Du Cubisme* [On Cubism]. Even though it never once mentioned Picasso and Braque, *Du Cubisme* was considered the cubist manifesto. Nor did Gleizes and Metzinger ever speak explicitly of the fourth dimension, although they write that "if we wished to tie the painter's space to some particular geometry, we should have to refer it to the non-Euclidean scientists."[144] In place of the artistic structures of Euclidean geometry, "Cubism substitutes an infinite liberty."[145]

While Gleizes and Metzinger's discussion of geometry and space parallels the one in Poincaré's *La Science et l'hypothèse,*[146] they deviate markedly from Poincaré's conventionalism, according to which all geometries are equally valid, and dogmatically state that "there is only one truth, ours, which we impose on everyone."[147] Their most widely quoted message is a variant on the one in Metzinger's article of 1911, that a cubist picture can be likened to "moving around an object to seize it from several successive appearances, which, fused into a single image, reconstitute it in time."[148] But this refers to the older cubism of Picasso and Braque.

Metzinger and his colleagues in the Puteaux circle attempted to explore the same ideas of space as Picasso and Braque before them, under the tutelage of the same man, Maurice Princet. But their commitment to figuration served as a brake on their imagination. Still, although ultimately a minor cubist painter, Metzinger was the first to appreciate intellectually what Picasso and Braque were trying to accomplish. He brought their cubism into the public domain and opened the field to others.[149]

Juan Gris was another story altogether. Gris arrived in Paris in September 1906 and through contacts in the Catalan community soon found his way to Picasso, who got him a flat in the Bateau Lavoir and a job as a cartoonist for *L'Assiette au beurre.* But the Picasso circle inspired Gris to more creative pursuits. By 1910 he was painting in the cubist style. Having developed an astounding facility with the new genre, Gris became the third man in cubism's inner circle. In addition, with his training in mathematics and physics, Gris was naturally attracted to Princet's informal lectures.[150]

Gris intellectual approach to cubism resulted in works of stunning construction, with highly intricate workmanship and an austere beauty that have led some critics to feel he approached art coldly, like a draftsman, with compass, measuring instruments and sharpened pencils in hand.[151] Close study reveals something entirely different. Although the finished works have an extreme precision, what remains of his preliminary sketches, as well as examination of alterations he did on the canvases themselves, tell a story of painstaking, tentative searches full of erasures and hand sketches. The appearance of certainty masks the exploration in Gris's preliminary work.[152] A Gris painting is not unlike a finished and pristine scientific paper, from which the angst of the creator has been removed.

In a letter to Amédée Ozenfant written in March 1921, Gris described his highly intellectual approach to cubism, in which any composition can be reduced to purely geometric forms: "I work with elements of the intellect, with the imagination. I try to make concrete that which is abstract. . . . I consider that the architectural element in painting is mathematics, the abstract side. . . . Cézanne turns a bottle into a cylinder, but I begin with a cylinder and create an individual of a special type. I make a bottle—a particular bottle—out of a cylinder."[153] But in practice, this idea of a pure synthetic cubism eluded even Gris.[154]

BEYOND CUBISM

Although Delaunay sometimes made analogies with the notion of simultaneity held by the futurists—one of fusing together space, time and motion—his own sense of simultaneity was based on the nineteenth-century color theories of Chevreul and Rood.[155] As Delaunay wrote in 1913, "simultaneity in light, it is the harmony, the rhythm of colours which created *Man's Vision*."[156] His distaste for the lack of color in the work of Picasso and Braque led Delaunay toward pure color and increased abstraction.

Through the Puteaux circle, Delaunay became friends with Princet, who wrote a nonmathematical catalog preface to Delaunay's 1912 exhibition at the Galerie Barbazanges. In a 1912 letter to Wassily Kandinsky, Delaunay hinted at a theory he had worked on for some time that concerned "transparency of colour, comparable to musical notes" whose laws were so new that "even my friend Princet has not seen them."[157] Princet, he reports, reacted positively, saying that these ideas followed "the lines of work that he has already started many years ago and that I have indeed alerted you to." As far as we know nothing ever emerged.

But Metzinger recounts the following story in his autobiography: In front of Metzinger and Gris, Princet discussed some recent work on reuniting "in a single system relations of colour . . . ; and, initiating us to the non-Euclidean geometries, he urged us to create a painter's geometry."[158] Could this work have been what Delaunay referred to? This is indicative of the role played by Princet both in the Picasso and Puteaux circles: a source and sounding board of fundamental theories. Princet died on 23 October, 1973, at the age of ninety-eight, having amassed a fortune in business ventures.

Apollinaire, though he may not have known a Rembrandt from a Rubens, had an excellent sense of what was new and different in art and was prepared to defend it. Recognizing that a new style was emerging from the studios of Delaunay, Léger, Picabia and Duchamp, Apollinaire gave it the name "Orphism" and defined it in *Les Peintres Cubistes*: "Orphic Cubism . . . is the art of painting from new ensembles with elements borrowed not from visual reality, but entirely created by the artist and endowed by him with a powerful reality."[159] Orphism was a source for pure abstraction. This trend must have been upsetting to Picasso and Braque, who were struggling to maintain a balance between abstraction and representation.

The Orphists "were influenced by contemporary science, technology, literature and philosophy as well as the actual experience of living in the contemporary world."[160] The principal Orphists, Frank Kupka, Duchamp, Delaunay and Léger, mixed these elements in different ways. Delaunay and Léger frankly reflected the technological world in which they lived; Duchamp's intellectual style is based in no small part on developments in mathematics and science;[161] and Kupka employed a mysticism deeply influenced by such scientific developments as X rays.[162] In the end, however, the Orphists' world is a fantasy. They have little interest in the program of artists such as Gris and Piet Mondrian, who considered themselves to be equivalent to scientists in a search for invariant laws of nature.[163]

For all of these artists, cubism set in motion a new manner of breaking down forms and reorganizing pictorial space. In their attempt to fuse art and science, Picasso and Braque thus set the stage for all of twentieth-century intellectual art. As Philippe Dagen, the editor of the Derain-Vlaminck correspondence, writes: "For these artists, there was no doubt at all that the scientific revolution and aesthetic revolution, technological modernity and pictorial modernity are indissociables."[164]

In 1912 Raynal commented on just this multidisciplinary effort by artists and scientists to reach beyond sense perceptions. On the dangers of trusting the appearances of nature, Raynal tells the story of Icarus trying to fly by imitating a bird. Yet the proper means was with a propeller: "The quest for truth has to be undertaken not merely with the aid of what we see, but of what we conceive."[165] He continues with a lesson already learned by the cubists and recently advocated by Einstein, that "judgements and reasonings based on perception only are for the most part erroneous [and] painting based solely on external perception is, therefore, inadequate."[166] Only the conceptualist method would bring the artist closer to the truth. The painter should rely on "only conceptions of objects, for these alone are created without the aid of those inexhaustible sources of error, the senses."

Another spin-off of cubism hit the public's eye in 1912 with Duchamp's *Nude Descending a Staircase.* Its kinetic style can be traced to Marey's photographic experiments, X-ray technology, futurism and Duchamp's desire to portray motion in time on a single canvas.[167]

Kupka's exploration of themes based on X rays ended in about 1912, when he joined with Kandinsky and Kazimir Malevich in a realm of complete abstraction. They could well have been motivated by the complete dematerialization of matter described in the new physics. For this they needed no awareness of Einstein's mass-energy equivalence, LeBon's writings were enough. Since all matter is at bottom amorphous, so should be its ultimate representation.

The cubist metaphor had an impact outside of art. Gertrude Stein recalled that at the beginning of the war she was standing with Picasso "on the boulevard Raspail when the first camouflaged truck passed. It was at night, we had heard of camouflage but we had not yet seen it and Picasso amazed looked at it and then cried out, 'yes it was we who made it, that is Cubism'."[168] Although the inventor of camouflage, Guirand de Scévola, had never met Picasso, he was aware of his work and is reported as saying: "In order to totally deform objects, I employed the means cubists used to represent them—later this permitted me, without giving reasons, to hire in my [camouflage] section some painters, who, because of their very special vision, had an aptitude for denaturing any kind of form whatsoever."[169]

The changing face of camouflage in the twenty-first century provides another link between art and science. It turns out that the optimum design to scatter radar so that an airplane as big as a barn looks like a speck in the sky is a collection of facets.[170] The United States Air Force's B117, the Stealth Bomber, is a flying cubist sculpture.

INTERMEZZO

The painter's studio should be a laboratory. There one does not make art in the manner of a monkey, one invents. Painting is a play of the mind.

—Pablo Picasso

Before turning to how Einstein discovered relativity theory in 1905, let us pause for a moment to recapitulate.

Almost immediately upon Picasso's arrival in Paris during May 1904, there was a sense that someone to be reckoned with had come on the scene. The highly charged Parisian artistic-literary arena found itself ignited by an astonishing artistic talent with breathtaking charisma. Picasso's piercing dark eyes radiated an absolute self-confidence that he would succeed on the grandest possible scale. Immediately and dramatically, he altered the personal lives and careers of his literary mentors, Apollinaire, Jacob and Salmon, rising stars in their own right who epitomized Parisian café society. The center of gravity of this core group shifted from the Left Bank to Montmartre and finally to Picasso's atelier in the Bateau Lavoir, on whose door hung the sign *Rendezvous des Poètes*.

La bande à Picasso was Picasso's think tank. Through articles in literary journals and in café conversations with such electric personalities as Jarry, they kept him informed of avant-garde intellectual fashions. These included whatever was new in literature, philosophy, or areas of science that touched on their attraction for the occult. The group significantly included

Maurice Princet, whose discussions of non-Euclidean geometry and the fourth dimension, based on Poincaré's writings, kept Picasso enthralled and set him thinking along radically new lines. To Poincaré's suggestion that the fourth dimension be represented as a succession of scenes, Picasso added a visually ingenious twist: Set down different views of a scene all at once, simultaneously. These were the essential elements in Picasso's discovery of a representation of nature that caught the enormous conceptual transformations occurring in art, science and technology at the beginning of the twentieth century. In late 1907 he joined forces with Braque in a collaboration, without precedent in art, that led eventually to the high formal phase of cubism. Here, too, Picasso's discussions with Princet, as well as his adventurous photographic experiments, were essential.

In the chapters to follow we will see how Albert Einstein, in Bern, Switzerland, also surfed the tidal wave of the avant-garde. The "Olympia Academy" was Einstein's think tank, in which he and his pals dissected classics in philosophy that also included Poincaré's *La Science et l'hypothèse.* It will turn out that simultaneity played a central role for Einstein, too, and that he also had to correct a suggestion by Poincaré. The relativity theory, like cubism, is a profound response to changes in the philosophical and scientific climate as well as to dramatic technological innovations.

Picasso's and Einstein's unique ways of confronting the tension between the entrenched classical and emerging nonclassical modes of thought at the beginning of the twentieth century were, at bottom, similar. Ultimately they were working on the same problem: How to represent space and time at just the moment in history when it became apparent that these entities are not what we intuitively perceive them to be. For Picasso the problem was framed by his rejection of such established art forms as impressionism, his elaborations on primitive conceptual art and then his realization of the key role played by new ideas in geometry. For Einstein it was his discovery of the wave/particle duality of light and then his unique use of thought experiments to go beyond the data of the laboratory, data that hindered the thinking of others. Both men sought representations of nature that transcend those of entrenched classical thought and reach beyond appearances. The intensity of their research, complications in their personal lives and the periods in which they were stuck forced them into a solitude and anxiety such as they had never before experienced.

Others who sensed the winds of change moved in what turned out to be the proper directions, yet drew back at critical moments. Derain's failing lay in not taking primitive conceptual art to its deepest conclusions—whereas Picasso succeeded with help from Princet; Poincaré's was his reliance on experimental data and sense perceptions in defining the concepts of time and simultaneity—whereas Einstein, as we will see, brought into play "data" from thought experimenters in order to go beyond both laboratory data and sense perceptions.

To my surprise, Henri Poincaré turns out to be a central figure for both Picasso and Einstein. In 1976, in Paris, I discovered Poincaré's letters and manuscripts, missing since his death in 1912, in the possession of his grandson, Monsieur François Poincaré. This led to me to reexamine how his philosophical views and scientific research reflected one another. Still, nothing prepared me for the impact of Poincaré's writings on Picasso and his circle. Since Poincaré will play an increasingly substantial role in what follows, some words of introduction are in order.

In 1873, at age nineteen, after a stellar lycée career, Poincaré entered l'École Polytechnique, from which he was graduated in 1875 and admitted to another of the Grandes Écoles, l'École des Mines. In 1879 he was awarded a doctorate from the University of Paris and immediately embarked on mathematical research of extraordinary originality.

From 1881 until his death in 1912, Poincaré held professorships at the University of Paris in mathematics and astronomy as well as similar positions at l'École Polytechnique. A testament of the breadth of his knowledge, and to his power in French academic circles, is that he was a member of both the Académie des Sciences (elected 1887, president 1906) and the Académie Française (elected 1908, director 1912). Poincaré occupied the apex of the pyramid that constituted French science.

Unlike Picasso's and Einstein's educational careers, Poincaré's was exemplary in every respect.[1] Like Picasso, at an early age he exhibited unusual ability in the subject in which he would do his most creative work, mathematics. Unlike Einstein's, Poincaré's talents were encouraged by his teachers. (Einstein, in contrast to Picasso and Poincaré, demonstrated no outstanding originality in his specialty until 1905.)

Poincaré's curriculum vitae is a 110-page book. He published on the order of 500 papers and 30 books, and received numerous honorary degrees as well as every major scientific prize except the Nobel—for which much lobbying took place on his behalf.

FIGURE I.1 Henri Poincaré, about 1910.

Besides being one of the greatest mathematicians who ever lived, Poincaré made significant contributions to all branches of physics and astronomy, in addition to formulating a unique philosophical viewpoint called conventionalism. Space vehicles are launched and their orbits calculated using Poincaré's hybrid version of Newton's mechanics, and his mathematical papers still promise new riches. It was a rereading of certain of Poincaré's classic papers from the late nineteenth century that, in part, revealed to twentieth-century mathematicians the basis of one of the most exciting areas of science today, chaos theory.[2] It is no exaggeration to say that Poincaré was the man who completed Newtonian mechanics, and that Einstein took the next step to widen its applicability and so truly extend it into the twentieth century.

Given his broad interests it is not surprising that Poincaré was sufficiently intrigued by creative thinking to agree to a detailed series of interviews in 1897 by the French psychologist Édouard Toulouse, as part of Toulouse's study of genius.[3] Toulouse was struck by the sharp difference between Poincaré's style of thinking and the stereotypical style of the scientist, as he reports comparing Poincaré with the writer Émile Zola:

> The one [Zola's] was an intelligence that was wilful, conscious, methodical, and seemingly made for mathematical deduction: it gave birth to a romantic world. The other [Poincaré's] was spontaneous, little conscious, more taken to dream than for the rational approach and seemingly throughout apt for works of pure imagination, without subordination to reality: it triumphed in mathematical research. And this is one of the surprises, which calls for direct studies touching on the deepest mechanisms.[4]

Poincaré himself believed in a commonality of creative faculties between artists and scientists. In a still widely read introspection from 1908, he described in some detail how he went about his own research. He spoke of the mathematician's "special aesthetic sensibility," which plays the role of the "delicate sieve" that filters out all but the few combinations that are "harmonious" and "beautiful."[5] This combined aesthetic and rational sense was, for Poincaré, the mind's highest faculty. "It is only through science and art," he remarked, "that civilisation is of value."[6] In *La Science et l'hypothèse,* he wrote with great passion about the scientist's quest for representations of nature that embody simplicity, symmetry and unity of foundations. To probe the deepest problem in philosophy and psychology—how we formulate exact knowledge from the potpourri of sensations that constantly bombard us—he went on to propose hypotheses about the mind's architecture. This is the sensibility that inspired young artists and scientists alike.

In previous chapters we analyzed how this inspiration was crucial to Picasso's discovery of the proper representation for *Les Demoiselles d'Avignon.* In what follows we will explore how aesthetics functioned in Einstein's discovery of relativity theory in 1905. The results will bear on the issue of why it was that only Picasso and Einstein succeeded in making these dramatic breakthroughs, and others who were more famous and experienced, such as Derain, Matisse, Lorentz and of course Poincaré, did not.

6

The Annus Mirabilis: How Einstein Discovered Relativity

I have been extremely interested by the fact that you are planning to return to the Patent Office, in that cloister among the men where I hatched my best ideas and where we spent such pleasant days together.

—Letter of Albert Einstein to Michele Besso, 12 December, 1919

At intervals of six to eight weeks starting on 17 March, 1905, Einstein sent off to the *Annalen der Physik* three papers that would appear in the journal's now famous Volume 17. Of these three masterpieces, the greatest is the special theory of relativity. This chapter explores how Einstein discovered it.

THE BEST PHYSICS DEPARTMENT IN EUROPE

From 29 October, 1903, to 15 May, 1905, the Einsteins lived in the old city center of Bern. Their two-room flat at 49 Kramgasse was a second-floor walk-up accessible by a steep staircase.

FIGURE 6.1 Einstein's living room at 49 Kramgasse, Bern, and the dining room table on which he may have written the relativity paper.

The apartment was a reasonable turn-of-the-century accommodation in Bern (Figure 6.1). There is one bedroom, a living room, part of which was partitioned off as a baby's room with a door that could be closed, a kitchen and a bathroom reached by a small stairway. Since there was no study, Einstein worked at the kitchen table or at the sitting room table. His financial situation was that of a lower-echelon civil servant with a wife and child, Hans Albert having been born 14 May, 1904.

Although Einstein's position was made permanent on 16 September, 1904, this carried no promotion and a pay raise of only 400 francs, to 3,900 Swiss francs.[1] In addition, he was helping to support his widowed mother. Mileva no doubt complained of the difficulty of making ends meet, and Einstein at one point considered applying for a better paid position at the Federal Postal and Telegraph Administration. No wonder that when he was out of the house and at the Patent Office he "was set free from everyday worries to produce [my] best creative work."[2]

The eight hours per day free for "mischief," plus Sundays, were the only periods available to Einstein for continuous concentrated thought. At home were a sulking wife and a baby son. But Einstein exhibited a stunning ability to remove himself from the "merely personal." Maja first noticed this during Einstein's extended periods at home in Milan, studying on his own after leaving the Gymnasium: "His work habits were quite odd—even in a large noisy group, he was able to withdraw to the sofa, take pen and paper in hand . . . and lose himself so completely in a problem that the polyphonic conversation stimulated rather than disturbed him."[3]

A similar description, from late 1910, comes from Einstein's graduate student Hans Tanner, at the University of Zurich. Tanner arrived one day to find Einstein smoking one of the cheap Swiss cigars that he always had at hand.

> [Einstein] was sitting in his study in front of a pile of papers full of mathematical formulae. Writing with his right hand, holding his younger son in his left arm, in the midst of all this he answered questions from older son Albert who was playing with bricks. With the words, "One minute, I'm almost finished," he handed over the children to me to tend while he went on working. From that I could see his power of concentration.[4]

It was in even worse conditions, at 49 Kramgasse, that Einstein completed two of the three papers for the 1905 Volume 17 of the *Annalen*, produced a draft of the relativity paper, wrote a Ph.D. thesis and, in March, published ten book reviews.[5]

On 15 May, 1905, the Einsteins moved to 28 Besenscheuerweg, on the outskirts of Bern, to be closer to the Bessos. At this point Besso and Einstein began their walks to and from the Patent Office. Mileva and Albert's social calendar was nowhere near as busy as Fernande and Pablo's. Mileva mostly cooked and took care of Hans Albert. Their closest friends were the Bessos. Mileva enjoyed visits from Habicht and Solovine but resented the lengthy Olympia Academy meetings, which sometimes lasted into the early hours of morning. When this camaraderie ended after 1905, the loss for Einstein is clear from his letter to Solovine of April 1906, in which he complained of having no social life anymore.[6]

He also enjoyed spending time with his son, at least as a child. Hans Albert fondly recalled his father's building toys for him out of matchboxes and string.[7] Later on he realized that Einstein's penchant for improvisation extended to his lecturing style. "For instance, when he had to give a talk he never knew ahead of time exactly what he was going to say. It would depend on the impression he got from the audience in which way he would express himself and into how much detail he would go. And so this improvisation was a very important part of his character and of his way of working."[8] For the immensely self-confident Einstein, this style was a way to impart to the audience that they were watching someone think on his feet. Tanner recalled Einstein giving his course lectures from notes written on a piece of paper the size of a visiting card.[9] He did not always accomplish this successfully, and so his students were able to observe how a physicist of his caliber works himself out of an impasse.[10] This made his lectures that much more exciting and inspirational.

The Einsteins took vacations, usually brief ones, and Einstein helped out around the house. As Hans Albert put it, "although he was not particularly clever with his hands to do more delicate things, he was always willing to help."[11] Around the time of the divorce in 1919, relations between father and son would become strained to the breaking point. Yet Hans Albert recalled of their infrequent meetings that his father always expressed interest in his engineering research.[12] He always felt that their talk about inventions and puzzles interested his father because of its reminder "of the happy, carefree and successful days at the Patent Office in Bern."[13]

At the Patent Office Einstein got along well with his thirty or so colleagues as well as his director, the brusque and hard-nosed Haller. By 1905 he had his close friend and sounding board Besso at his elbow. Along with his top desk drawer, this constituted his department of theoretical physics. It turned out to be the best one in Europe.

HENRI POINCARÉ AND THE PHYSICS OF 1905

Discussing Henri Poincaré's work with the Olympia Academy, Einstein regarded with fascination how the master philosopher dissected Lorentz's electromagnetic theory. Poincaré had followed Lorentz's theory with great interest ever since its appearance in 1892. Uniquely among competing electromagnetic theories, Lorentz's was compatible with Poincaré's quest for unity of the sciences.[14] But although Lorentz's local time could explain away the failure of the less exact ether-drift experiments, Poincaré could rate the theory only satisfactory because of its inadequate explanation for the failure of the Michelson-Morley experiment. Lorentz's contraction hypothesis, inserted to account for this failure, seemed to Poincaré an arbitrary fix, stuck in by hand with no physical justification. In *La Science et l'Hypothèse*, wondering sardonically whether more such fixes would be needed whenever someone performed a new ether-drift experiment of higher accuracy, he commented, "Hypotheses are what we lack least."[15] The situation was coming to resemble Ptolemy's Earth-centered universe, which required the continual introduction of new epicycles whenever new sightings were made.

In 1905, the ether was an integral part of electromagnetic theory. The repeated failures to find any variation in the velocity of light measured on the moving Earth thus posed an awkward and confusing situation for physicists.[16] Even if it was undetectable, ether had to be there in order to support light waves in transit. Yet even if, like Poincaré, one was not a strict positivist, experimental data were nevertheless paramount because "experiment is the sole source of truth."[17]

In Chapter 3 this situation was developed in some detail. Let me reiterate here. The measured velocity of light always turned out to be the same as if the Earth were at rest in the ether. But the Earth undergoes several motions. Hypotheses were proposed to cancel out predicted effects on optical phenomena caused by the Earth's motion through the ether, and these worked for experiments of lowest-order accuracy.[18] Lorentz's hypothesis of a local time coordinate, in particular, acted like a magic wand. "Local time" comprises two terms. One of them is the "usual time coordinate" used in physics, while the other depends on the clock's relative motion. The usual time coordinate, the time we read off our clocks and wristwatches, always seems to be the same regardless of any relative motion between ourselves and others. The term that depends on relative motion, on the other hand, is much too small to be observed and so could have no claim to physical reality.[19]

Lorentz, Poincaré and all other physicists considered local time to be useful solely for the mathematical purpose of providing terms to cancel out predicted effects that were not observed—in short, to explain the failure of ether-drift experiments. The real time was the one that was the same for all observers regardless of their relative motion. This would turn out to be wrong: They were misled by their intuition, which was based on sense perceptions. Einstein would realize that Lorentz's local time is correct mathematically *and* physically. But this required the courage to go beyond sense perceptions.

The accuracy that Michelson and Morley claimed for their experiment was a serious blow to Lorentz's electromagnetic theory. The concern it aroused has led some commentators to suppose that Einstein discovered relativity theory in 1905 in direct response to the failure of this experiment.[20] Einstein himself made comments that have been marshaled to support this view.[21] The issue is a deep one in the foundations of science, because it concerns the link between experimental data and scientific discovery.

What actually happened was more complex. Sometimes Einstein's comments depended on the occasion at which he was speaking or how the question was asked.[22] The objective historical fact is that Einstein had studied the Michelson-Morley experiment prior to 1905. His own assessment, made almost fifty years later, is probably the most accurate: "The influence of the famous Michelson-Morley experiment upon my own deliberations has been rather indirect."[23] Thorough and impartial historical research supports Einstein's statement made almost fifty years later. Besides, Einstein could only have been chagrined that anyone could believe that so complex and beautiful a theory as special relativity could have its roots in a single experiment.

In 1981 I systematically set up a classification of the books, monographs and journals Einstein had definitely read, very probably read, and possibly read by the end of June 1905, when he sent the relativity paper to the *Annalen*. The upshot is that Einstein was never out of touch. Between 1896 and 1900 he could use the library facilities at the Swiss Polytechnic. In the bleak years 1900–1902, libraries were usually, though not always, open to him in Zurich. Starting in 1902, his work at the Patent Office involved use of its library as well as the one at the University of Bern, where he also perused literature for his own research. For someone out of the academic mainstream, Einstein read selectively but enough to be aware of the main trends in physics research.[24] This information is important because the relativity paper has no citations to the scientific literature. He was surely not unaware of Michelson-Morley, but there were wider concerns.

Although Poincaré believed in a close connection between experimental data and theory, his handling of data was extremely complex. Whereas Ernst Mach and his followers of the positivist school saw science as a compendium of facts and mathematics as a mere means of classification, Poincaré disagreed: "an accumulation of facts is no more a science than a pile of stones is a house. . . . Most of all the scientist must predict."[25]

The changing character of laboratory data, particularly effects such as radioactivity, X rays and cathode rays that might be caused by unseen entities such as atoms and electrons, was intriguing to Poincaré. In 1902, however, he still insisted that atoms, the *cause célèbre* of the positivist-antipositivist debate, could not be granted any degree of physical reality. "Hypotheses of this genre," he wrote, "have . . . only a metaphoric sense."[26]

Yet unbeknownst to Einstein, by 1905 had Poincaré changed his mind. The astounding successes of Lorentz's theory had by then led physicists to suggest that it should be the fundamental theory of all matter in motion. This programmatic research effort was called the "electromagnetic world-picture."[27] Its goal was to explain how the electron's mass is generated by its own radiation reacting back upon it, and then ultimately to deduce Newton's theory of mechanics from Lorentz's electromagnetic theory.[28] The procedure was to add hypotheses onto the electromagnetic theory to describe a particular sort of electron, and then conduct experiments to test whether the electron really behaved as described. Lorentz opted for representing the electron as an elastic sphere—like a balloon—smeared with charge. When moving, it undergoes a Lorentz contraction and becomes flattened like a pillow, with its longer axis at right angles to the direction of motion. This hybrid version of the electromagnetic theory is Lorentz's theory of the electron. It produced predictions for the electron's mass that were then compared with experimental data on high-speed electrons obtained by the German scientist Walter Kaufmann, whose experiments were praised as nothing less than classic—ringing confirmation of the dominant theory of their time. Kaufmann is virtually unknown today; his data misled.[29]

Though aware of this research toward an electromagnetic world-picture, Einstein was more intrigued by Poincaré's overview of physics in *La Science et l'Hypothèse*. In Poincaré's opinion, physics faced three pressing fundamental problems: the ether-drift experiments; the peculiar way in which ultraviolet light liberates electrons from metals, known as the "photoelectric effect";[30] and the erratic microscopic dance of pollen grains and dust particles, known as "Brownian motion."[31] The reason for Einstein's interest in these three problems is that in early 1905, he alone knew that the electromagnetic world-picture was a chimera. By the end of June he had solved all three in a novel and surprising way.

How this came about turns out to have unexpected roots.

PHYSICS AND MUSIC

Besides the Patent Office, physics and the Olympia Academy activities, Einstein had his beloved music. Hans Albert recalled his father trying to educate him in a wider sense: "He often told me that one of the most important things in his life was music. Whenever he felt that he had come to the end of the road or into a difficult situation in his work he would take refuge in music that would usually resolve all his difficulties."[32]

From his adolescence through his student days and into the 1940s, music was at the core of Einstein's creative life, and Mozart and the violin were at the core of his music. At age sixteen Einstein had an epiphany in the school cafeteria at Aarau. Reflecting on Bismarck's famous line that "beer makes one stupid and lazy," Einstein vowed that he would be a theoretical physicist and henceforth become intoxicated instead on physics and Kant's *Critique of Pure Reason*.[33] To celebrate, he invited his friend Hans Byland to accompany him on the piano in a Mozart sonata. What happened next, Byland never forgot: "When his violin began to sing, the walls of the room seemed to recede—for the first time Mozart in all his purity appeared before me, bathed in Hellenic beauty with its pure lines, roguishly playful, mightily sublime."[34] Einstein's passionate musical rendering mirrors the conviction so indigenous to youth that everything is possible, while his musical taste shows a preference for classical purity that would also be reflected in his physics. Music was the window into the place where Einstein sealed all his emotions in order to avoid dealing with interpersonal relationships.

Einstein preferred the highly structured, deterministic music of Bach and Mozart. He imagined Mozart plucking melodies out of the air as if they were ever present in the universe,[35] and he thought of himself as working like Mozart, not merely spinning theories but responding to Nature, in tune with the cosmos. About composers he had some very definite opinions.[36] To an insistent reporter asking his opinion of Bach, Einstein replied brusquely, "This is what I have to say about Bach's life work: listen, play, love, revere—and keep your mouth shut."[37] He found Handel interesting but somewhat shallow; Beethoven too melodramatic and, besides, he "created" his music; some of Brahms he considered significant, "but most of his works have for me no inner persuasiveness [and] I do not understand why it was necessary to write them"; he thought Richard Strauss gifted but "concerned only with outside effects"; and Debussy was "delicately colourful but shows a poverty of structure." After a performance at the Bern Opera, in 1908, of Richard Wagner's *Götterdämmerung*, Einstein commented to his companion, "Wagner is, God forgive me, not to my taste."[38]

One seldom speaks of "tastes" in science, but Einstein's tastes in music and physics were linked. He thought of both musical and physical truths as Platonic forms that the mind must intuit. Great music cannot be "created" any more than great physics can be deduced strictly from experimental data. Some aesthetic sense of the universe is necessary for both.

Einstein's phenomenal creative output in the spring of 1905 could only have occurred if he had the main themes of the three papers well thought out beforehand. His results were not the sort that emerge from reams of calculations, nor did he have time for that approach. They were fruits of the most difficult kind of thinking—conceptual thought, a mode that Einstein deliberately cultivated.

To his biographer Carl Seelig's question as to whether "relativity theory has a definite birthday," Einstein replied,

> Four or six weeks elapsed between the conception of the idea of special relativity and completion of the corresponding publication. But it would be hardly correct to designate this as the birth date, because earlier arguments and building blocks were being prepared over a period of years, although without bringing about the fundamental decision.[39]

I will discuss these building blocks later. Here I want to emphasize Einstein's persistence in thinking about the problem of relative motion and the ether. He thought about these issues in a manner strikingly different from other physicists. He based his ideas on concepts from our daily movement about the world, he later noted, because "scientific thought is a development of pre-scientific thought."[40] This valuable insight into his own thinking was elaborated in an essay with the captivating title "Physics and Reality," in which Einstein wrote, "The whole of science is nothing more than a refinement of everyday thinking. [The scientist] cannot proceed without considering critically a much more difficult problem, the problem of analyzing the nature of everyday thinking."[41]

The origin of notions such as time was always at the heart of his research. At the Olympia Academy, Solovine recalled Einstein's curiosity about how children develop their knowledge of time, which was allied with an interest in creative thinking.[42] Allied with this concern was Einstein's interest in the nature of creative thinking. The Gestalt psychologist Max Wertheimer, a colleague of Einstein's at the University of Berlin, recalled the "hours and hours" they discussed this topic in 1916.[43] Relativity's radical revision of the scientific concept of space and time emerged from a deep understanding of—and thus an ability to question—the way those entities were intuitively conceived.

Einstein often began his most esoteric scientific and philosophical lectures by reminding his audience of the importance of conceptual analysis: "The theory of relativity is intimately connected with the theory of space and time. I shall therefore begin with a brief investigation of the origin of our ideas of space and time."[44] Almost always, he paid homage to Poincaré: "It is essential here also to pay strict attention to the relation of experience to our concepts. It seems to me that Poincaré clearly recognised the truth in the account he gave in his book, *La Science et l'hypothèse.*"[45]

Just as Poincaré began his famous 1908 introspection with the problem, "What, in fact, is mathematical invention,"[46] Einstein began his by asking "What, precisely, is 'thinking'?"[47] Like Poincaré he was deeply interested in how ideas metamorphose. Our first impressions of a world external to ourselves, he wrote in 1946, are "sense impressions" from which "memory-pictures" emerge.[48] Certain memory pictures form series. A memory picture that occurs a great many times in several different series can serve as an "ordering element" for those series. Einstein referred to this ordering element as a "concept." Thinking is "operations with concepts . . . the creation and use of definite functional relations between concepts and co-ordination of sense experiences to these concepts."[49] Concepts are organizing principles that enable us to turn sense perceptions into exact knowledge.[50] Subconscious thinking is a "free play with concepts."[51] When Einstein told Seelig about "building blocks . . . being prepared over a period of years," he meant a continual process of "free play," and then kept in the unconscious.

That creative thinking is essentially nonverbal seemed clear to Einstein: How else could "we 'wonder' quite spontaneously about some experience"?[52] (Two of Einstein's childhood "wonders," as we saw, were the compass needle and the geometry booklet.) Wondering "quite spontaneously" is at the root of his highly visual thought experiments. For Einstein, creative thinking occurred in visual imagery, and words "were sought after laboriously only in a secondary stage."[53]

In summary, as a musician and a physicist Einstein was an antipositivist. In music, beyond notes and instruments was the sublime realm where melodies floated. In physics, beyond observations and theory lay the music of the spheres, where laws of nature waited to be plucked out of the cosmos. His great breakthrough was to use organizing principles and the visual imagery of thought experiments to go beyond sense perceptions and its associated form of intuition.

THREE PAPERS, ONE THEME

By the spring of 1905, the twenty-six-year-old Einstein had decided that physicists were "out of [their] depth."[54] From calculations based on Planck's radiation law, Einstein drew the astounding "general conclusion" that light can be a particle and a wave, and in fact both at once, a wave/particle duality.[55] Therefore the electromagnetic world-picture could not succeed, because Lorentz's theory could represent radiation, or light, only as a wave, and so could never provide a way to explain how the electron's mass is generated by its own radiation. Whereas Planck had discovered certain peculiarities about the energy of radiation, Einstein set out to explore the structure of radiation itself. Einstein's particles of light differed fundamentally from Newton's in ways that even he did not yet fully realize.

Around the third week of May 1905, Einstein sent his friend Habicht what are surely some of the greatest understatements in the history of science. He wrote that he had only some "inconsequential babble" for his friend, whom he lambastes for neither writing nor visiting him during Easter: "So what are you up to, you frozen whale, you smoked, dried, canned piece of soul . . . . I promise you four papers." The first paper is the light quantum paper that Einstein referred to as "very revolutionary." The second suggested a means to measure the size of atoms using diffusion and viscosity of liquids. The third one explored Brownian motion using methods of the molecular theory of heat. "The fourth paper is only a rough draft at this point, and is an electrodynamics of moving bodies which employs a modification of the theory of space and time; the purely kinematic part of this paper will surely interest you."[56] What is so incredible about this outburst of creativity is that by late May two papers were completed and the third was in draft form.

The first, third and fourth papers were meant for the *Annalen*, while the second was his Ph.D. thesis.[57] The first of the papers destined for the *Annalen der Physik* was entitled "On a Heuristic View Concerning the Production and Transformation of Light."[58] It concerned processes where light impinging on matter is absorbed and there is a subsequent emission of electrons, for example the photoelectric effect, the ability of light to generate an electric current in some metals. As a heuristic hypothesis—or theoretical expedient—he suggested that in such circumstances it is useful to represent light as a particle, or "light quantum."[59] Einstein did not mention the baffling wave/particle duality of light but rather focused on the particle mode. In proposing such an extreme approach Einstein astoundingly brought up no experimental data. Instead he argued for the particles' existence on aesthetic grounds, thus introducing into twentieth-century physics an entirely new method of reasoning.

Straightaway the paper called attention to something that offended Einstein's sensibility: "There exists a profound formal distinction between the theoretical conceptions physicists have formed about gases [and] Maxwell's theory of electromagnetic processes."[60] What Einstein is driving at is the unnatural distinction electromagnetic theory drew between light and its source. Imagine dropping a stone into a pond and seeing spherical waves spread out from the point of impact. This imagery is the one assumed by Maxwell's electromagnetic theory; an accelerated electron is the source of spherical electromagnetic waves, or light, in the ether. To Einstein it was unnatural, or unaesthetic, to have wave and particle representations, that is, continuity and discontinuity, side by side. Only by linking particulate electrons with particulate light could such important phenomena as the photoelectric effect be made explainable. It was this paper, not relativity, that won Einstein the 1921 Nobel Prize.[61]

Although there existed no experimental evidence to refute Einstein's hypothesis of light quanta, physicists resisted them on intuitive grounds. Max Planck argued that no visual imagery based on particles of light could explain the phenomenon of interference, whereas a perfectly satisfactory imagery existed that was abstracted from how water waves interfere.[62] This criticism had been leveled, as well, at Newton's particles regarding Young's double-slit experiment (see Chapter 3).

The second paper was Einstein's third (and finally successful) attempt at a doctoral dissertation, "On a New Determination of Molecular Dimensions."[63] He had decided to continue the academic "comedy" both to advance himself at the Patent Office and to try again to obtain a university position in the normal way. This doctoral thesis and the third paper, "On the Movement of Small Particles Suspended in Stationary Liquids Required by the Molecular-Kinetic Theory of Heat,"[64] provided a way of determining the size of atoms and explained the mystery of Brownian motion, the random jiggling seen in a microscope, as resulting from impacts of atoms and molecules. The importance of these two papers, at a time when the reality of atoms was not universally accepted, cannot be overstated.[65]

By March 1905 Einstein had realized that Poincaré's three problems addressed a common theme, which was the nature of light and its relation to the limits of physical theory. This was the theme of Einstein's three papers in the *Annalen.* The paper on light quanta demonstrated the limits of electromagnetic theory, and the one on Brownian motion probed the limits of thermodynamics and mechanics. These two papers convinced Einstein, as he recalled in 1907, of the field's need for fundamental changes.[66] The third paper was to be the phoenix rising from the ashes of nineteenth-century physics. It would not only overturn certain assumptions regarding the propagation of light but would propose a new means for specifying the proper form of a physical theory. The nature of radiation, its properties and structure, was the principal problem that all three papers addressed.[67]

Einstein's startling insight was sparked by his courageous acceptance of certain results as axiomatic, such as Planck's radiation law. But the key element in his intellectual toolbox came from outside science per se and concerned notions of aesthetics and philosophy. Einstein was able to grasp the unity of Poincaré's three problems and the means to solve them because he was willing and able to intuit a reality beyond perceptions.

THE RELATIVITY PAPER

The corpus of Einstein's fourth paper entitled "On the Electrodynamics of Moving Bodies," the so-called relativity paper, is at first glance no different from other scientific papers of that era.[68]

Yet first glance deceives: It was daring in both style and content. Today no leading physics journal would publish it because of its complete lack of citations to the literature. Einstein's two other papers in the *Annalen*'s Volume 17 at least contained citations, however sparse. The *Annalen*'s policy was that an author's address was the city from which the paper was sent. Einstein signed off with "Bern, June 1905." Most people went to look for him at the university.

The format of a journal paper in theoretical physics has hardly changed since 1905. There are essentially three parts: statement of a problem concerning empirical data; proposal to modify an already existing theory to explain these data; and deduction of further predictions.

Now let us consider Einstein's paper (Figure 6.2). Imagine that you are on the editorial board of a prestigious physics journal and that you receive a paper from a little-known author that is unorthodox in style and format; whose title has little to do with most of its content;[69] that has no citations to the literature; whose first half is largely devoted to philosophical ban-

3. ***Zur Elektrodynamik bewegter Körper;***
von A. Einstein.

Daß die Elektrodynamik Maxwells — wie dieselbe gegenwärtig aufgefaßt zu werden pflegt — in ihrer Anwendung auf bewegte Körper zu Asymmetrien führt, welche den Phänomenen nicht anzuhaften scheinen, ist bekannt. Man denke z. B. an die elektrodynamische Wechselwirkung zwischen einem Magneten und einem Leiter. Das beobachtbare Phänomen hängt hier nur ab von der Relativbewegung von Leiter und Magnet, während nach der üblichen Auffassung die beiden Fälle, daß der eine oder der andere dieser Körper der bewegte sei, streng voneinander zu trennen sind. Bewegt sich nämlich der Magnet und ruht der Leiter, so entsteht in der Umgebung des Magneten ein elektrisches Feld von gewissem Energiewerte, welches an den Orten, wo sich Teile des Leiters befinden, einen Strom erzeugt. Ruht aber der Magnet und bewegt sich der Leiter, so entsteht in der Umgebung des Magneten kein elektrisches Feld, dagegen im Leiter eine elektromotorische Kraft, welcher an sich keine Energie entspricht, die aber — Gleichheit der Relativbewegung bei den beiden ins Auge gefaßten Fällen vorausgesetzt — zu elektrischen Strömen von derselben Größe und demselben Verlaufe Veranlassung gibt, wie im ersten Falle die elektrischen Kräfte.

Beispiele ähnlicher Art, sowie die mißlungenen Versuche, eine Bewegung der Erde relativ zum „Lichtmedium" zu konstatieren, führen zu der Vermutung, daß dem Begriffe der absoluten Ruhe nicht nur in der Mechanik, sondern auch in der Elektrodynamik keine Eigenschaften der Erscheinungen entsprechen, sondern daß vielmehr für alle Koordinatensysteme, für welche die mechanischen Gleichungen gelten, auch die gleichen elektrodynamischen und optischen Gesetze gelten, wie dies für die Größen erster Ordnung bereits erwiesen ist. Wir wollen diese Vermutung (deren Inhalt im folgenden „Prinzip der Relativität" genannt werden wird) zur Voraussetzung erheben und außerdem die mit ihm nur scheinbar unverträgliche

FIGURE 6.2
To the right is the first page of Einstein's relativity paper, with a translation from Miller (1998a), pp. 370–371 (opposite).

ter on the nature of certain physical concepts taken for granted by everyone; that discusses only one experiment (the generation of current in a wire loop in motion relative to a magnet) and one that is adequately explainable using Lorentz's electromagnetic theory and is not even considered to be fundamentally important; and whose author audaciously declares the very core of electromagnetic theory, "luminiferous ether," to be "superfluous."[70] The article concludes with certain results concerning electrons that, in papers where electrons are discussed, are generally at the beginning. To the reader of 1905, the paper was written backwards.

This is how Einstein's submission probably struck the editor of the *Annalen der Physik,* Paul Drude, who had just moved to the University of Berlin from the backwater of Giessen University. Recall that in his letter to Mileva of 7 July, 1901, Einstein reported pointing out errors in Drude's work in what may not have been the most diplomatic tone. Luckily Drude was in Berlin, where most of the Curatorium, or editorial board, sat. Else he might have taken it upon himself to bin the young upstart's paper.

ON THE ELECTRODYNAMICS OF MOVING BODIES

By A. Einstein

That Maxwell's electrodynamics – the way in which it is usually understood – when applied to moving bodies, leads to asymmetries that do not appear to be inherent in the phenomena is well known. Consider, for example, the reciprocal electrodynamic interaction of a magnet and a conductor. The observable phenomenon here depends only on the relative motion of the conductor and the magnet, whereas the customary conception draws a sharp distinction between the two cases in which either the one or the other of these bodies is in motion. For if the magnet is in motion and the conductor at rest, there arises in the neighborhood of the magnet an electric field with a certain definite energy, producing a current at the places where parts of the conductor are situated. But if the magnet is at rest and the conductor in motion, no electric field arises in the neighborhood of the magnet. In the conductor, however, we find an electromotive force, to which in itself there is no corresponding energy, but which gives rise – assuming equality of relative motion in the two cases discussed – to electric currents of the same path and intensity as those produced by the electric forces in the former case.

Examples of this sort, together with the unsuccessful attempts to discover any motion of the earth relatively to the "light medium," lead to the conjecture that to the concept of absolute rest there correspond no properties of the phenomena, neither in mechanics, nor in electrodynamics, but rather that as has already been shown to quantities of the first order, for every reference system in which the laws of mechanics are valid*, the laws of electrodynamics and optics are also valid.

We will raise this conjecture (whose intent will from now on be referred to as the "Principle of Relativity") to a postulate, and moreover introduce another postulate, which is only apparently irreconcilable with the former: light is

The *Annalen*'s editorial policy was that an author's initial publications were scrutinized either by the editor or a member of the Curatorium.[71] Subsequent papers were published without further refereeing. Since he had already published five times in the *Annalen*, Einstein's paper could have been accepted on receipt. His earlier papers were considered good enough for the *Annalen* to have invited him to contribute to their book review journal.[72] Still, Drude must have had some doubts, because he gave this new work to the theorist closest to hand, Max Planck. Planck immediately recognized its prospects.

Page for page, Einstein's relativity paper is unparalleled in the history of science in its depth, breadth and sheer intellectual virtuosity. Einstein developed one of the most far-reaching theories in physics in a literary and scientific style that was parsimonious yet not lacking in essentials; in a pace that, where necessary, possessed a properly slow cadence yet was not devoid of crescendos and *tours de force*. The 1905 theory of relativity, presented in thirty pages of print, is developed almost like an essay. Written in white heat in about five weeks, it is pristine in form, and yet in its own way as complete as Newton's book-length *Principia*.[73] It remains an excellent source from which to learn relativity theory.

Its thrust was literally antithetical to what physicists on the cutting edge were doing. Then as now, they were trying to formulate a Theory of Everything, except that life was "simpler" then because there were only two known forces, electromagnetism and gravitation. How could such a grandiose research effort be critiqued from a simple observation about electrical currents generated in dynamos?

EINSTEIN'S APPROACH TO ELECTRODYNAMICS

Several channels of thought came together for Einstein in spring 1905: scientific, philosophical and engineering. In his "Autobiographical Notes" Einstein recalled his mindset after the bombshell of his 1904 paper, in which he became convinced of the insufficiency of Lorentz's theory for light and electrons:

> By and by I despaired of the possibility of discovering the true laws by means of constructive efforts based on the known facts. The longer and the more despairingly I tried, the more I came to the conviction that only the discovery of a universal formal principle could lead us to assured results. The example I saw before me was thermodynamics. The general principle was there given in the theorem: the laws of nature are such that it is impossible to construct a *perpetuum mobile*.[74]

By "constructive efforts" Einstein meant trying to understand phenomena with a theory based on atoms of electricity—electrons. The "known facts," or laboratory data, so essential to Lorentz and Poincaré as to every physicist of that era, were the ether-drift experiments and Kaufmann's data on the electron's mass. The problem was that Lorentz's theory was incomplete in its description of light and so could not lead to "discovering the true laws."

What to do? At this point Einstein recalled how he discovered the fatal shortcoming in Lorentz's theory, which was by accepting Planck's radiation law as axiomatic and then deducing its consequences. Perhaps he could try this method with problems of relative motion, too, problems that, as we know from his letters to Mileva, had been on his mind since 1899. In seeking a "universal formal principle" Einstein saw before him the example set by thermodynamics, the science of heat, on which he was an expert.[75]

Thermodynamics fascinated Einstein because its fundamental principles are axiomatic as well as being independent of the material qualities of the systems they govern.[76] Some years later Einstein referred to such theories as "theories of principle," and theories such as Lorentz's, "constructive theories."[77] One of the cornerstones of thermodynamics is the principle of the conservation of energy, which simply demands that every physical theory conserve energy, no questions asked.

Such principles are essential to theory building. Scientists apply them with great assurance because they solve the difficult problem of when to stop asking why. Einstein said of this problem: "Scientific endeavours are quite extraordinary; often nothing is of more importance than seeing where it is not advisable to expend time and effort."[78]

But it is one thing to recognize the need for a formal principle and quite another to find the proper one. *La Science et l'hypothèse* could only have helped.

POINCARÉ'S PRINCIPLE OF RELATIVITY

In Newtonian science there is a class of preferred observation platforms called "inertial reference systems." These platforms move relative to one another in straight lines at constant velocities. Inertial reference systems are special because Newton's laws of motion are the same in all of them. Consequently, no mechanical experiment can reveal the system's motion; you might as well be at rest. This statement is known as the "principle of relativity." In Newton's theory of motion it is axiomatic.[79]

Being interested in the cognitive basis of science, Poincaré gave as an important reason for this principle's validity that "the contrary hypothesis would be singularly repugnant to the mind."[80] The principle of relativity, in Poincaré's view, was connected to the origins of geometry and of knowledge itself. Basically, we formulate geometry by inspecting the relations between displacements of material objects. Then we abstract to the realization that the goal of science "is not the things themselves . . . it is the relations between things; outside of these relations there is no reality knowable."[81] Science should restrict itself to exploring the relative motions of material objects.

Yet the very basis of electromagnetic theory, the ether, violated the principle of relativity. Inertial reference systems that were at rest in the ether were theoretically different from those in motion. All you had to do to reveal your system's motion was measure the velocity of light (on, say, the moving Earth) and compare it to the value postulated for measurement in a laboratory fixed in the ether.[82] This was the goal of the ether-drift experiments. Poincaré did "not believe . . . that more precise observations can ever reveal anything other than the relative displacements of material bodies"—that is, they would continue to fail.[83] Yet they were undertaken because "we were not certain of [their result] in advance."[84] Consistent with his emphasis on laboratory data, Poincaré required more evidence before he could promote the principle of relativity in Lorentz's electromagnetic theory to the axiomatic status it held in Newtonian science.

Einstein saw how Poincaré struggled with the failure of the ether-drift experiments. He noticed that Poincaré drew back from crossing the Rubicon and leaving these failures behind. Poincaré's allegiance was to the ether and to a constructive theory couched in experimental data. After all, light from a distant star "must be somewhere and supported, so to speak, by some material medium."[85]

But if the principle of relativity became an axiom in Lorentz's electromagnetic theory, then only effects due to the relative motion of material objects could be measured. What about the ether? Poincaré blandly assured readers that "we assume that the ether is within our grasp."[86] In *La Science et l'hypothèse* he hinted at experiments, explained in more detail in a 1900 publication, that aimed to measure the ether's recoil on an apparatus that fired light pulses.[87]

In summary, despite his belief in the importance of a principle of relativity, Poincaré maintained the contradictory need for the ether, and he was sure that experiments other than the ether-drift experiments would measure its effects. The ether was the very core of Poincaré's definition of what is real, because it provided "relations between things." From readings in Poincaré's *La Science et l'hypothèse* and in Lorentz's 1895 treatise, Einstein was well aware that physicists were attempting to *construct* a principle of relativity for electromagnetic theory whose status would be the same as the one in Newton's mechanics, but whose underlying support would be a myriad of hypotheses like the local time. This seemed to be going well. But Einstein knew it could never succeed because of the particle nature of light, which most other physicists ignored.[88]

EINSTEIN'S PRINCIPLE OF RELATIVITY

Given the physics of 1905, the opening sentence of Einstein's relativity paper was totally unexpected: "That Maxwell's electrodynamics—the way

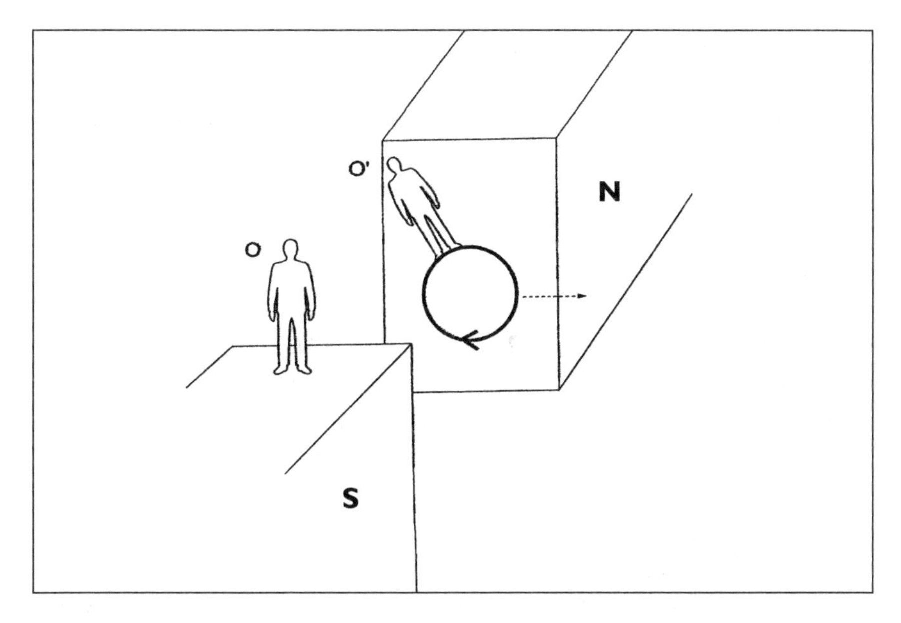

FIGURE 6.3 N and S designate pole faces of a permanent magnet between which a wire loop travels to the right as indicated by the dashed arrow. Observer O' rides on the conducting loop and O stands on the magnet's south pole. Both observers measure the same quantity—the current generated in the wire loop as a result of its motion relative to the magnet. The current's direction is clockwise, as indicated by the arrow on the loop. Although the measured quantity (electric current) depends only on the relative motion between loop and magnet, the two observers explained the current as arising in two fundamentally different ways. Einstein found this duality of explanations to be "unbearable."

in which it is usually understood—when applied to moving bodies, leads to asymmetries that do not appear to be inherent in the phenomena is well known."[89] This is strikingly like the first sentence of the light quantum paper, where Einstein wrote of the "profound formal distinction" made in electromagnetic theory between the continuous (wave) of radiation and its discontinuous (particle) source. In both instances he is pointing out something artificial, redundant and therefore unaesthetic. Yet to everyone else it was natural. Contrary to what Einstein wrote, the asymmetry in Maxwell's electrodynamics was not "well known."

All Einstein needed to expose it was the most elementary example of electromagnetic induction, a phenomenon discovered in 1831 by the English scientist Michael Faraday and which was at the heart of the second industrial revolution. It involves a conducting loop of wire and a magnet in relative motion (Figure 6.3).

The observer measures an electrical current in the wire. Through his work in the Patent Office, Einstein was well acquainted with subtle problems in dynamo design. In his view, the root of these problems resided in this straightforward example, as he wrote in the relativity paper.[90] In Einstein's careful way of discussing the genesis of his thinking, in a 1919 unpublished manuscript he recalled that the "Faraday [experiment] on electromagnetic induction played for me a leading role."[91]

Einstein alone was disturbed about the following situation: "The observable phenomenon [measured current] here depends only on the relative motion of the conductor and the magnet, whereas the customary conception draws a sharp distinction between the two cases in which either the one or the other of these bodies is in motion."[92] He is drawing the reader's attention to a profound formal distinction between interpretations of how the current arises, depending on whether the observer is sitting on the conducting loop or on the magnet. An observer on the conducting loop interpreted the current as arising from the loop moving through the magnet's magnetic field, while the one riding along with the magnet attributed the current as arising from an electric force whose basis is in an electric field produced by the magnet.[93] Yet the measured current depends on only their relative velocity. Einstein was vexed by the need for two basically different interpretations of a phenomenon dependent on only a single quantity, relative motion. One was in terms of a *field* and the other a *force*. This is redundant and, so, asymmetric. Here Einstein has introduced a new meaning for *asymmetry*, which differs from the one in art.

In 1919, Einstein recalled that he found this redundancy so "unbearable" that it "forced me to postulate the (special) relativity theory [because] the difference could not be a real difference between these two cases [but] only a difference in the choice of reference point."[94] A single phenomenon (electromagnetic induction) can be "viewed" from different vantage points, but it can have only one explanation. For Einstein an asymmetry can be redundancy in explanation. He will wield this version of Ockham's razor to whittle down physical theory to a satisfactory minimalist form.

In the next paragraph of the relativity paper Einstein writes that "examples of this sort" (unnamed), together with the "unsuccessful" ether-drift experiments, "lead to the conjecture" that to the same low order of accuracy the principle of relativity from Newton's mechanics holds also in electrodynamics and optics. At this point the physicists of 1905 must have asked themselves, "How does this guy link optics and mechanics with electromagnetic induction?"

The answer turns out to be fiendishly simple. Calculating the electrical current arising in the wire requires knowing the wire's velocity relative to the magnet, which entails the laws of mechanics. Optics enters because electromagnetic theory is also a theory of light. Then, to lowest-order accuracy, Newton's principle of relativity also covers electromagnetic theory and optics.[95] Einstein's great insight is that electromagnetic induction is the arena in which optics, electromagnetic theory and mechanics all overlap.

Einstein's next step is pure bravado: He raises this inexact principle of relativity to an axiom. In Einstein's terminology it is also a concept, because it reorganizes hitherto misunderstood data. And so, in one stroke of the pen, Einstein declares all ether-drift experiments foregone failures. Right from the start he eliminates unmeasurable quantities, such as velocities relative to the ether, and insists that all fundamental notions such as space, time and velocity be measurable—if not by real experiments then at least in principle by thought experimenters. Notions like "observers in the ether" became meaningless, and so the "introduction of a 'luminiferous ether' [is] superfluous"[96]—exit metaphysical obscurities. Mach's insistence on the measurability of quantities was of help here. So was Poincaré's analysis of the foundations of mechanics in *La Science et l'hypothèse,* where he cast a critical eye over the way physics dealt with absolute quantities: "too often we enunciate mechanical facts as if there were an absolute space to which we could refer them."[97]

Next Einstein proposes a second principle: In inertial reference systems, the velocity of light is always the same and has the value that only observers at rest in the ether were assumed to measure, 186,000 miles per second. The light quantum, already counterintuitive, becomes even more so: It travels at the same velocity no matter how fast the laboratory is moving. This further distances Einstein's light particle from Newton's.

If this were not enough to make any physicist's blood boil, the closing paragraph of Einstein's introduction surely did:

> The theory to be developed is based—like all electrodynamics—on the kinematics of the rigid body, since the assertions of any such theory concern the relationships between rigid bodies (coordinate systems), clocks and electromagnetic processes. Insufficient consideration of this circumstance is the root of the difficulties with which the electrodynamics of moving bodies presently has to contend.[98]

By consensus, in June of 1905 the cutting edge of theoretical physics concerned complex mathematical formulations for a theory of the electron. Scientists assumed that spatial intervals were measured on rulers and time on clocks and that was that.[99] Einstein knew otherwise.

His first inkling came from his 1895 thought experiment, which he had always kept in mind. In this scenario (see Figure 3.2, Chapter 3), someone is trying to catch up with a point on a light wave whose source is at rest in the ether. While chasing this point, the experimenter measures changes in its velocity relative to his platform. By increasing the platform's velocity, the experimenter ought to be able to catch up with the point, in agreement with Newton's mechanics and our daily experience. It is as if you are trying to catch up with another car. As you pull even, the relative velocity between the two cars decreases and eventually becomes zero. The "paradox," as Einstein recalled in 1946, was that when the thought experimenter caught up with the point on the light wave, the light wave would be frozen in space.[100] But such a phenomenon had neither been observed, nor did it have a place in electromagnetic theory.

By this time Einstein realized that the Aarau thought experiment encompassed every possible ether-drift experiment. Electromagnetic induction had "forced me to postulate the (special) relativity principle,"[101] which supported what the thought experimenter would find "intuitively clear": that "from the standpoint of [the thought experimenter] everything would have to happen according to the same laws as for an observer who, relative to the Earth, was at rest."[102] By *intuitive* Einstein had something very definite in mind—the Newtonian principle of relativity must remain inviolate.

Yet even this development did not completely resolve the paradox. Put yourself in the place of the thought experimenter. No matter how fast you travel, the point on the light wave always moves at the same speed relative to you—the speed of light.

All of these results rested ultimately on the nature of time. How Einstein came to realize this is a fascinating story that involves his job as a patent clerk and again brings in our friend Henri Poincaré, this time in one of his more practical pursuits.

POINCARÉ, EINSTEIN AND THE NATURE OF TIME

Consider the following situation. Two people are located so far apart that they cannot contact one another by voice. They have decided beforehand that at 3:00 P.M. they will point their flashlights skyward and turn them on. How can they be sure their clocks are synchronized? In the first section of the relativity paper, Einstein argued that a foolproof way to accomplish this is for observers with clocks to exchange light signals.

Synchronized clocks are at the very core of Einstein's exposition: All results follow from their behavior. He structured his argument this way because he felt that "insufficient consideration" of the nature of time was the Achilles' heel of physics. How Einstein conceived of an analysis of time and simultaneity by synchronizing clocks with light signals has always been a mystery akin to Picasso's realization that geometry was a key to a new representation in art. Let us seek precedents.

While a student at the Swiss Polytechnic, Einstein read about the local time in Lorentz's 1895 treatise. But Lorentz wrote nothing about clocks and simultaneity. In *La Science et l'hypothèse*, however, Poincaré prefaced a penetrating discussion on the foundations of Newton's mechanics with two astounding passages in which he separated general problems concerning time into two: absolute time and simultaneity. In the first passage he wrote:

> There is no absolute time. To say that two durations are equal is an assertion which has by itself no meaning at all and which can only acquire one by convention.[103]

By "absolute time" Poincaré meant Newton's concept of a time relative to an absolute reference system, with theological implications. Mach attacked Newton's absolute time as a metaphysical obscurity. Yet Poincaré goes on to admit into the study of motion, as is customary, an absoluteness of time. This time does not depend on the relative motion of the clocks that register it but is the same for everyone.

Poincaré continues by reminding us of something taken for granted: "Not only have we no direct intuition of the equality of two durations, but we have not even one of the simultaneity of two events occurring in different places. I have explained this in an article entitled 'The Measurement of Time.'"[104] A distinction between the problems of time and simultaneity had never appeared anywhere else. This passage could only have jolted the young men in Bern munching cheese, sipping tea and smoking cheap cigars. Perhaps Einstein raised an eyebrow and said something like, "Well, this is interesting. Let's see what else Monsieur Professor Poincaré has to say about this." Although Poincaré went no further in the work at hand he took the unusual step of providing the exact reference to his article "The Measurement of Time" as "*Revue de Métaphysique et de Morale,* t. vi, pp. 1–13, January 1898."[105] It is the only literature citation in the entire book. Might Einstein and his friends have accepted Poincaré's invitation and looked up his paper in a local library? Einstein's French was not bad, while Chavan and Solovine were fluent.[106]

Had they done this, they would have read an erudite article on the relation between the velocity of light and clock synchronization, to the best of my knowledge the only such exposition of its day. When, in 1976, I had the good fortune to discover Poincaré's missing correspondence and manuscripts, I half expected to find a letter from Einstein in some way acknowledging his debt to Poincaré's writings and particularly to the 1898 essay. I found no such letter, nor has anything like it emerged from the Einstein archives. Probably Einstein never read "The Measurement of Time." More likely Chavan or Solovine reported on it at an Olympia Academy session.

Any discussion of Poincaré's approach to clock synchronization using light signals must mention a scientific paper he published in 1900 entitled "Lorentz's Theory and the Principle of Reaction." We can build a solid case for Einstein's having read this paper before writing his relativity paper. Poincaré's paper appeared in a volume dedicated to Lorentz on the twenty-fifth anniversary of his doctorate, which included several papers that would have interested Einstein;[107] Einstein's reasoning in 1905 regarding synchronization of clocks using light signals is incredibly similar to Poincaré's; and in 1906, Einstein cited this paper.[108] Whereas in the 1898 essay Poincaré wrote broadly on the nature of time without explicit mention of local time, the 1900 paper explores local time with full mathematical regalia and adds a splendid discussion of the foundations of Newtonian mechanics. It is one of Poincaré's classics. Still, while this treatment suited Einstein, the 1898 essay expressed, more accessibly, many of the same concerns.

"The Measurement of Time" is a window through which we may glimpse how avant-garde technological thought influenced both Poincaré and Einstein. Poincaré's choice of title was intentional: by the late nineteenth century, standardization of time was paramount for technology and commerce. A passenger changing trains at a large station in Germany could find the railroad clocks set according to times in five different cities, while his own watch kept the time of his own town.[109] This was a source of chagrin to Germany's most ardent supporter of standardized time, Count Helmuth von Moltke of the German General Staff. In France, clocks inside railroad stations were set at *l'heure de la gare*, the railroad time, five minutes slower than Paris time. Clocks outside the station were set at *l'heure de la ville*, the local time. The United States maintained some eighty different railroad times, which differed from the local times,[110] so that a traveler going from Eastport, Maine, to San Francisco had to change his watch as often as he changed trains—some twenty times. And this was just to keep up with railroad times.[111] In the end, however, it was not the passengers' convenience that brought a uniform time to all railroads, but the need for efficient, reliable freight transport.[112]

Among the industrialized nations, Great Britain had taken the lead in standardizing time, legislating a standard railroad time in 1854. Beginning in the 1850s, time signals were sent out by telegraph from a master clock in Greenwich. Time delays in transmission due to the number and type of relays, as well as to the velocity of light, were taken into account.[113] The United States followed suit. Starting in 1865 telegraphic time signals were sent out daily from the U.S. National Observatory in Washington, D.C., which promoted the notion of a uniform time. By October 1866 longitude differences in the United States were ascertained relative to Greenwich by means of the transatlantic cable that had been laid the year before. In this case additional time delays accrued due to transmission over long insulated underwater cables.[114]

After consultation with the railroads, in 1870 Charles Ferdinand Dowd suggested the scheme for time standardization in the United States that is essentially the one in use today, with four standard meridians spaced 15 degrees apart forming the middles of four time zones.[115] The easternmost passed through Washington, D.C., so the U.S. Naval Observatory defined a prime meridian.[116]

In addition to railroad scheduling, weather forecasting also required time standardization.[117] For the purpose of systematically collating weather reports from across the United States, in 1870, Cleveland Abbe, in charge of the U.S. Signal Service's meteorological functions, urged a standardization of time. Among the problems Abbe had to contend with was observations of the aurora borealis, or northern lights, by widely separated observers. It was very hard to tell whether events reported by far-flung observers were simultaneous. This is the problem of distant simultaneity dealt with by Poincaré in 1898 and Einstein in 1905. Both men, in similar fashion, distinguished between events occurring so closely together that analysis cannot separate them, and those occurring far enough apart to be amenable to scientific analysis.[118]

Through the urging of Abbe and the good sense of Dowd's suggestion, in 1883 the United States and Canada adopted a time standard based on the Greenwich meridian, and Dowd's time zones. The next step was international standardization. In October 1884, the United States government invited twenty-five nations to Washington to discuss this situation. The result of the Prime Meridian Conference was to divide the Earth into twenty-four one-hour time zones, spaced 15 degrees apart, with Greenwich as the zero longitude line or prime meridian.[119]

Despite their early and essentially chauvinistic resistance to this plan, by the beginning of the twentieth century the French became supporters, at least indirectly.[120] The next problem was to establish worldwide synchronized clocks that would serve also for determination of longitudes in sparsely explored parts of the globe. Throughout their history, the French were especially keen on this sort of exploration, which also served to better map the Earth's surface, thereby improving determination of the Earth's shape.[121]

Such problems were at the core of Poincaré's research interests. As the only member of l'Académie des Sciences who qualified for membership in each of its five sections—geometry, mechanics, astronomy, physics, geography and navigation—Poincaré could not have been unaware of the 1884 Prime Meridian Conference. Most likely he was informed by J. Janssen himself, Director of the Paris Observatory and one of France's two delegates.

Longitude determination depends on clocks synchronized with a standard clock at an agreed-on prime meridian. A major problem was how to improve on the accuracy of hand-carried chronometers. Poincaré's 1898 essay is meant to inform us of the "definition implicitly assumed by the savants [which we can observe if we watch] them at work and look for the rules by which they investigate simultaneity."[122] The key word here is the word "implicitly." Poincaré was a master at rooting out implicit assumptions.

His first example is the way astronomers measure the velocity of light by observing the eclipses of Jupiter's satellites. This entails the standard definition of velocity, which is distance divided by time. There are two nonsimultaneous events here: reflection of sunlight by Jupiter's satellites and its reception on Earth.[123] The implicit assumption is that the velocity of light is the same in every direction. Poincaré pointed out that research in astronomy would be impossible without this assumption, even if it could never be verified directly.

What Poincaré left unsaid is that astronomers and electrodynamicists treated the velocity of light in very different ways. While the astronomer's universe was empty, the electrodynamicist's was filled with ether. Consequently what was axiomatic for the astronomer—the constancy of the velocity of light in every direction—for the electrodynamicist emerged only from a confluence of hypotheses, including Lorentz's local time.

Poincaré next explored longitude measurement. The problem is how to set up clocks synchronous with one in Greenwich or Paris. This can be accomplished as follows: Carry a clock that has been synchronized with one at a prime meridian (but this is unreliable in practice); or coordinate observations of astronomical phenomena (but this is laborious and extremely difficult). Poincaré opts for electromagnetic signals, but he treats this case differently from the astronomical one. As an example, Poincaré considers someone in Paris telephoning a friend in Berlin. In the course of the conversation the person in Paris asks the time. The friend in Berlin replies that it is 5:00 P.M. and the Parisian sets his clock to 5:00 P.M. To be rigorous, says Poincaré, one must account for the time delay in transmission owing to the enormous but finite velocity of light. For this situation, however, such corrections are unnecessary. So, "in general, we neglect duration of the transmission and consider the two events as simultaneous."[124]

Poincaré considered simultaneity a psychological concept, whose definition could depend on sense perceptions. In the case of Jupiter's satellites there is a perceptible difference in the two events involved. Although there is no perceptible time delay for longitude measurement made with electromagnetic signals, precise corrections have to be made.[125] But this does not apply to every circumstance of telegraphic or wireless communication. Since there "is no general rule, no rigorous rule [but] a multitude of little rules applicable to each particular case," any analysis of simultaneity boiled down to time measurements.[126] Consequently, "the qualitative problem of simultaneity is reduced to the quantitative measurement of time."[127]

But in the final analysis, Poincaré had to conclude that even time measurements were somewhat qualitative. While it is all right, if you wish, to make corrections for the finite velocity of light or other delays in transmission over transatlantic cables, any other corrections would greatly complicate the "enunciation of the laws of physics, mechanics and astronomy."[128] Here he could only have meant complications involved with taking Lorentz's local time as the real time, which conflicts with what we expect from our sense perceptions.[129] In the end, for Poincaré there remained an absoluteness of time. His argument for it centered on his belief that scientific theory and technological practice should mirror concepts from the world we encounter daily.

Poincaré provided some quantitative bite to these deliberations in his 1900 paper, where he explores how to synchronize two clocks at relative rest while moving through the ether. Observers with the clocks decide to synchronize them by exchanging light signals, rather than use cumbersome telegraphic devices. In this case "signals are transmitted with the same velocity in both directions."[130] This can be accomplished if each clock reads Lorentz's local time so that observers moving with them are unaware of their motion, as it should for inertial reference systems.[131] But Poincaré's result is approximate because the mathematics of local time were not yet completely worked out.[132]

The preferred method for clock synchronization was, of course, wireless telegraphy. In 1905, wireless telegraphy on ships was a major topic: It promised to improve safety and keep passengers informed about the latest news, and of course it had military implications. Einstein and his coworkers may also have discussed how, in 1905, the United States Navy began to send time signals out to the fleet from Washington, D.C.[133] By then the problem had arisen of how to standardize transmission and reception devices.[134] Lieutenant Commander Edward Everett Hayden of the United States Navy anticipated "the day when wireless telegraphy will perhaps allow of a daily international time signal that will reach every continent and ocean in a small fraction of a second."[135]

In the Patent Office, issues such as time standardization and transmission by cable and wireless telegraphy were discussed daily. Haller, the office's director, had been trained as a railroad engineer and may have encouraged such discussions by adding his opinions. Einstein could not have failed to be struck by the link between the velocity of light and clock synchronization. His friend Chavan, however, was the more likely candidate to have reported on Poincaré's 1898 paper, since he worked for the Federal Postal and Telegraph Administration, an organization that had a vested interest in standardization of time as well as sharing a building with the Patent Office. Although Einstein could have acquired knowledge of the relation between clock synchronization and light from discussions at work, the similarity between Einstein's deliberations in 1905 and Poincaré's in his papers of 1898 and 1900 is too striking to be ignored. Both men explored how clocks are synchronized by exchanging light signals and emphasized the importance of analyzing the simultaneity of spatially separated "events."[136]

In his "physical (thought) experiments"[137] on synchronizing clocks, Einstein's description of time and simultaneity strikingly combines elements from Poincaré's statement of the problem of simultaneity in *La Science et l'hypothèse* with elements of railroad time and distant simultaneity. But unlike Poincaré, Einstein insists that issues of simultaneity *and* time are quantitative issues with no element of subjectivity.

> If we wish to describe the *motion* of a material point, we give the values of its coordinates as functions of the time. Now we must bear carefully in mind that a mathematical description of this kind has no physical meaning unless we are quite clear as to what we will understand by "time." We have to take into account that all our judgements in which time plays a role are always judgements of *simultaneous events*. If, for instance, I say, "That train arrives here at 7 o'clock," I mean something like this: "The pointing of the small hand of my watch to 7 and the arrival of the train are simultaneous events."[138]

Einstein's essential thought experiment for clock synchronization is straightforward compared to Poincaré's in 1900. This is because, in relativity theory, the equality of the to-and-fro light velocities is axiomatic. Consequently, the time on the clock to be synchronized is just the average of the times of the emission and reception of a light ray by a "master" clock at relative rest. The problem requires only grade-school arithmetic. Synchronization is then spread throughout the other clocks at relative rest in this inertial reference system.

The sleekness of Einstein's definition for synchronized clocks, however, carried a high price for physicists: the shattering of their conceptual framework. Assuming a principle of relativity that was exact and covered all branches of physics, and assuming that the velocity of light has the same value in every inertial reference system, amounted to rejecting the ether.

In the second section of the relativity paper, Einstein deduced from this definition of clock synchronization yet another terribly counterintuitive result. Two events that are simultaneous in one inertial reference system are not necessarily simultaneous when measured in another. There is no preferred viewpoint from which you can say that they are "really" simultaneous or not.

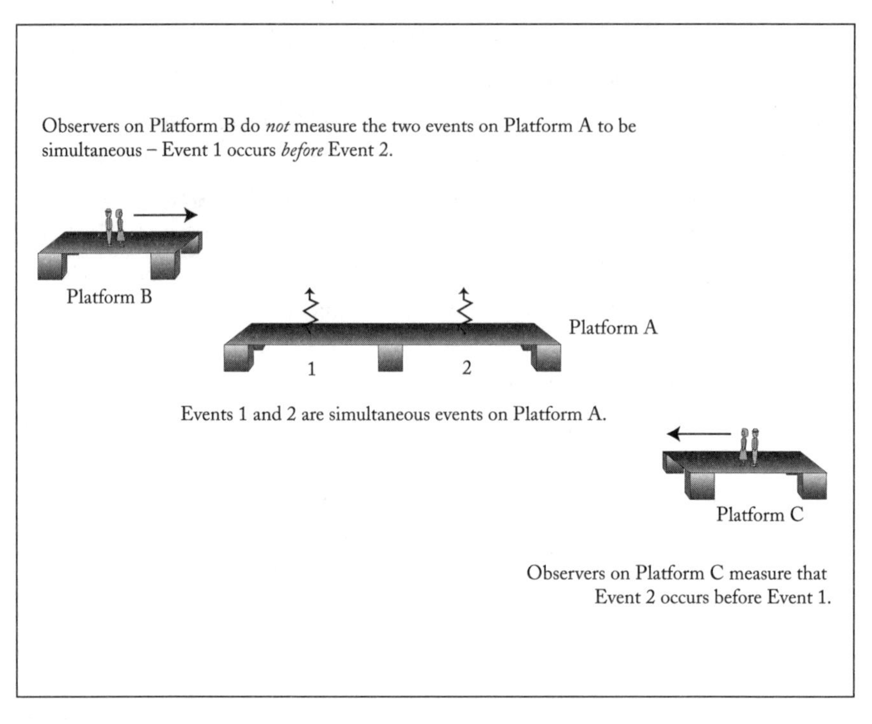

FIGURE 6.4 Simultaneity according to special relativity

In Figure 6.4, two observers at rest on Platform A have agreed that at 3:00 P.M. they will both point their flashlights skyward and flash them. These two events are simultaneous. When viewed, however, by someone on another platform in motion relative to Platform A, they are not simultaneous. Simultaneity is a relative concept. Someone on Platform B sees event 1 occur *before* event 2, while someone on Platform C sees event 1 occur *after* event 2. There is no true simultaneity, just as there is no true shape to an object. Cubism and relativity both require one to pick out from nature particular aspects of it. Einstein's temporal simultaneity matches Picasso's spatial one. Both amounted to representing nature from several viewpoints at once. How you measure or view a scene, that is what it is.

Counterintuitivities begin to pile up. Among them is Einstein's demonstration that this result entails the slowing down of a clock in inertial motion as measured by clocks in another inertial reference system. These results go beyond the time order and spatial order of events that Poincaré alluded to in *La Science et l'hypothèse* when he discussed how to view the fourth dimension.

The earliest testimony from Einstein himself of how he discovered that time and simultaneity are relative is a "reminiscence" written in 1907. After reviewing the status of Lorentz's theory and his ad hoc contraction postulate, Einstein went on to describe what his own view of the situation had been in 1905: "Surprisingly, however, it turned out that a sufficiently sharpened conception of time was all that was needed to overcome the difficulty discussed. One had only to realise that an auxiliary quantity introduced by Lorentz and named by him the 'local time' could be defined simply as the 'time.'"[139] Whether conscious or not, Einstein's phrasing is a nice turn on Poincaré's statement, in his 1900 paper, that when synchronizing clocks using light signals one must employ "not the true time t, but a certain *local time* t'."[140] Recall, however, that neither Lorentz, Poincaré, nor any other physicist was willing to grant Lorentz's local time any physical reality, because that would have meant that time depends on the relative motion of clocks, and this result clashes with our sense perceptions. Only Einstein was willing to go beyond appearances.

First, however, he required more support to assure himself that local time was the real physical time. Someone like him could not be convinced solely by its usefulness for explaining "known facts," such as the failure of ether-drift experiments or how the electron's mass depends on its velocity. The crucial insight came from the Aarau thought experiment. After ten years of deliberation, Einstein finally understood its role: It went beyond the "known facts."

At this juncture we can imagine Einstein doing the following elementary calculation. In the definition of velocity, distance divided by time, instead of using Newton's time he inserted Lorentz's local time with its two terms, the "usual" commonsensical time and a second term that depends on the clock's relative motion. The result must have astounded him. All measurements of the velocity of light made in inertial reference systems now yielded the same result, 186,000 miles per second.[141] He found a new way of combining velocities in which no measurement of the velocity of light can reveal the motion of an inertial reference system. This provided the mathematical support for the thought experimenter's intuition that Newton's exact principle of relativity could be widened to cover optics and electromagnetic theory.

Here Einstein required some cognitive guidelines. Despite its far-reaching consequences, the difference between the local time and the time of our sense perceptions is imperceptible. Lorentz and Poincaré were fond of producing calculations to demonstrate that the difference between the local time and the "real" physical time was much too small to be measured, not to say perceived.[142] For help Einstein turned to the eighteenth-century Scottish philosopher David Hume. As he wrote to Besso on 6 January, 1948, "So far as I can recall, David Hume had a greater direct influence on me [than Mach]; I read him in the company of Conrad Habicht and Solovine" at the Olympia Academy.[143] Whereas Mach emphasized the need to maintain a healthy skepticism, the more incisive Hume cut deeper. With Hume's classic conclusion that exact laws of nature cannot be obtained from sense perceptions or empirical data, he provided the confidence Einstein needed to deal with Lorentz's local time, because it turned out, contrary to Poincaré, that simultaneity *and* time are both quantitative problems.

And why is Besso acknowledged in the relativity paper? As far as I know, Einstein's only comment on Besso's contribution came in an impromptu lecture he delivered on 14 December, 1922, in Kyoto, Japan. Einstein spoke of a session with Besso during which they discussed aspects of a "problem," which we may assume was time. Evidently Besso's critical turn unlocked something within Einstein, because the next day, "without even saying hello," Einstein greeted Besso with "Thank you. I've completely solved the problem. An analysis of the concept of time was my solution. Time cannot be absolutely defined and there is an inseparable relation between time and signal velocity."[144] Alas, there are no details.

Whatever brainstorming went on, it brought results quickly. In the letter to Habicht written sometime between 18 and 25 May, 1905, Einstein mentions a paper that "is only a rough draft at this point [on the] electrodynamics of moving bodies which employs a modification of the theory of space and time; the purely kinematic part of this paper will surely interest you."[145] Since Einstein moved house on 15 May to be closer to Besso, one of their very first conversations on the way from work must have led to Einstein's solving the "problem" on which he was stuck. The conversation with Besso triggered what he had learned from Poincaré's 1898 and 1900 papers—or perhaps what they had heard at work about synchronizing clocks using wireless telegraphy.

Taking into account Einstein's interview with Seelig some decades later, the draft version he described to Habicht contained the "conception of the idea of special relativity," which was setting clocks using light signals and the relativity of time. Einstein's recollection to Seelig that "four to six weeks elapsed [before] completion of the corresponding publication" squares with the *Annalen*'s receiving the paper on 30 June. We may further assume that the pre-Besso drafts were the ones on which Einstein was stuck. The conversations with Besso were decisive, and Einstein was properly generous. In his conclusion to the relativity paper he writes: "In conclusion I wish to say that in working at the problem here dealt with I have had the loyal assistance of my friend and colleague M. Besso, and I am indebted to him for several valuable suggestions."[146]

Einstein never wavered on the importance of the Olympia Academy and his friendship with Besso. Forty years later he expressed concern to Besso that his biographer Carl Seelig did not realize this. Seelig, Einstein complained, was preoccupied with his childhood, which the biographer justified by claiming that

> my existence is known in detail. Which is not the case, and it is precisely the years spent in Switzerland. That gives a wrong idea, as if, so to say, my life started in Berlin! It is important to add that we discussed scientific questions, every day, while returning home from the office. He should mention the friendship with Maurice Solovine (now in Paris). . . . In Bern I regularly had evenings of reading philosophy and discussion with C. Habicht and Solovine, in the course of which we occupied ourselves especially with Hume (in a very good German edition). This reading had a considerable influence on my development, along with Poincaré and Mach. It was you who recommended [Mach] at the time of my years of study [at the Swiss Polytechnic], after we met [at a musical soirée] at Mme. Caprotti.[147]

Although Apollinaire wrote of Picasso's creativity soon after *Les Demoiselles d'Avignon*, Besso, who was Einstein's Salmon, never reminisced about what happened in Einstein's "atelier" in spring of 1905. This is not surprising, as Besso tended toward constipation whenever it came to writing anything with the remotest chance of publication. Nevertheless, we have enough historical data to understand, at least potentially, the intellectual supernova that flared during the incredible Bern spring of 1905.

EARLY RECOGNITION

The eager Einstein did not have to wait long for reactions to his published papers. The first one came quickly, on 30 November, 1905. It was from the experimentalist Kaufmann who commented on a "short publication by Mr. A. Einstein" (the relativity paper) whose results "are formally identical to those of Lorentz's [electron] theory." The "formal" identity that Kaufmann mentioned was that of the electron's mass.[148] Kaufmann referred to a "Lorentz-Einstein" theory that did not fare well against his recent data.[149] In a longer paper in the *Annalen*, in early 1906, Kaufmann pulled no punches, stating bluntly that his experimental results were "not compatible with the Lorentz-Einstein" theory.[150]

Nevertheless, Einstein's reputation began to grow and he was happy about this: "My papers are much appreciated and are giving rise to further investigations. Professor Planck (Berlin) has recently written to me about that."[151] Planck had early on recognized that Einstein was after bigger game in the relativity paper than an electron theory. Ever in search of absolute laws in physics, Planck was impressed by the universality of Einstein's principle of relativity. At the annual September meeting of German Scientists and Physicians, the major scientific event in Germany, Planck rose to the defense of the Lorentz-Einstein theory against Kaufmann's data.[152]

Einstein announced his last publication of 1905 to Habicht in a letter written somewhere between 30 June and 22 September in which he suggested that he might be able "to smuggle you in among the patent slaves." It concerned a result that he had overlooked in the relativity paper, that "mass [is] a direct measure of the energy contained in a body. . . . The consideration is amusing and seductive."[153] On the world stage this result turned out be anything but amusing. Forty years later its fundamental scientific content became overshadowed by premonitions of an absolute destructive power, revealed through the equivalence of mass and energy, $E = mc^2$.[154]

Perhaps given at this point to resting on his laurels, or maybe momentarily spent, Einstein complained to Solovine, "As for my science, I am not all that successful at present. Soon I will reach the age of stagnation and sterility when one laments the revolutionary spirit of the young."[155] Not quite.

By September 1906 Einstein received requests for his relativity paper from the Nobel Prize–winner Röntgen, and even Drude. On 2 June, 1906, Max Laue, who would become a lifelong friend, wrote Einstein commenting on certain points in the light quantum paper.[156] The same age as Einstein, Laue had been Planck's *Assistent* at the University of Berlin since autumn of 1905. The first physics lecture he heard Planck give was on Einstein's relativity theory.[157]

In the summer of 1907, Planck sent his *Assistent* to seek out the man in Bern who was writing such fascinating and provocative papers. Of course Laue went to the university and luckily was redirected to the Patent Office. Laue described the scene when he arrived there: "In the general waiting room an official said to me: 'Follow the corridor and Einstein will come out to meet you.' I followed his instructions but the young man who met me made such an unexpected impression on me that I could not believe he could be the father of the relativity theory. So I let him pass and only when he returned from the waiting room did we actually become acquainted."[158]

"During the first two hours of our conversation he overthrew the entire mechanics and electrodynamics," wrote Laue on 2 September, 1907, to his friend Jakob Laub, a physics graduate student at the University of Würzburg.[159] On their walk from the Patent Office Einstein offered Laue one of his favorite cheap Swiss cigars. Laue found it so horrible that at the first opportunity he tossed it into the river Aare.

Laub was already acquainted with Einstein's relativity theory, having been assigned to give a colloquium on it by Würzburg's most esteemed professor, Wilhelm Wien, who would be awarded the Nobel Prize in 1911. Laub recalled that "there was a lively discussion from which it was clear that it was not too easy to get inside the new concepts of time and space."[160] In July 1907 Einstein began a correspondence with Wien on the velocity of light. In 1912 Wien nominated Einstein and Lorentz jointly for the Nobel Prize.[161]

The year 1905 did not exhaust Einstein's creative outpourings. Before he left his "secular cloister," in 1909, he would produce one more great thought experiment.

7

I Really Would Not Have Thought Einstein Capable of That!

After a dinner at the house of Émile Borel, Paul Valéry asked [Einstein]: when an idea comes to you, how do you make arrangements to remember it? A notebook, a scrap of paper . . . Einstein responded: Oh! An idea, it is so rare!

—Émile Borel, 1922

Imagine that you are a university science graduate who found it impossible to obtain an academic position and so went into a civil service post. Yet you persist in your research and succeed in formulating a bold new theory of space and time. Others, however, interpret it as merely giving a firmer foundation to an already existing theory whose emphasis lies elsewhere. Even so, your name has become linked with one of the great scientists of the day. You are elated and even dare to dream of a university position. Then an experiment disconfirms the dual-named theory. Everything collapses.

PERSEVERANCE

This was precisely what Einstein faced in 1906, when his theory of space and time was interpreted as providing an underpinning to the esteemed

Lorentz's theory of the electron. The Lorentz-Einstein electron theory, however, was judged by most physicists to have been disconfirmed by the renowned experimentalist Walter Kaufmann. Kaufmann measured a variation of the electron's mass with velocity that was other than the one predicted by the Lorentz-Einstein theory. Lorentz, devastated, wrote to Poincaré on 8 March, 1906, that his theory "is in contradiction with Kaufmann's results, and I must abandon it. I am at the end of my wits."[1]

In the spring of 1907, the portion of the physics community who supported the Lorentz-Einstein theory was in a state of collective despair. If this was the mind-set of people like Lorentz, what about the patent clerk? Einstein should have been shattered. But perseverance under the most severe conditions is a hallmark of genius. As the psychologist Howard Gardner has found, "The extraordinary individual is . . . perennially at risk for pain, rejection, and loneliness. [Yet he or she has the] capacity to see not so much the bright side of a setback as the learning opportunity it offers."[2]

We know Einstein's view of the situation from a review paper on relativity that he commenced writing in September 1907 for the prestigious *Yearbook for Radioactivity and Electronics*.[3] Einstein first assessed Kaufmann's experiments on the mass of moving electrons and found them by and large satisfactory; however, there was the possibility of systematic errors from unknown sources. Second, he offered his "opinion" that since Kaufmann's data supported electron theories lacking in generality, they ought not to be taken as definitive.[4] Einstein's third step was even more extraordinary: ignoring Kaufmann's data, he went on to propose a sweeping generalization of his supposedly disconfirmed theory of space and time to include gravity. That's chutzpah! Eight years later this proposal culminated in one of the most beautiful theories in the history of scientific thought—the general theory of relativity. A precursor to all of this, however, was the second and last major thought experiment of Einstein's scientific career, which emerged in 1907. Its resolution required new notions of aesthetics.

THE SECOND THOUGHT EXPERIMENT

A scientist writing a review paper often finds that the integrated material takes on a new meaning; once gathered into one place, a large body of research can be better understood, perhaps even at a deeper level. Previously unconnected aspects can connect in unexpected ways that lead to new insights. This is what happened to Einstein between September and December 1907. The insight came in the form of a thought experiment, of which Einstein provided detailed recollections soon afterward. This is in contrast to the 1895 thought experiment, about whose origins we know next to nothing.

Thought experiments occur to the prepared mind—there are no bolts from the blue.[5] They must be preceded by periods of conscious work. Psychological studies reveal that highly creative problem solving often runs through a cycle of conscious thought, unconscious thought, illumination (hopefully!) and verification. Let us try to reconstruct the period leading up to Einstein's 1907 thought experiment to see if it follows this pattern.[6]

In 1919, Einstein recalled that in 1907, while he "was working on a summary essay concerning the special theory of relativity,"[7] he put a great deal of effort into generalizing Newton's theory of gravitation in a way consistent with relativity, and failed. Yet instead of becoming disillusioned, Einstein used his troubles as learning opportunities. While mulling over the failure to generalize Newton's theory of gravity, it occurred to him to think again about the relativity of electric and magnetic fields from 1905. This is the conscious part of problem solving.

The illumination from unconscious thought surfaced unexpectedly, as he related in an address given in Kyoto, Japan, on 14 December, 1922: "I was sitting in my chair in the Patent Office in Bern when all of a sudden a thought occurred to me: 'If a person falls freely he will not feel his own weight.' I was startled. This simple thought experiment made a deep impression on me. It led me toward a theory of gravitation."[8]

What was behind this moment when "all of a sudden a thought occurred to me"? Clues emerge from Einstein's unpublished manuscript of 1919, where he recalled this thought experiment in more detail:

> At this point, there occurred to me the happiest thought of my life. Just as is the case with the electric field produced by electromagnetic induction, the gravitational field has similarly only a relative existence. *For if one considers an observer in free fall, e.g., from the roof of a house, there exists for him during this fall no gravitational field—at least not in his immediate vicinity.* Indeed, if the observer drops some bodies, then these remain relative to him in a state of rest or in uniform motion, independent of their particular chemical or physical nature . . . . The observer therefore has the right to interpret his state as "at

> rest." . . . The experimentally known matter independence of the acceleration of fall is therefore a powerful argument for the fact that the relativity postulate has to be extended to coordinate systems that, relative to each other, are in nonuniform motion.[9]

To understand what Einstein means by the "relative existence" of the electric field and the gravitational field, let us recall the relative nature of simultaneity. Figure 6.4 of Chapter 6 shows how two events that are simultaneous in one reference system need not be simultaneous to observers in relative motion: There is no "true" description of simultaneity.

The situation with the magnet and conductor in Figure 6.3 of Chapter 6 is similar. While both observers measure the same current in the conducting loop when there is relative motion between magnet and conductor, their descriptions differ. The observer on the magnet attributes the current to the conductor moving through a magnetic field, while the one on the magnet explains the current resulting from the conductor moving through an electric field whose source is the moving magnet. The observer's state of motion determines whether electromagnetic induction is caused by a magnetic or an electric field, and so these fields have a "relative existence." In his 1919 recollection, Einstein rhapsodizes about this in a Mozartian vein: "and a kind of objective reality could be granted only to the *electric and magnetic* field together, quite apart from the state of relative motion" between observers.[10] Both fields are "out there" to be plucked out of the cosmos. Consequently, the "asymmetry mentioned in the introduction as arising when we consider the currents produced by the relative motion of a magnet and a conductor, now disappears."[11] According to Einstein's relativity theory, electromagnetic induction can be discussed in terms only of fields and not of forces and fields, as Lorentz's theory had done.

His 1907 illumination led him to discover that the "gravitational field has similarly only a relative existence." Two observers, one on the ground and the other falling with the stone, agree on the single measurable quantity—how long it takes a stone to hit the ground—but give different interpretations. The observer on the ground sees the stone accelerating, while the other falls alongside it at relative rest.

Einstein's first attempts at exploring this situation had to go something like this.[12] He considers the problem first from the viewpoint (reference system) of the observer standing on the ground, for whom the only force acting on the freely falling stone is its weight. Using straightforward methods, Einstein then moves the mathematics into the reference system of the observer who is freely falling with the stone, while both are at relative rest, and so for whom the stone is weightless. Whereas scores of scientists have "seen" or imagined this elementary textbook situation, Einstein alone saw its "deep structure": With a certain proviso (to be explained in a moment), the equations of Newton's mechanics reflect the actual situation of the falling observer and the co-moving weightless stone at relative rest only under one condition—if the relative acceleration between the two observers, as measured by the falling one, is equal and opposite to the gravitational field experienced by the observer on the ground. In this case "*there exists for* [the freely falling observer] *no gravitational field*. . . . The observer therefore has the right to interpret his state as 'at rest.'"

This result has a deep connotation: Acceleration and gravity are equivalent. Einstein would call this the "equivalence principle."[13] The existence of the gravitational field is *relative* to acceleration. The classic illustration is what you would experience in a descending elevator. If you happened also to be standing on a scale, you would notice that you weigh less than if you were standing on the ground. Alternatively, you can imagine that the falling elevator is actually at rest and you are in a gravitational field less than that of Earth—less by precisely the amount of the elevator's acceleration. If the scale suddenly registers zero you are in trouble, because it means the cables have broken and you are in free fall. In this case you are weightless, like the stone in Einstein's thought experiment, and so there is no gravitational field in your immediate vicinity.

Once he was able to consider the falling observer's reference system "at rest," the laws of physics in this reference system could not differ from those in any other. So, as Einstein wrote in 1919, "the relativity postulate has to be extended" to accelerating reference systems.

Einstein's discovery of the equivalence principle was a first step in generalizing special relativity. An immediate offshoot was that every reference system can be used as a measurement platform. They all are on an equal footing, which removes an asymmetry in special relativity. As an extension of Newtonian mechanics to systems moving at velocities comparable to that of light, it too had given a preferred status to inertial reference systems. Special relativity also inherits from Newtonian mechanics confusions over what observers would experience in accelerating reference systems, which the equivalence principle would eventually alleviate.[14]

Among the astonishing results of the equivalence principle is that time depends on the gravitational field in which the clock happens to be, and so does the velocity of light.[15] Consequently, in seeking a general theory of relativity, Einstein had to drop the second axiom of special relativity: Light does not travel in a straight line at the same velocity in every reference system.

The direct relation between acceleration and gravity required that Einstein drop, as a redundancy, a time-honored distinction in Newton's mechanics. This brings me to the proviso I mentioned earlier. Newtonian science assumes that objects have two different sorts of masses, gravitational and inertial. Gravitational mass is a measure of a body's gravitational attraction, while inertial mass is a measure of how the object responds to pushes and pulls. Einstein declared these two masses to be the same thing. Critical to his calculation toward the equivalence principle was that they in fact cancel each other out. He thus removed another asymmetry that was not inherent in the phenomena: Why have two masses when one will do?

Were any clues available in 1907 to point Einstein to the equality of gravitational and inertial masses?[16] Once again we are brought to *La Science et l'hypothèse,* where Poincaré ferreted out yet another tacit hypothesis.[17] This one was concealed in the customary way that astronomers used Newton's laws of motion to "weigh" planets with satellites, such as the Earth. In the mathematics, as a matter of course, they canceled the Moon's gravitational mass with its inertial mass. So, wrote Poincaré, we tacitly assume their exact equality. But as he did with his observation about simultaneity in the same book, Poincaré went no further.

Einstein saw that this tacit assumption, or asymmetry, masks a deep law of nature: the equivalence between gravity and acceleration. Given that *La Science et l'hypothèse* was a best-seller, what could have piqued Einstein to see a deeper meaning in this passage? I believe it was his minimalist aesthetic that he wielded to remove asymmetries.

What is astonishing is that even in its historical context, the mathematical details of Einstein's 1907 thought experiment are so incredibly straightforward. No complicated mathematical considerations are necessary because Einstein went right to the conceptual core of the problem. A hallmark of high creativity is having such complete command of all technicalities that one can soar over inessential details and go right to the heart of the problem. Mozart did it in music, Picasso in art, and Einstein in physics.

THE GEOMETRIZATION OF SPACE AND TIME

Faced with the same empirical data as Einstein—failed ether-drift experiments and Kaufmann's measurements on the mass of high-speed electrons—Poincaré produced, in two elegant papers published in 1905 and 1906,[18] a mathematical formalism identical to Einstein's special relativity. Consistent with his insistence on considering only empirical data and, when it came to space and time, only sense impressions, Poincaré did not deign to analyze time and simultaneity. This would have required abandoning empirical data for the thought experimenter's lofty realm of visual imagery. Poincaré remained earthbound. He meant these two papers only to provide a secure mathematical foundation for Lorentz's theory of the electron.

Among the results in the 1906 paper that outlived its stated purpose are certain mathematical techniques that Poincaré brought into play for the first time in order to facilitate analysis of Lorentz's electron theory. They involve setting up a mathematical formalism in a four-dimensional space, with time as the fourth dimension. Surprisingly, Poincaré did not mention any visual geometrical aspects of this process.[19]

Another mathematician did. In 1907, the forty-three-year-old Hermann Minkowski realized the power of Poincaré's methods. Late in life, Einstein recalled Minkowski as one of the "excellent teachers" at the Swiss Polytechnic from whom "I really could have gotten a sound mathematical education."[20] In 1902 Minkowski received the call to Göttingen where, during the semester of 1907–1908, he organized a seminar on recent developments in electrodynamics. Papers by Einstein and Poincaré were among those discussed. By then Minkowski had achieved some fame as a mathematician, but his wider renown among physicists would come from the linking of his name to that of a former student who regularly cut his classes. On reading Einstein's relativity paper, Minkowski's immediate reaction was, "I really would not have thought Einstein capable of that!"[21] At least he remembered him.

Minkowski discovered that Poincaré's mathematical method provided the basis for a geometrical formulation of "Lorentz-Einstein electron theory" in four-dimensional space, complete with visual imagery. He realized that the velocity of light is what connects the three spatial dimensions with time (Figure 7.1).

Minkowski went further by generalizing his mathematical analysis into non-Euclidean geometry.[22] His ultimate dream, a twentieth-century

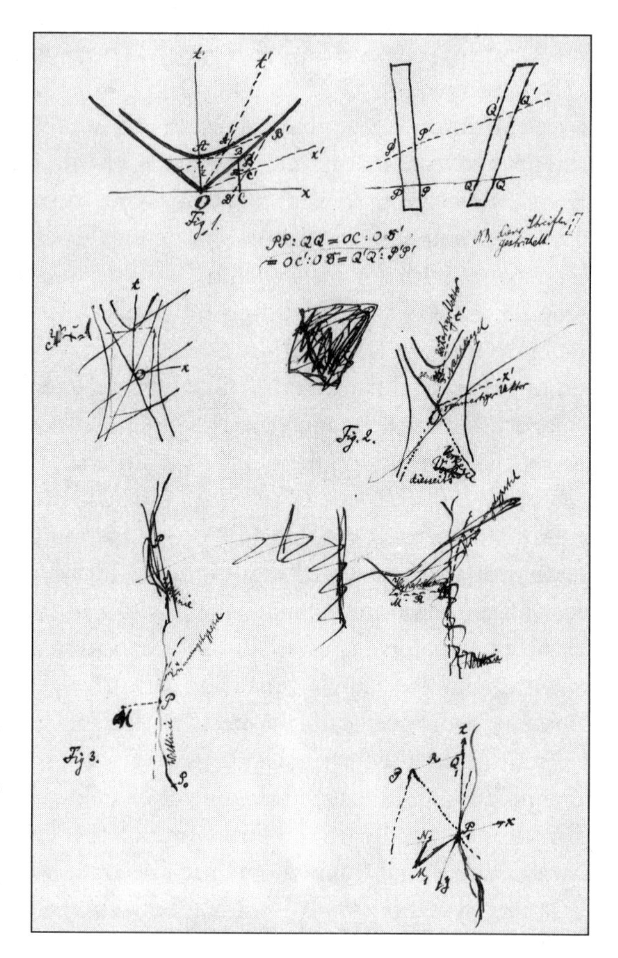

FIGURE 7.1 Hermann Minkowski's sketches of his space-time diagrams meant for his address "Space and Time," delivered in 1908. The numbers refer to the address's text.

version of René Descartes, was to fill up space with points in a space-time. Although Poincaré too clearly saw that in his mathematical formalism space and time were connected into a space-time, he chose to ignore its deeper ramifications. In 1912 Poincaré wrote that to take this union as anything but mathematical was to admit a relativity of time.[23] For Minkowski, however, there were profound physical consequences to be explored. As he eloquently put it in 1908: "Henceforth space by itself, and time by itself, are doomed to fade away into mere shadows, and only a kind of union of the two will preserve an independent reality."[24] Passing away in 1909, at age forty-five, Minkowski truly died before his time.

Einstein described Professor Carl Friedrich Geiser's courses in higher geometries at the Swiss Polytechnic as "true masterstrokes of pedagogical art."[25] He recalled Geiser's lectures while working on generalizing the special relativity theory.[26] But it was in *La Science et l'hypothèse* where Einstein read about the foundations of geometry. For example, in the relativity paper Einstein pointed out the importance "of the properties of homogeneity which we attribute to space and time."[27] Poincaré likewise attributed the property of "homogeneity" to the pristine space of mathematics in which all points are equivalent.[28] Without this property, it would be impossible to establish relations between objects, which is essential to Poincaré's theory of the origins of geometry. Einstein moved Poincaré's requirement of homogeneity into the space of physics, which, now that ether has been declared "superfluous," is likewise empty. Because homogeneity guarantees a democracy of points in space and the inertial reference systems that occupy them, the mathematical equations for moving between reference systems could be reduced to their simplest possible form. Additionally, like Poincaré in *La Science et l'hypothèse,* Einstein carefully stated right at the start that to analyze mechanics he would employ "the methods of Euclidean geometry."[29]

At first Einstein was unimpressed with Minkowski's geometrical presentation, considering it to be mere erudition. Starting in 1912, however, Einstein began to realize that special relativity could be fully widened to include gravity only if it was highly geometrized.[30] Here his friend Grossmann, since 1907 a professor of mathematics at the Swiss Polytechnic, came to his aid, for the third time.[31]

Even physicists who were more mathematically astute than Einstein did not immediately grasp what Minkowski had proposed. They continued to explore Einstein's relativity theory with methods that avoided non-Euclidean geometry.[32] Mathematicians who had studied Poincaré's comments on relations between geometry and physics were quicker off the mark. But it was Einstein's own research, starting in 1912, that demonstrated the deeper meaning of Minkowski's suggestions. By 1916 the preeminent mathematician of the time, David Hilbert, could declare "*the physicist must become a geometer.*"[33] Artists had grasped the importance of a geometrical representation of nature some years earlier.

Einstein completed the general theory of relativity in 1915. In this theory space-time is a four-dimensional structure whose shape is distorted by massive objects within it and in which light rays glide over its curvature. It is the intellectualization of Lorentz's ether: Its properties are given by the theory's mathematics in a more systematic and less ambiguous manner than in Lorentz's electromagnetic theory, with its unmeasurable quantities.

THE ACADEMIC COMEDY CONCLUDES

After having studied Einstein's papers on relativity, Jakob Laub wrote to Einstein on 1 March, 1908, "I must confess to you that I was amazed to read that you have to sit in an office for eight hours a day! But history is full of bad jokes."[34]

Another bad joke was the comedy of Einstein's quest to enter academia. The previous year, disenchanted with the Patent Office, he had decided to try again for the post of *Privatdozent.* To a friend with whom he played music Einstein wrote, sometime between 5 January and 11 May, 1907, "All goes well with me; I am a respectable Federal ink shitter with a decent salary. I am riding my mathematical-physical hobby horse and I fiddle my violin—both within the constraints my two-year-old little boy has imposed on me for superfluous things of this kind."[35] He needed more time for "mischief" than the Patent Office allowed.

Having now earned a Ph.D., Einstein decided to look into the traditional route to becoming a *Privatdozent.* By then he was corresponding with such prominent German scientists as Wien, Laue, Minkowski and Planck and had just completed a major review paper on relativity. He felt he could avoid writing a *Habilitationschrift*—the formal and original piece of work, beyond a doctoral dissertation, that such a position required.

So, feeling his oats, Einstein bound seventeen of his papers and submitted the package to the University of Bern with a brief curriculum vitae. Published papers were acceptable under the university's application guidelines as "other outstanding achievements," the same sort of escape clause as at the University of Zurich.[36] But one of the physics faculty—most likely the very elderly and possibly senile professor of experimental physics, Aimé Forster—considered Einstein's relativity theory "very problematical," and his application was denied.[37] But in February 1908, when Einstein submitted a formal *Habilitationschrift* based on his current work on radiation, the university dared not refuse, owing in large part to Einstein's recognition in major German universities. By spring 1908 he was a *Privatdozent.* Attending his first course were Chavan and two Patent Office colleagues, Besso and Heinrich Schenk. Two semesters later, only one student registered and Einstein cancelled the class.[38]

FAREWELL TO THE PATENT OFFICE

Something big on the academic front finally broke in 1908, when Einstein was considered for a position as Associate Professor at the University of Zurich. Before that came about, however, yet another academic comedy ensued. Alfred Kleiner, the professor of experimental physics at Zurich who had delayed reading Einstein's 1901 Ph.D. dissertation, had actually been following Einstein's career with interest for a number of years.[39] The time seemed propitious to think seriously about a position for him. To assess Einstein's teaching ability, Kleiner set up a visit to Einstein's class in Bern during the summer semester of 1908 which, by this time, he was presenting to an audience of one. Kleiner arrived to "size up the beast" and found him wanting.[40] Not standing for this, Einstein responded immediately, "On that occasion I did not lecture divinely—partly because I was not prepared very well, partly because the state of having-to-be-investigated got on my nerves a bit."[41]

Einstein also resented that Kleiner had told others of his poor assessment. Nevertheless, Kleiner insisted on satisfying himself as to Einstein's teaching ability. A second assessment was arranged and Einstein succeeded.[42] "I was lucky," he wrote to Laub. "Contrary to my habit, I lectured well on that occasion—and so it came to pass."

"It came to pass" on 6 May, 1909, when Einstein's position was approved at Zurich, to commence 15 October, 1909. On hearing the news Laub wrote to Einstein, "It was high time for you to get out of the Patent Office."[43] "So now I am an official of the guild of whores," replied Einstein, somewhat resentful of the effort it had taken and of the research he was still expected to produce.[44] In July 1909, the University of Geneva awarded him his first honorary degree. Among other recipients were Marie Curie and Wilhelm Ostwald, the Leipzig physicist who had never replied to Einstein or his father in the dark days of 1901. The following year Ostwald nominated Einstein for a Nobel Prize.

On 6 July, 1909, Einstein informed the Patent Office of his intention to resign. Haller wrote a touching final report on him, saying "his departure is a loss to the Patent Office."[45] Einstein had also been invited to his first major physics conference, the annual meeting of German Scientists and Physicians, 21–25 September, 1909, where he presented a keynote lecture.[46] He left Bern for Salzburg two weeks before the meeting was scheduled to begin. In a well-received and provocative lecture entitled "On the Development of our Visualisations [*Anschauungen*] of the Existence and Constitution of Radiation," he explored the clash between the time-honored wave representation of light with that of the light quantum.

To this day physicists puzzle over how light can be simultaneously a wave and a particle. As Einstein had written to Habicht sometime between 18 and 25 May, 1905, he considered light quanta "very revolutionary," and from his use of the term *Anschauung* in the title of his lecture we can understand why. Light quanta are a "wonder" because they clash with our customary representation of light as a wave phenomenon. On the other hand, Einstein never considered relativity theory to be revolutionary, but a natural extension of Newtonian science.

At the University of Zurich Einstein taught seven hours per week and took it very seriously, receiving high evaluations from his students. But he was not as happy as he expected. "I have less *really* free time than at Bern."[47]

SOME RECEIVED OPINIONS OF SPECIAL RELATIVITY AND ITS AUTHOR

Einstein's great discoveries of 1905 and 1907 were made virtually alone. His only colleagues were not professional scientists but enthusiastic young men whose precocity and opinions stimulated his thinking. His home life was frugal and unhappy.

The early recognition that came to Einstein could not have been completely fulfilling. The paper on the photoelectric effect was considered bizarre.[48] Soon after Einstein's theory of Brownian motion appeared, another theory was published whose methods, while less deep, most scientists found more comprehensible.[49] And his electrodynamics work was interpreted as a generalization of Lorentz's electron theory. The unity and meaning of this triad of papers were for a long time missed. But what could only have been of lasting personal regret to Einstein was that the physicists whose opinion he most cared about—Lorentz, Poincaré and Planck—never fully accepted special relativity.

His results differ fundamentally from Lorentz's and Poincaré's. For example, in Einstein's theory, different observers in relative motion to one another measure different lengths for rods at rest in other reference systems. A rod has no such thing as a true length. Length is a relative quantity. In Lorentz's theory, a rod's true length is its length when at rest in the ether—but the failure of the ether-drift experiments prevented measurement of this dimension. By eliminating from physics the notion of a preferred reference point, Einstein also eliminated unknown quantities that served no purpose.

For Poincaré the conflicts between Lorentz's theory and experimental data meant that he could never raise the principle of relativity to the level of a statement beyond experimental proof.[50] Even after Kaufmann's data were shown to be wrong, there remained the incorrect predictions of the extension of Lorentz's theory to gravitational phenomena.[51] The patent clerk's "opinion" about Kaufmann's data turned out to be right. After this, despite their emphasis on the importance of experimental data, Lorentz and Poincaré stopped demanding further experimental proof.

Poincaré and Einstein met at the 1911 Solvay Conference, convened in Brussels from 30 October to 3 November. The topic was the structure of light, relativity theory having been accepted by most physicists.[52] Although there are no recorded exchanges between them, Einstein informed Zangger that "Poincaré was in general simply negative and, for all his acuity showed little understanding of the situation."[53]

Despite their disagreements, shortly after Solvay, Poincaré wrote a letter on Einstein's behalf for a professorship at his old undergraduate university, the Swiss Polytechnic: "M. Einstein is one of the most original thinkers I have ever met. . . . Since he seeks in all directions, one must, on the contrary, expect the majority of the paths on which he embarks to be blind alleys."[54] No doubt relativity was one of these. Poincaré never referenced or even mentioned Einstein's relativity papers in print, nor did he ever associate Einstein's name with the principle of relativity. To the end of his life, in 1912 at age fifty-eight, Poincaré considered Einstein's theory as just one view of the situation; he preferred Lorentz's.[55]

Had Poincaré lived to see Einstein's general theory of relativity I believe he would have completely changed his opinion. What must have impressed Einstein about Poincaré, although possibly not immediately, was his honesty as a true philosopher-scientist. By this I mean his philosophical view was highly sensitive to advances in science. Poincaré drastically altered his view at least once in his career, when he switched from believing that atoms are perhaps only metaphors to thinking them real entities. Only through such openness could he fully participate in frontier research toward an electromagnetic world-picture. General relativity would have finally convinced Poincaré of the relativity of time and simultaneity and swayed him from his conviction that three-dimensional Euclidean geometry is privileged because it is the simplest or most convenient form. Einstein's special and general relativity theories redefined the notion of simplicity and common sense. Even more, Poincaré would have had to abandon the sharp distinction he always maintained between mathematical and representative or physical space, and acknowledge that the geometry of physical space is experimentally testable. On this issue, wrote Einstein in 1921, "*sub specie aeterni* [in the eternal sense], in my opinion, [Poincaré] is right."[56] But *sub specie aeterni* is not the point. The question is which geometry pertains to representative space. The general theory of relativity was the first theory in which it actually matters what geometry is used. It makes definite statements about the structure of space and time.

Max Planck made two great discoveries in physics—the quantum of energy and Albert Einstein. Despite being relativity theory's earliest major supporter, however, Planck never accepted that Lorentz's ether was superfluous, nor did he cease to ask for further experimental evidence to support the principle of relativity. He could not accept it as an axiom. In his November 1911 letter to Zangger, Einstein remarked that "Planck is blocked by some undoubtedly false preconceptions" regarding relativity theory, and the quantum as well.[57]

Despite these fundamental disagreements, Planck was immensely impressed with Einstein as a physicist. So much so that in the summer of 1913 Planck and the eminent German physical chemist Walther Nernst made Einstein an extraordinary offer at the University of Berlin that included a highly paid professorial position with no teaching duties; the directorship of an Institute of Theoretical Physics, of which Einstein would be its only member and whose headquarters could be in his flat; and a salaried membership in the Prussian Academy of Sciences. Einstein accepted.

But before departing for Berlin in late 1913, he wrote to Besso: "In the meantime I shall have to seek out Lorentz in order to discuss fundamental matters. He exhibits interest, and so does Langevin. Laue is not accessible when it concerns matters of principle, also not Planck. . . . The free and impartial view is hardly characteristic of the German adult (blinkers!)."[58] Einstein had already had a taste, during Planck's recent visit to Zurich, of this attitude. When Einstein discussed his recent work on generalizing relativity to include gravitation, he recalled Planck's saying, "As an older friend I must advise you against it for in the first place you will not succeed; and even if you succeed, no one will believe you."[59]

As for H. A. Lorentz, Einstein described him in his 15 November, 1911, letter to Zangger: "H. A. Lorentz is a marvel of intelligence and tact. He is a living work of art. In my opinion he was the most intelligent among the theoreticians present [at Solvay]."[60] He had been corresponding with Lorentz since March 1909 on radiation theory. The tone of the exchange affected Einstein deeply: "I admire this man like no other; I might say, I love him.[61] . . . For me personally, he meant more than all the others I met on my life's journey."[62] On 22 November, 1911, after turning down an offer of a professorship from Utrecht, Einstein expressed his deepest emotions to Lorentz: "I write this letter with a heavy heart, like a person who has done some kind of injustice to his father. . . . You have probably sensed that I revere you beyond measure."[63]

Fluency in many languages, superb diplomatic skills and an encyclopedic grasp of science made Lorentz the perfect person to chair international colloquia such as the Solvay Conference. This was no mean feat in such a gathering of prima donnas. Yet to Einstein's great disappointment, Lorentz viewed relativity theory as merely an alternative to his own electron theory, in which there was no relativity of time. Although the theories were indeed mathematically equivalent, they were not at all equivalent physically.[64] Still, Lorentz supported and even inspired Einstein's work toward generalizing special relativity. In general relativity Lorentz saw his ether return in the guise of a structured space-time that supported light in transit.

Lorentz alone became a personal inspiration to Einstein, and he only after 1907. During Einstein's major annus mirabilis of 1905 and the minor one of 1907, he needed no muse other than himself. Certainly Mileva no longer played that role. In this Einstein resembled Picasso in 1907, when Cézanne became his inspiration, but only after the completion of *Les Demoiselles d'Avignon*.

THE "MERELY PERSONAL"

Einstein's troubled domestic situation is clear from a letter he wrote to Laub on 19 May, 1909, in which he declined Laub's invitation to visit him at the university in Heidelberg. "As for my coming to Heidelberg for Pentecost, I can't do that to my wife. She doesn't get much out of me as it is."[65] Later that year, after they moved to Zurich, Mileva wrote her friend Helen Savic: "You see with such fame, not much time remains for his wife. . . . You see I am starved for love."[66] And well she was. Mileva's jealousy of Albert's fame and the way he went about his life peaked in spring of 1909.

The incident that finally broke their marriage for good has its roots in a vacation that Einstein took with his family in August 1899 at the Hotel-Pension Paradies in Mettmenstetten, Switzerland. He struck up a friendship with the hotel owner and was introduced to the owner's sister-in-law, Anna Schmid, with whom Einstein evidently more than flirted. He even composed a little ditty for her, in the familiar style of his letters to Marie and Mileva. "Little girl small and fine/What should I write for you here/I could think of many things/Including also a kiss/Upon your little mouth."[67] This romance occurred at the time that Einstein was writing his most passionate letters to Mileva. It was not a case of out of sight, out of mind.[68]

The summer romance with Anna Schmid ran its course. A decade later, however, in spring 1909, Zurich newspapers announced that the up-and-coming Albert Einstein was about to join the university's faculty. Anna Schmid, now Anna Meyer-Schmid, saw the announcement and sent Einstein a congratulatory note addressed to the Patent Office. He responded immediately about how "immeasurably happy" he was to receive her postcard and how much he cherished the "memory of the lovely weeks I was allowed to spend near you"[69] in Mettmenstetten: "[I] am sure that you have become as exquisite and cheerful a woman today as you were a lovely and joyful young girl in those days."[70] Einstein continued in a vein that caught the tenor of his personal life at the time: "So, now I've become such a big schoolmaster that my name is even mentioned in the newspapers. But I have remained a simple fellow who asks nothing of the world—only my youth is gone, the enchanting youth that walks forever on air."[71] Almost as an afterthought he added: "Miss Maritsch has indeed become my wife." If it was not clear already that an old spark had been rekindled, Einstein took the plunge in the postscript: "P.S. As of 15 October I will be in Zurich, mostly in the Institute for Physics, Rämistr[asse]. If you ever happen to be in Zurich and have time, look me up there; it would give me great pleasure."[72]

Anna's response was intercepted by Mileva, who wrote immediately to Anna's husband. Mileva constructed her letter to the effect that, like herself, Albert was mortified by Anna's "somewhat inappropriate letter" and that to prevent any further problems Albert had returned it with a note saying that he did not understand it.[73] Albert, of course, had done no such thing. Greatly embarrassed, Einstein replied to Anna's husband that her letter merely "reawakened the old affection we had for each other," and no harm was meant.[74] He promised no more contact between Anna and himself.

In the Einstein household, all hell broke loose.

This fragile home situation was probably the reason for Einstein's early departure from Bern for the Salzburg meeting. Almost five months later, from Zurich, in the midst of a strictly scientific letter to Besso, Einstein abruptly interjected a personal note—"Mental balance lost because of M.

not regained."[75] Einstein remained deeply embarrassed; his moodiness over this episode was still noticeable a year later.[76]

Four years after Mileva and Albert separated, Einstein wrote to Besso that "I would have remained true to Mileva . . . if it had been bearable with her. . . . But Mileva was absolutely intolerable to me."[77] No doubt the Meyer-Schmid episode played a major role. Forty years later the memory was still painful. When Anna's daughter contacted Einstein about her mother's relationship with him and why he terminated it so abruptly, he replied that it was due to his wife's jealousy, a trait that is often linked "with an uncommon ugliness."[78]

It seems safe to say that by 1909 the Dollie-Johnnie days were long gone. The couple that could not live apart proved unable to live together. Einstein may well have tried to save the marriage. He continued to buy Mileva Christmas presents, and a second son, Eduard, was born 28 July, 1910.[79]

To add to their problems, the financial strain continued. Einstein's salary in Zurich turned out to be barely higher than it was in the Patent Office. He once recalled jokingly, "In my relativity theory, to each point of space I attach a clock, but in reality I find it difficult to provide even one clock."[80]

A *Privatdozent* at Zurich, David Reichenstein, described the Einsteins at home in terms strikingly similar to Hans Tanner's: "I entered Einstein's room; calmly philosophic he was with one hand rocking the bassinet in which the child was lying (his wife was at work in the kitchen). In his mouth Einstein had a bad, a very bad cigar, and in his other hand an open book. The stove was smoking horribly. How in the world could he bear it."[81] Reichenstein went on to recall that one day after lying down to take a nap Einstein was overcome by the stove's fumes. Luckily Zangger happened by in time to revive him.[82]

After lectures Einstein often continued his discussions with students at a nearby café. This extra attention, in addition to his informal demeanor, made Einstein a popular teacher. Tanner recalled him as a quintessential absent-minded professor, coming into lectures in threadbare clothing with trousers too short, placing on the lectern a pocket watch with a long iron chain.[83] Tanner also noted Einstein's sharp cynical manner, first manifested on the international scene in the directness of his replies at Solvay in 1911. These remarks were sometimes softened by his beloved mentor Lorentz.

Einstein's flirtations continued. Reichenstein recalled an incident when the two of them attended a lecture on psychoanalysis and then accompanied the lecturer and other interested persons to a café. The lecturer was especially keen to hear Einstein's opinion about the scientific basis of his ideas. In the middle of a long and careful explanation he suddenly realized that Einstein's attention had drifted toward two sisters "of rare beauty and charm."[84] The lecturer became extremely indignant. Sheepishly looking to Reichenstein for support, Einstein mumbled that science was more important to him than flirtation.[85]

By this time his career was meteoric. In March 1910 the kingpin of German science, Walther Nernst, visited him in Zurich. Nernst had become interested in Einstein's work on the heat-conducting properties of matter, which Einstein based on Planck's quantum hypothesis. Einstein's results lent support to Nernst's recent work in thermodynamics and he came away extremely impressed.[86] In 1910 Planck referred to Einstein's relativity theory as changing our notion of the world in a way not experienced since Copernicus.[87]

As the editors of the Einstein Collected Papers have astutely observed: "Coupled with Einstein's ambitions as a scientist and his cavalier attitude toward women was a strong and active concern for his academic career."[88] Scarred from past experiences, Einstein was ruthless when it came to career moves. In September 1910 he accepted a significant salary increase at the University of Zurich, in the face of overwhelming student support regarding rumors that he might go elsewhere, and assured university authorities he would stay. Barely four months later, in January 1911, Einstein accepted a position as a full professor at the German University in Prague.[89] It is still a mystery why he did. There was no one there for him to discuss fundamentals. That bane of university life also unleashed itself on him: "The paper shit is endless."[90] Einstein went out of his way to clash with the stuffed shirts who ran the place. To cap things off, his office window looked out on the courtyard of an insane asylum.

Of the offers that were starting to pour in, one was of particular interest. It was from his alma mater, the Swiss Polytechnic, which that year had been elevated to the Eidgenössische Technische Hochschule, Swiss Federal Technical University, or ETH for short. Good friends such as Zangger and Grossmann had argued strongly that the ETH should hire a highly regarded professor of theoretical physics. It worked and Einstein was elated. But before that he used the offer of a position at Utrecht to hurry the ETH in 1912 to crystallize (and sweeten) their offer. Although such ploys have become standard fare in academia, this one disappointed Lorentz and ended up embarrassing Einstein.[91] At the time, however, he was extremely happy. "Two days ago I was appointed to the Polytechnic in Zurich (hallelujah!) and I have already submitted my imperial royal resignation [in Prague]."[92] The circle closed. Einstein returned to the university that had all but kicked him out only twelve years earlier. Now he was its most eminent professor and well paid, too. He did not remain there for long.

Switzerland then had the reputation as a "first-class waiting room" for German academics hoping to return home.[93] Einstein was one of them, even though he had renounced his German citizenship in 1896 and been a Swiss citizen since 1901. After a great deal of preliminary groundwork, in July of 1913 Nernst and Planck smoothed the way to his plush appointment at the University of Berlin. A mark of Einstein's high prestige was that his renunciation of German citizenship, his lack of Prussian military service and his Jewishness were all conveniently forgotten so that the Prussians might add the new Copernicus to their intellectual elite.

But there was another reason why Einstein was eager for the Berlin offer. In 1912 Elsa Löwenthal (née Einstein), divorced and with two daughters, looked up her now well-known cousin in Zurich. Elsa was the antithesis of Mileva. A warm, heavyset woman, typically bourgeois, without thoughts of any career, her goal in life was to take care of Albert, or Albertle as she fondly called him. They began an affair. To prevent the inevitable problems, however, Einstein decided to break it off,[94] but Elsa, seeing a good thing, persevered until Albert surrendered. After he accepted the Berlin appointment he wrote her that "now we can be together."[95] The love letters began in October 1913: "I now have someone of whom I can think with unalloyed pleasure and for whom I can live,"[96] he wrote to her on the tenth. On 7 November, 1913, Einstein signed off "Kisses from your Albert."[97] Meanwhile, Elsa was doing a makeover on Albert, particularly with regard to his personal hygiene. In Mozartian jest, Einstein wrote on 2 December, 1913, "So, a foul profanity and a hand kiss from a hygienic distance from your really filthy Albert."[98]

The acrimonious divorce proceedings dragged on for four years. Finally, on 14 February, 1919, Einstein's marriage to Mileva was dissolved. As part of the divorce settlement, Einstein awarded Mileva 8,000 Swiss francs per year against the payment of his yet-to-be awarded Nobel Prize. His confidence now knew no bounds. Despite the Swiss ban against his marrying for two years, on 2 June, 1919, Einstein and Elsa were married at the registry office in Berlin.

Cracks in the relationship had already appeared, however. We know this from an astounding letter that one of Elsa's daughters, Ilse, wrote on 22 May, 1918, to a male confidant, Georg Nicolai, with instructions to "destroy this letter immediately after reading it!"[99] Naturally he did not. Ilse was in dire need of his advice and concluded the letter with "Help me!"

Apparently in the course of a humorous conversation, Nicolai had suggested to Ilse that perhaps she ought to marry Einstein. Jokingly, Ilse related this remark to Elsa and Einstein. To her shock, Einstein thought it was not such a bad idea. One thing led to another, and at a certain point Einstein confided to Ilse "how difficult it is for him to keep himself in check" in her presence.[100] The two women left it up to Einstein to choose between them. Elsa was willing to step aside for Einstein's happiness. Ilse, on the other hand, was beginning to wonder whether the passion of an affair with Einstein could be sustained in a marriage.

In the end it made little difference. "As long as I was here in the house, it would not (in A[lbert]'s opinion) be a great difference to me whether or not I were married, at most it would be merely convenient." We know that Albert and Elsa had separate bedrooms at opposite ends of the house and that Ilse lived there with them. No wonder Ilse ended the letter with "Help me!"

ABSOLUTELY FAMOUS

On 6 November, 1919, an English eclipse expedition to the tiny island of Principe, off the west coast of Africa, verified general relativity's most spectacular prediction, the deviation of starlight in the vicinity of a massive body. Einstein was deified. The war-weary world rushed to embrace a man who, sitting in his study, had seemingly divined laws of the universe. The stars were not where they were supposed to be, but everything is all right, because Einstein understands it. Space and time are welded together into a warped four-dimensional structure on which light rays travel. This esoterica was hawked in just about every newspaper in the Western world, with photographs of the man who seemed to have glimpsed the creation itself.[101] Einstein became the first media icon of the

FIGURE 7.2 Einstein, Charlie Chaplin and Elsa, at the world premiere of *City Lights,* at the Los Angeles Theater, Los Angeles, 30 January, 1931

twentieth century. Hollywood stars befriended him. At the opening of *City Lights* in Los Angeles on 31 January, 1931, Charlie Chaplin joked to Einstein about the crowds cheering them: "They cheer me because they all understand me, and they cheer you because no one understands you."[102] Awed by the whole thing, Einstein turned to Chaplin and asked him what it all meant. "Nothing," replied Chaplin.[103]

8

Creativity in Art and Science

Thinking for its own sake! . . . When I have no special problem to occupy my mind, I love to reconstruct proofs of mathematical and physical theorems that have long been known to me. There is no goal in this, merely an opportunity to indulge in the pleasant occupation of thinking.

—Albert Einstein

The important thing is to create. Nothing else matters; creation is all.

—Pablo Picasso

The two creations whose genesis we have explored here, Albert Einstein's special relativity theory and Pablo Picasso's *Les Demoiselles d'Avignon*, are the works that brought science and art into the twentieth century. But beyond their historical moment—their shared response to the tension between classical and nonclassical thinking—these masterpieces share a deeper connection. At the creative moment boundaries between disciplines dissolve. Aesthetics becomes paramount.

The psychologist Howard Gardner has argued persuasively "for the existence of several *relatively autonomous* human intellectual competences," which he refers to as "multiple intelligences."[1] It would be too easy, however, simply to say that Einstein displayed logical-mathematical intelligence and Picasso, spatial intelligence. As Gardner rightly observes, "one might underestimate the component of spatial thinking in the sciences."[2] Conversely, in Picasso's discovery of cubism, logical-mathematical thinking has likewise been underestimated. Einstein was a scientist who relied heavily on spatial thinking and Picasso an artist for whom logical-mathematical thought played a crucial role.

Such a confluence of "intelligences" had not appeared since the sixteenth and early seventeenth centuries, the days of Albrecht Dürer, Leonardo da Vinci and Galileo. That was a time when science was struggling to free itself from the Earth-centered Aristotelian cosmogony and aesthetics offered a way out. Elegance lay at the core of the Polish astronomer Nicolaus Copernicus's arguments in favor of a Sun-centered universe. Having sacrificed the geometric perfection of circular orbits in favor of elliptical ones, the Copernican view gained a deeper, more satisfying symmetry: it placed the planets in orbital paths that are related to how fast they move around the central sun. To the sixteenth-century mind, a Sun-centered universe displayed the hand of the Deity as embodied in the light and warmth flowing from the central source.[3] These reasons were of great importance to Galileo who, like Copernicus, founded his scientific theories on aesthetics. There would be no data to support a Sun-centered universe for over two hundred years.

In Einstein's 1905 papers, aesthetic arguments reappeared with a force unseen for centuries. Einstein's raison d'être for the light quantum was the "profound formal distinction" in current physics between particle and wave, and the resulting clash between continuity and discontinuity.[4] He declared this situation unaesthetic. His argument for an exact principle of relativity was that it was needed in order to remove "asymmetries that do not appear to be inherent in the phenomena," as he began the 1905 relativity paper.[5] For Einstein, as for Copernicus and Galileo, aesthetics were data.

So sensitive was he to aesthetics that he internalized clashes with it. In 1905, the difference in explanations for electromagnetic induction became so "unbearable" that it "forced" him to propose a widened principle of relativity. In 1907, he opted for one unified definition of mass instead of the two postulated by Newton's theory of motion. For Einstein, minimalism was an aesthetic principle—an axiom. Nature was not redundant; therefore one had to pare away redundancies in order to reveal laws of nature. Redundancy in explanation hid the principle of relativity, while redundancy in the concept of mass obscured the equivalence between gravity and acceleration. There is a link between aesthetics and fundamental laws of nature.

For Dürer the perspective point was of the essence. It is the point from which light emanates, the point at infinity; it became an aesthetic. Artists in the late nineteenth and early twentieth centuries, particularly Cézanne, felt otherwise, and Picasso and Braque obliterated it altogether. For their new art they conceived a new aesthetic: reduction to geometrical forms.

These aesthetic goals required Einstein and Picasso to confront the concept of simultaneity. Picasso's spatial simultaneity goes beyond Bergson's, which is typified in the art of Cézanne. Cézanne set down on canvas the sum of views of a scene that had been stored in his unconscious over a long period. Picasso's notion of spatial simultaneity was something drastically new in art: the simultaneous representation of *entirely different* viewpoints whose sum constitutes the depicted object. The squatting demoiselle, represented simultaneously in full frontal and profile views, was interpreted as a projection from the fourth dimension. In the fashionable occult terms of the time, it is as if Picasso found a way to sit on the "astral plane."

Picasso's concept of time in painting transcends that of impressionist art, as represented by Claude Monet's temporally static representations of haystacks or of Rouen Cathedral. The treatment of time in *Les Demoiselles d'Avignon* is quite complex. We can see the painting as a series of five motion picture frames tending toward increased geometrization, and the squatting demoiselle, meanwhile, as a sequence of snapshots superimposed on each other. Picasso continued his research into spatial and temporal simultaneity at Horta de Ebro when he experimented with printing superpositions of photographic negatives. Einstein's temporal simultaneity shares with Picasso's the notion that there is no single preferred view of events.

Both men's epiphanies grew out of a sense that something was missing in the way science and art were then understood. For Einstein, redundant explanations and clashing representations pointed to incorrect notions of time and simultaneity. For Picasso, seeing African objets d'art at the Trocadéro snapped him into realizing that the inconsistencies in the present state of *Les Demoiselles d'Avignon* could be alleviated only by moving toward a more radically conceptual style than anyone had ever attempted.

Geometry had to become the language of the new art. This was a stunning discovery that formalized the formerly informal language of art and brought it closer to science. Geometry, especially in its newer forms, provided the common ground.[6] Yet neither Einstein, nor most other physicists at first for that matter, appreciated the non-Euclidean framework for relativity proposed by Minkowski in 1907. It would take five more years for Einstein to realize that special relativity could be fully generalized only through geometrization.

We can now return to the issue I raised in the Intermezzo: Why were Picasso and Einstein the ones to make these breakthroughs, and not such better-established figures as Derain, Matisse, Lorentz and of course Poincaré?

First, both men were opportunists, willing to take advantage of every hint offered by the intellectual currents in which they swam. Picasso was the only artist with the courage to explore Iberian and African conceptual art to their very depths, which required taking heed of Princet's discussions on geometry and the fourth dimension. This played a large role in convincing Picasso that geometry is the language of the new art. Derain remained entangled in doubts about how to proceed with primitive art, while Matisse simply refused to see beyond its conceptual message into the necessary mathematical means for abstraction.

Lorentz, for his part, was unable to discover relativity because he was absolutely stuck in nineteenth-century modes of thought. He could not abandon the concept of the ether or of absolute time, even when his own mathematical formulations made them unnecessary.[7]

Poincaré, however, almost had it all. Despite his courageous willingness to shift his philosophical view with developments in physics, experimental data of the laboratory remained for him the sole source of truth. This, in conjunction with his emphasis on sense impressions, led Poincaré to conclude that no systematic definition can be given for distant simultaneity, which he wrongly concluded was merely a qualitative problem. Time, on the other hand, was a quantitative problem, which he resolved in favor of the commonsense intuition of a time independent of a clock's motion, rather than Lorentz's local time.

Still, of the candidates other than Einstein, Poincaré came the closest to discovering special relativity.[8] By June 1905 Einstein and Poincaré had at their disposal the same experimental data and went on to propose identical mathematical formalisms to explain them.[9] But Einstein inferred a meaning Poincaré did not. His thought experiments enabled him to *inter-*

pret the mathematical formalism as a new theory of space and time, whereas for Poincaré it was a generalized version of Lorentz's electron theory.[10] Poincaré was after bigger game than Einstein: He sought nothing less than a unified theory of the then-known forces, electromagnetism and gravity.[11] Einstein alone realized that physics was not ripe for such an ambitious program, principally because it had not come to terms with the wave/particle duality of light.

The view of simultaneity Poincaré set out in *La Science et l'hypothèse* turned out to be unsuited for either art or science. But his analysis was of such depth and clarity that it inspired Picasso and Einstein to rethink this concept and discover its proper statement. In art, spatial simultaneity is different points of view presented all at once instead of as a succession of perspectives. In physics, simultaneity is described by Lorentz's local time and depends on a clock's motion. It is not absolute—a notion, seemingly so clear, that on deeper inspection turns out to be unacceptably vague. Both new definitions of simultaneity are counterintuitive to everyday common sense because they do away with preferred viewpoints and so go beyond sense perceptions.

ANXIETY, EGO AND EMOTION

Einstein and Picasso were men of immense egos and irresistible force and charm, yet who preferred emotional detachment. Their creative drives were the guiding forces in their lives—"nothing else matters; creation is all," Picasso put it bluntly.[12] As young men both vowed to devote their lives to creativity. In Picasso this occurred in 1894, when his younger sister Conchita fell ill with diphtheria. He swore an oath to God that he would give up painting if she recovered.[13] When she died, Picasso interpreted this as God's will, and a calling he felt bound to honor for the rest of his life.[14] Einstein, too, as a teenager declared it was his ultimate devotion to "strenuous intellectual work," as he wrote Marie Winteler's mother in May 1897.[15] Their single-minded quests required an emotional aloofness and severity that many acquaintances noted throughout their lives.

During Einstein's stay in Prague, Max Brod, a young writer, introduced him into intellectual circles that included Brod's close friend Franz Kafka. Brod had a reputation for writing novels with acute psychological insights. Among them is *Tycho Brahe's Path to God*, in which Brod styled the seventeenth-century astronomer Johannes Kepler after Einstein. In a striking passage, the Danish astronomer Tycho Brahe is thinking about his young assistant Kepler: "There was something incomprehensible in his absence of emotion, like a breath from a distant region of ice. . . . He had no heart and therefore had nothing to fear from the world. He was not capable of emotion or love."[16] Philipp Frank, Einstein's successor at Prague, as well as his lifelong friend and astute biographer, recalled that everyone was astounded at how well Brod caught Einstein's psyche. Frank recalled Walther Nernst remarking to Einstein, "You are this man Kepler."[17]

Françoise Gilot, the only one of Picasso's mistresses to escape with her dignity intact, described him as reducing women from "goddesses to doormats."[18] Neither man made any effort to conceal his habit of using people to his own ends, or his ultimate preference for emotional isolation. Einstein, however, was never as sadistic as Picasso. Picasso's wives and mistresses generally ended up as bitter women, while male friends were sometimes abandoned in times of dire need. In 1911, when Apollinaire was falsely accused of an art theft, Picasso pretended not to know him;[19] and he offered no help when Max Jacob was arrested by the Gestapo and sent to Drancy concentration camp in 1943, where he died. Picasso heaped public embarrassments upon his select group of Catalan pals, which they were forced to endure in order to prove their fealty.[20] Neither he nor Einstein was much good as a father. Yet they did not hesitate to play prominent roles in international organizations promoting world peace, prosperity and understanding. Their humanism was better in the abstract.

Despite their extreme self-confidence, at great creative crossroads, both men suffered from extreme anxiety. Einstein's torment over the explanation of electromagnetic induction is matched by Picasso's restless striving when he became stuck during the first campaign on the *Demoiselles*. These periods of tension can be seen as catalysts for critical epiphanies. Yet unlike such thinkers as T. S. Eliot and Mahatma Gandhi, they were not wracked throughout their lives by self-doubt.[21]

At this point we may be tempted to turn to Freudian psychoanalysis for further insights. Yet we would learn little because the role Freud assigns to the unconscious is the arena in which the drives of aggression and sexuality play out their tensions. While sex, wealth, power and adulation are undoubtedly important as fuel for the creative process, they cannot be the sole dynamics of genius. Such a hypothesis, by explaining too much, in the end explains too little.[22]

Owing to its effect on Picasso studies, it is apropos to mention the first psychoanalytic commentary on Picasso, which was by Freud's former friend and collaborator Carl Jung. Jung's piece was occasioned by a large Picasso retrospective held at the Zurich Kunsthaus in September and October 1932, and appeared in the *Neue Züricher Zeitung* of 13 November, 1932. That the great Swiss psychoanalyst agreed to embark on this work is a sign of the importance of Picasso's art. Zervos immediately translated it into French, excerpted it, and then gave it a critical working over in *Cahiers d'Art.*[23]

Jung constructed his analysis by comparing certain of Picasso's paintings with drawings made by neurotic and schizophrenic patients. While drawings made by neurotics exhibit some sentiment and elements of symmetry, even though they might be completely abstract, those by schizophrenics are almost devoid of sentiment and full of broken lines that, in Jung's interpretation, reveal psychological fissures. Jung fit Picasso's Blue Period paintings, cubist paintings and later works into the schizophrenic context. Picasso's entry into the Blue Period, he contended, marked the first stage of schizophrenia. These paintings represent a farewell to the world about him in preparation for a descent into hell. In this way Jung tried to attribute Picasso's creativity exclusively to psychological problems.

The gist of Zervos's criticism is that Jung is completely ignorant of history. For example, Jung finds his "descent into hell" in Picasso's choice of subjects among the Montmartoise underworld. Rather, Zervos emphasizes, Picasso was following in the footsteps of El Greco and Lautrec. Even worse, in Zervos's eyes, Jung never cites Cézanne's influence on Picasso. Still, Jung's analysis of Picasso's creativity as stemming from psychic crisis was not ignored, and could not be. Eunice Lipton finds Zervos's criticism of Jung's analysis, however flawed it is in its gross historical oversights, an overreaction. She points out that the psychoanalytic analyses of the 1930s opened the way for studies of Picasso as a human being.

Lipton provides an interesting account of the immediate post-Jungian analyses, many of which focused on Picasso's anxiety. The French art historian and critic Germain Bazin, an avid follower of psychoanalytic trends, followed up Jung's analysis in 1935. With some poetic license, Bazin attributed Picasso's anxiety to the general anxiety at the beginning of the twentieth century, which led in Picasso's case to psychological nomadism, a fleeing from himself, that was perhaps ultimately rooted in a general aversion to bourgeois ideals and satisfaction. Bazin goes on, without any substantiation, to connect Picasso's nomadic turnings with his Semitic soul. Waldemar George, another French art critic who like Bazin had been writing on Picasso's anxiety since 1932, considered Picasso to be the paradigmatic example of modern anxiety, the anxious and egocentric modern man. Without the slightest evidence, both critics made frequent references to Picasso's Jewish blood. As Lipton points out, "one wonders how it relates to mounting anti-Semitism in France in the 1930s."[24] In the end, however, Bazin's analysis is itself rooted in the 1930s existential crisis in France, a haunting sense of emptiness and irrelevance, out of which, in 1938, Jean-Paul Sartre composed *La Nausée.*

What then differentiates certain people from others? After all, no matter how many hours a day the overwhelming majority of physicists or artists ply their trade, they will never approach the league of an Einstein or a Picasso. This goes for every field of endeavor that has its supergeniuses, including sports. To understand what constitutes high creativity, we need a theory of the dynamics of the unconscious: how concepts are moved about in the mind toward finding incredibly novel combinations.

Toward this I will use a model I have been developing that incorporates certain notions from cognitive science, including Gestalt psychological theory of creativity.[25] This model will enable us to incorporate the information we have discussed regarding the discoveries of Einstein and Picasso, while providing clues as to how they utilized it.

A MODEL FOR CREATIVITY

Consistent with reminiscences of Einstein, Poincaré and Picasso, among others, and in conjunction with a host of psychological experiments, we know that seminal ideas emerge not in any real time sequence but as an explosion of thought.[26] The results are so different from anything before, and the culmination of such extreme interdisciplinary thought, that any model for creative thinking based on digital encoding is tainted with a linear and overly rational underpinning. Typically in computer simulations of thinking, the mind is caricatured as working down the branches of a decision tree within a space of possibilities that is of necessity restricted, simply so that software can be written that can be run on existing hardware.

So-called "discovery programs" for discovering laws of physics are an interesting first step. They inevitably give the result to be discovered and use a definition of scientific creativity as a search for patterns in laboratory data—data that are selected and treated more carefully than the discoverer ever did.[27] Moreover, that necessary key to scientific discovery—proper problem choice—is not taken into account, nor is the proper historical sequence of events.[28]

All this directly contradicts the reports of great discoverers themselves, who tell us that the solution to a problem emerges suddenly into conscious thought. Here I want to present a model of creative thinking that, in addition to having none of the above faults, has the important attribute of bringing into play aesthetics, thought experiments and the notion of parallel lines of unconscious thought, as well as emphasizing the role of intuition as more than a deus ex machina. My goal is both to summarize the detailed analysis in previous chapters and to explore how Einstein and Picasso processed information in order to make their momentous breakthroughs. Creativity in art can be explored like creativity in science because artists and scientists use many of the same strategies toward discovering new representations of nature. Just like scientists, artists solve problems.

The model's basis is a hypothesis that has emerged from studies in highly creative problem solving—that creativity occurs in a cycle of conscious thought, unconscious thought, illumination (hopefully!) and verification.[29] The historical precedent for this sort of analysis is in Poincaré's introspection of 1908, where he made it abundantly clear that he actually depended on unconscious thought for his discoveries. This assertion in turn is supported by studies on Poincaré done in 1897 by the French psychologist Toulouse. Toulouse noticed that Poincaré had learned when to cease work on a problem, because "during the intervals he assumes . . . that his unconscious continues the work of reflection."[30] Poincaré believed that in the unconscious "there reigns what I would call liberty, if one could give this name to the simple absence of discipline and to the disorder born of chance. Only this disorder permits of unexpected connections."[31] Einstein, too, believed in "free play with concepts" in the unconscious.[32] Poincaré insisted, however, that "unconscious work is not possible and in any case is unfruitful, unless it is preceded . . . by a period of conscious work."[33] Neither Toulouse nor Poincaré nor Einstein used the term "unconscious" in the Freudian sense, and neither do I. Rather it will be used in the more neutral sense of cognitive science, as a part of the mind to which the consciousness has no access but which does not have a particular set of emotional agendas separate from the conscious one.

Data from modern psychology support Poincaré's insights and have expanded on them with such new knowledge as how information is stored in long-term memory. While consciousness plays the important role of setting boundaries on our everyday actions, in the unconscious we can activate complexes of information in long-term memory without boundary. Information held in long-term memory can be processed in parallel in the unconscious and then find its way into conscious thought.[34] The *unconscious-thought* part of this cycle is sometimes called "incubation." Psychological experiments indicate that after a false start on a problem, subjects can develop a block to further progress. During the incubation period, these unsuccessful attempts can fade away, after possibly activating relevant portions of long-term memory, to leave the way open for fresh attempts.[35]

We have already studied, in Chapter 7, Einstein's discovery in 1907 of the equivalence principle that led to the general theory of relativity. To further compare the creative thinking of Einstein and Picasso, I will use the previous chapters' accounts of Einstein's discovery of special relativity theory and Picasso's discovery of *Les Demoiselles d'Avignon* as "data" that are grist for the model's mill and in this way inform us further about Einstein's and Picasso's creativity.

Conscious Thought

For Einstein, *conscious thought* involved the choice of a problem: He undertook to reformulate Lorentz's electromagnetic theory so that its fundamental statements, or axioms, were independent of the constitution of matter. From his letters to Mileva and his recollection of the 1895 thought experiment, we know that Einstein had unsuccessfully worked on problems of the ether and relative motion for ten years. During this period he became adept at the accepted methods of theoretical physics. By early 1905 he had realized the inadequacy of Lorentz's theory to serve as the fundamental theory because it could not account for the wave/particle duality of light. He was also aware of Kaufmann's data on the mass of high-speed electrons and had in hand the "data" of the 1895 thought experimenter. Then there was what he had learned at the Swiss Polytechnic as well as his philosophical readings. Einstein recalled that he took from the theory of thermodynamics the idea that he should seek axiomatic statements that are independent of the atomic nature of matter.

We must not forget that at the Patent Office Einstein became well versed in design problems of electrical dynamos and spoke often with friends from the Federal Postal and Telegraph Administration about issues in wireless telegraphy and synchronization of clocks. His readings at the Olympia Academy, which included *La Science et l'Hypothèse,* cannot be discounted. Then there was Einstein's passionate interest in the violin compositions of Bach and Mozart, composers whose specialty was a structured motif with sublime melodies. Einstein came to believe he could create this in physics, too.

All of this information is taken from long-term memory and mapped onto, or worked on, by a dedicated representational medium capable of dealing with logical reasoning based in symbols or elements not reducible to logical symbols, such as visual imagery. Conscious and unconscious thought occur in this medium. All of this affected Einstein's problem choice, which was reformulating Lorentz's electromagnetic theory. Activation is maintained in the unconscious by the intense desire to solve the problem at hand. By the beginning of May 1905, however, he became stuck. Problem choice will be dealt with again later under "Intuition."

For Picasso, *conscious thought* concerned his choice of subject matter, the bordello theme. He was out to shock in a way that went far beyond current art, and certainly beyond warnings about venereal disease. But current art gave only a hint from Ingres and some others. The primitive objets d'art and the Iberian reliefs from Osuna told him he should take a conceptual turn, and he was actively thinking about geometry and the representation of space. Picasso was concerned as well about the new direction taken by Derain and Matisse, and he was out to compete with them. By the end of May 1907, however, Picasso was stuck. All five *Demoiselles* were in the Iberian motif when he turned "his paintings to the wall."[36]

Unconscious Thought

Unconscious thought can be encapsulated in what I have called "network thinking."[37] In network thinking, concepts from apparently disparate disciplines are combined via the proper choice of mental image or metaphor. The intense desire to solve a problem can produce stresses that in turn cause associations that are not possible in conscious thought—in other words, the mind brings up the problem in all sorts of unlikely contexts. Eventually, however, it must focus on particular approaches. It is this choice—finding the proper image or metaphor—that catalyzes the illumination, the nascent moment of creativity.

Wertheimer's Gestalt theory of creativity asserts that the mind has an irresistible urge to create structures, or arrangements of facts into patterns that possess maximal symmetry. These are good *Gestalts.*[38] Many cognitive scientists believe that Gestalt principles of perception are hard-wired into the brain and that their function is to make sense of the perceptions with which we are continually bombarded.[39] People for whom this drive toward good form is especially strong are most likely to be personally offended by inconsistencies and unaesthetic representations. Thus Einstein used words like "unbearable" to express his distaste for the way electromagnetic induction was interpreted by physicists and engineers in 1905.

Einstein's *unconscious thought* involved parallel yet intersecting streams that contain, for example, the 1895 thought experiment, which he kept in mind for ten years; his repeated attempts to understand problems of relative motion; issues concerning how electrical dynamos function; Mach's resounding message always to question accepted foundations; and Poincaré's struggles to understand the failure of ether-drift experiments and their relation to Lorentz's local time.

Picasso's *unconscious thought* entailed his involvement with the Parisian visual culture, including Méliès's highly edited movies, the experimental photography of Marey and Muybridge, and his own photographic research; X rays; the occult with its astral plane; symbolist literature; philosophical currents; and William James's ideas on perception. Picasso's drive to succeed was fueled, inter alia, by his competition with Derain and Matisse, his oedipal relationship with his father, Fortier's picture postcards, dread of venereal disease, his Spanish-tinged attitude toward women and his sexual predicament with Fernande. The mix included concepts of geometry and Poincaré's intriguing discussion of how to view the fourth dimension.

Fernande recalled how Picasso virtually secluded himself, for the first time in his life, in order to solve the new problems posed by the *Demoiselles.* Apollinaire and Kahnweiler remembered that he did it virtually alone, with little discussion and almost no encouragement. When Picasso stopped working on the *Demoiselles* in late May 1907, his unconscious remained highly activated, awaiting vital input. In the Gestalt theory of creativity, this crucial moment, the vital input, serves to bring to equilibrium knowledge already in the field of thought, which is itself in great turmoil. Picasso's visit to the Trocadéro did just that. He recalled this moment as one of "shock" and "revelation."[40]

Illumination

The flash of revelation is not a *what* but a *how*. It does not provide new facts; rather, finding the right image or metaphor tells one how to think about facts already at hand. Insights or implications come later, as guidelines emerge that serve to constrain and direct new combinations of facts. These guidelines, in both art and science, both maintain logical continuity with previous work and are at once visual, aesthetic and intuitive. Such a structure for thinking is a powerful motivator and ingredient of creativity.[41] Poincaré referred to it as a "delicate sieve" and stressed the "mathematician's aesthetic sensibility."[42] By this he meant instances such as when certain mathematical equations retain their form under, say, inversion of all spatial coordinates. Mathematicians look for such solutions to problems because they usually turn out to be "at once useful and beautiful."[43] In science this is particularly important because, for example, not every mathematical equation can represent a theory. Rather an equation, beyond being mathematically justified, must agree with certain even more stringent guidelines such as the principle of relativity. Illumination both arises from and gives form to the elements of creativity.

Einstein's creative speed after his walk with Besso can be likened to Picasso's producing *Les Demoiselles d'Avignon* after his visit to the Trocadéro,[44] to Mozart's compositional powers and to the way Poincaré wrote up his scientific papers—the first draft was it.[45] They all dealt in concepts that they set down in wholes, or *Gestalten*. This habit contrasts with Beethoven's scores, which were laboriously worked over, Charles Darwin's evolutionary theory, which took over twenty years of gestation, and the marathon sessions of writing, verbal exchanges and rewriting for which the atomic physicist Niels Bohr was famous.

Recall Einstein's reply to his biographer Seelig's question whether "relativity theory has a definite birthday": He answered that the final arguments and building blocks were "prepared over a period of years."[46] Continual activation of the problem in Einstein's unconscious kept vibrant a "free play with concepts."[47] An absolutely free play with information, however, can be unproductive and especially so in science because only certain combinations are permissible.

So there is the need for guidelines for creativity that act like Poincaré's "sieve" in order to filter out the most productive results. For our purpose the key ones are aesthetics, visual imagery, continuity in theories and intuition. These four elements were all active for Einstein and Picasso. I will briefly note a few examples in the first three categories, and then give rather more attention to intuition.

Aesthetics. Einstein applied his minimalist aesthetic to light in his abhorrence of continuity and discontinuity side by side, and to electromagnetic induction in his refusal to accept a redundancy in explanation. Picasso's aesthetics were, of course, extremely complex, but they included seeking a conceptual, rather than naturalistic, depiction of objects in order to probe the "deep structure" of representations.[48]

Visual Imagery. Einstein used visual imagery as a means to frame his thought experiments of 1895 and 1905. His power of visual thinking was to frame a problem as a thought experiment in which he could "see" its "deep structure."[49] In 1895 the problems that were encapsulated in the thought experiment of that year boiled down to Einstein's discovering that time is a relative quantity; in 1905 it was the "unbearable" situation in electromagnetic theory that called for a widened principle of relativity. Picasso's visual imagery in *Les Demoiselles d'Avignon* is again very complex, but one aspect of it is that it is expressed in the language of geometry.

Continuity. Continuity in theoretical development was essential to Einstein's efforts to widen Newton's principle of relativity to include electromagnetic theory and light. In 1907 Einstein further enlarged this principle to include measurements from accelerating platforms. This step removed an asymmetry in special relativity theory, which was that measurements could be made only on platforms undergoing inertial motion. But however jarring these ideas were to others, Einstein always saw himself as a classicist, and his theories as extending, rather than breaking, the tradition of Newtonian physics.

Continuity is also of the essence for artists. Picasso worked and reworked earlier versions of the *Demoiselles* along the lines of Ingres, Cézanne and El Greco, while bearing in mind the bold conceptual advances made by Derain and Matisse in 1906. While perhaps more consciously revolutionary than Einstein, Picasso too acknowledged his debts to tradition.

Intuition. The term "intuition" is often used as a catchall to describe how Einstein's and Picasso's dazzling discoveries literally seemed to burst forth. This, however, masks the way these men used their deep insights to see issues and nuances hidden from those unable to penetrate beyond technical difficulties. An Einstein or a Mozart or a Picasso has such a high degree of technical expertise that he can soar over technicalities and take in the underlying structure of a problem.

Both men had a penchant not for details but generalities. This is clear, for instance, in Einstein's thinking after his first two fledgling papers. To some extent Picasso was ahead of Einstein in technical expertise in that he had, so to speak, painted his way through art forms that prevailed at the turn of the twentieth century, as well as those of his ancestors. Einstein, who despite his university training was really an autodidact, was less widely read in physics than Picasso was widely versed in painting. Yet he could discern what was a fundamental problem.

Commenting on why he chose not to become a mathematician, the subject in which he always scored highest, Einstein recalled that he might have suffered the fate of "Buridan's ass which was unable to decide upon any specific bundle of hay" and starved.[50] In other words, he did not trust his nose for problems. In physics, on the other hand, "I soon learned to scent out that which might lead to fundamentals and so turn aside from everything else, from the multitude of things that clutter up the mind and divert it from the essential."[51] So, for example, he gave up trying to derive Planck's radiation law, considering this approach unfruitful and incorrect, and instead accepted it as a law of nature and turned to deducing its consequences.

For Picasso in 1907, problem selection meant choosing a subject that he could transform into a conceptual style. As today's artists sometimes seem to forget, to work conceptually one must actually have a concept, as well as a theme and a way to merge them into a whole. The concept was a new depiction of space and time; "choosing a problem" meant choosing a scene in which this concept could be fully realized in complex and aesthetically satisfying ways.

Howard Gardner draws a distinction among the great geniuses between "masters" and "makers." Mozart was a master who worked within a "domain."[52] Except for some of his operas, he produced no drastically new style of music. But through musical scores of superb structure and sublime melody, he brought the Baroque period to conclusion and paved the way for the Romantic era, in which Beethoven broke new ground. Einstein and Picasso were makers who shattered domains, and created new ones. Even for makers, however, it is a misuse of the idea of intuition to suppose that their insights came from out of the blue.

Einstein's intuition included his quest for generality, which was dependent on his ability to sense when to raise an hypothesis to an axiom. His widening of Newton's principle of relativity satisfied the 1895 thought experimenter's intuition, his minimalist aesthetic and his faith in visual imagery—in the Kantian notion of *Anschauung* as a means of insight.

It is difficult to pin down Picasso's intuition in the same way as Einstein's, which was an outgrowth of both culture and science. What I can say, however, is that they were both interested in expanding the concept of perception out of its common-sense basis. The notion of a new common sense that includes the relative nature of space and time extends perception beyond the ken of our senses. In art, this new perception emerged from a geometrical substrate, a revealing of forms that again extends beyond our immediate vision.

Verification

I will discuss three sorts of verification. The first two refer specifically to scientific theories, while the third refers to art as well.

Looking back on his experiences as a scientist, Einstein was able to succinctly express the first two: "The first point is obvious: the theory must not contradict empirical facts."[53] If a scientific theory cannot be verified in the laboratory, it comes under question and may have to be discarded. But not so fast, Einstein continues: "However evident this demand may in the first place appear, its application turns out to be quite delicate."[54] This is just what occurred with the Lorentz-Einstein theory when it was contradicted by Kaufmann's data, whereas other theories specialized to the electron fared better. While Lorentz panicked, Einstein confidently proceeded to generalize his relativity theory. The reason lay in Einstein's second means of assessing a scientific theory, which is to opt for theories "whose object is the totality of physical appearances."[55] This was the goal of the Lorentz-Einstein theory, and more specifically, in Einstein's mind, his relativity theory. These theories pertained to more than electrons. Einstein refused to let the issue be decided by just one set of empirical data.

The third means of assessment, which applies to both art and science, is even more subtle. It is verification by influence: Does the new idea lead anywhere? Does it inspire others to produce useful science or important art? Does it become part of a worldview? Clearly Einstein's relativity theory and Picasso's *Les Demoiselles d'Avignon* satisfy all of these criteria. Their creative drives became inspirations and their personas the stuff of which movies are made and novels written. Their great works, spun during the most intensely creative period of their lives, were at first spurned, then accorded accolades, then incorporated into the intellectual milieu that they themselves had spawned, and finally superseded. But they cannot ever be forgotten, because they are now part of the very rock on which all of science and art will be forever built. During their lives, their creators' influence waned, no doubt to their own regret. Einstein and Picasso could no longer produce the works they did as young men, when nothing mattered for them but ideas. They longed for the days of the Patent Office and the Bateau Lavoir. Genius burns brightly, but only for a short time, and then begins the slow process of burning out.

Perhaps Einstein was thinking of himself as much as Newton when, toward the end of his life, he wrote so poignantly:

> Newton, forgive me; you found the only way which, in your age, was just about possible for a man of highest thought—and creative power. The concepts you created, are even today still guiding our thinking in physics, although we now know that they will have to be replaced by others farther removed from the sphere of immediate experience, if we aim at a profounder understanding of relationships.[56]

While it took until 1905 for this to happen to Newton, it happened to Einstein in his own lifetime.

THE END OF THE CLASSICAL WORLD

The trend toward abstraction quickly left its two prime movers behind. Neither Einstein nor Picasso crossed the Rubicon into the extreme abstractions their works inspired in others. In a sense it's a sign of the depth of their discoveries that they led where the discoverers themselves could not follow.

Advances in atomic physics, dependent on methods Einstein pioneered for theory building as well as on his relativity theory, led to abstractions in visual imagery and a concomitant break with classical causality with which he could never come to terms.[57] Picasso could never come to terms with an abstract art devoid of any figurative content whatsoever. Both men had the same notion of visualization.

The visual imagery in special and general relativity is the same as in Lorentz's electromagnetic theory and Newton's mechanics. It is based on phenomena that we have actually witnessed with our senses and is *imposed*

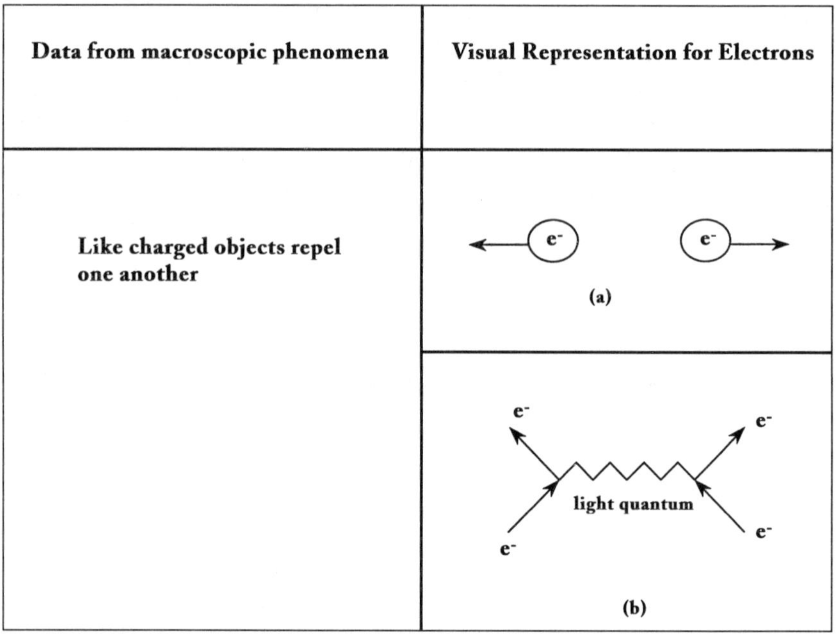

Data from macroscopic phenomena	Visual Representation for Electrons
Like charged objects repel one another	e^- e^- (a)
	e^- e^- light quantum e^- e^- (b)

FIGURE 8.1 Different visual representations for the data that two like-charged objects, in this case electrons, repel one another. Figure (a) is the extrapolation into the atomic domain of how this process occurs in our daily world with, say, two charged billiard balls. Figure (b) is the proper visual representation, a Feynman diagram, that displays the "deep structure" of how two electrons repel one another by exchanging a light quantum.

upon these theories. Consider, for example, how two electrons interact, as in Figure 8.1. The macroscopic world tells us that like-charged bodies repel one another. The representation of this observation in Lorentz's theory is in Figure 8.1(a), where repulsion is indicated by the oppositely directed arrows. Electrons are assumed to act like charged billiard balls, an extrapolation of the observation that like-charged bodies repel one another.

Similarly in atomic physics, Niels Bohr's highly successful theory of 1913, formulated along the lines of Einstein's axiomatic approach to theory building, was based on the visual imagery of the atom as a minuscule solar system (Figure 8.2).[58]

So astounding was this visual imagery, and so successful was the theory, that such a sober-minded and austere physicist as the young Max Born was moved to write in a fit of ecstasy: "The thought that the laws of the macrocosmos in the small reflect the terrestrial world obviously exercises

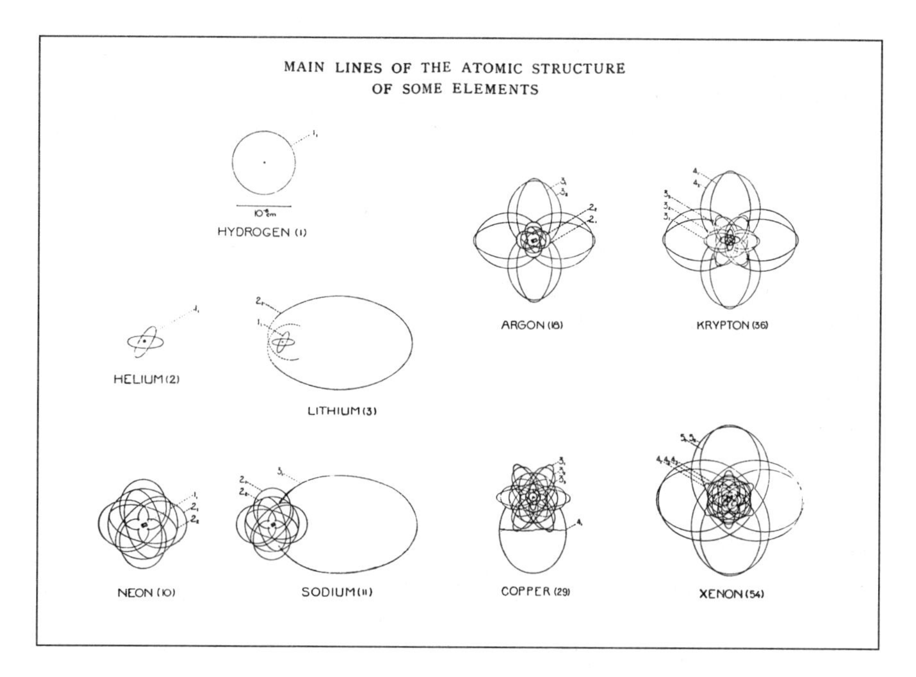

FIGURE 8.2 Depictions of the atom from Bohr's atomic theory (Kramers and Holst, 1923)

a great magic on mankind's mind; indeed its form is rooted in the superstition (which is as old as the history of thought) that the destiny of men could be read from the stars."[59] Einstein was particularly impressed by Bohr's accomplishment.

But in 1923, physicists were forced to abandon the image of the atom as a solar system.[60] That year, by combining special relativity with quantum theory, the French physicist Louis de Broglie discovered that like light, the electron also has a wave/particle duality. Consequently, the electron is unimaginable and diagrams such as the ones in Figures 8.1 and 8.2 were jettisoned. At this point successive reformulations of Bohr's theory emerged, none of which had any accompanying visual imagery, until that theory totally collapsed at the beginning of 1925. In mid-1925 Werner Heisenberg formulated a new atomic physics, called quantum mechanics, predicated on unvisualizable particles. He wished specifically to avoid the old imagery, since it had led to confusion and inconsistency.

Physicists divided themselves into two camps. One side clung passionately to a classical figurative mode of representation, seeking to understand the atomic world through imagery abstracted from phenomena in the daily world, suitably reinterpreted for the atomic realm. The German word for such imagery, *Anschauung*, appears often in the German-language journals where most cutting-edge research in atomic physics appeared. The principal members of this group were Einstein, the former radical now fighting a rearguard action against followers more radical than he, and his Berlin colleague, the Austrian physicist Erwin Schrödinger.

Foremost on the other side were Bohr and, most strongly, Heisenberg, who rebelled with great fervor against the old mode of visual representation.[61] The clash was an aesthetic one. Einstein and Schrödinger preferred a wave representation for electrons, with its visual imagery and associated continuous and hopefully causal description of atomic physics, while Heisenberg preferred a particle representation with its inherent discontinuities, lack of visualization and questionable causality. Heisenberg replaced Einstein as the leader of the avant-garde. His subsequent research laid the groundwork for increasingly abstract representations of the atomic and subatomic world. The breakthrough occurred in 1949 when, using a version of quantum mechanics consistent with relativity theory, Richard Feynman formulated the diagrams that bear his name.[62]

As an introduction to Feynman's imagery, let us return to Figure 8.1. The proper visual representation of two electrons repelling one another turns out to be the one in Figure 8.1(b), which is a Feynman diagram in which the two electrons exchange a light quantum.[63] The details of this interaction are unimportant. The point is that we would not have known how to draw the Feynman diagrams without the mathematics of quantum mechanics. Whereas the visual imagery of pre-quantum theories was *imposed* upon them and turned out to be confusing and incorrect, the visual imagery of quantum theory is *generated* by the theory's mathematics. Physics came full circle back to Plato, who argued that mathematics is the key to visualizing nature.

Figure 8.3 illustrates this spectacular transformation in representation in atomic physics. All of the figures depict the same thing, a hydrogen atom interacting with light. Figure 8.3(a) is the familiar solar-system version of the atom from Bohr's 1913 atomic theory. This visual imagery is imposed on the theory by mathematical symbols taken from Newton's celestial mechanics. (Despite being almost a century out of date and completely unworkable, this image of the electron still shows up in modern books.) The abstraction increases until we reach Figure 8.3(d), a Feynman diagram for the same process.

The earliest direct interaction of the new art with the new physics came with Niels Bohr's interest in cubism. As one of Denmark's most prominent citizens, in the 1930s Bohr moved into a mansion owned by the Carlsberg

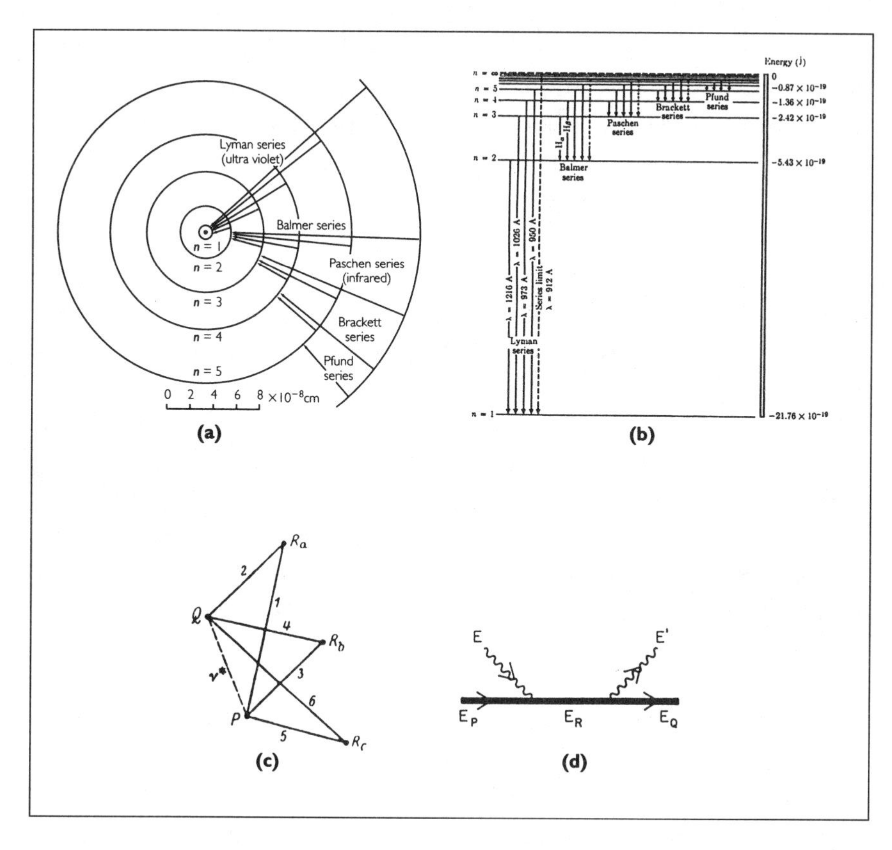

FIGURE 8.3 Representations of the hydrogen atom interacting with light. Figure (a) is from Bohr's 1913 atomic theory, which represents the atom as a minuscule solar system. Figure (b) is a more abstract representation that survived the demise of Bohr's theory. It is the visual representation of one of the theory's basic assumptions about the atom's energy levels. Figure (c) is from work Heisenberg produced just before he discovered the new atomic physics, or quantum mechanics. Although this representation is based on the theory's mathematics, it is closely connected with Figure (b). Figure (d) is totally new and was unexpected. It is a Feynman diagram representation that is completely generated from the mathematics of an advanced version of Heisenberg's quantum mechanics. The horizontal line represents the atom as it proceeds through space and time, while the wiggly lines represent light quanta. Note the contrast between Figures (a) and (d).

Foundation and was given carte blanche to furnish it. For inspiration he hung in his study a large cubist painting. But instead of choosing one of the acknowledged masters of the genre such as Braque, Gris or Picasso, he bought Metzinger's *La Femme au Cheval* (circa 1911).[64] This choice suggests that Bohr's interest in cubism was rooted more in Metzinger's writings than his art, and particularly in the book Metzinger wrote with Alfred

Gleizes, *Du Cubisme.* This book provided what has become the standard description of cubism as representing a scene as if the observer were "moving around an object [in order to] seize it from several successive appearances."[65] How you look at it, that is what it is.

According to the Danish artist Mogens Andersen, Bohr was "vitally interested in the new ground so swiftly broken by modern painting during his lifetime."[66] He took great pleasure in explaining Metzinger's painting to visitors, thus disabusing them of any preconceived ideas of what art should be.[67] The lecture was meant to parallel the lessons painfully learned by atomic physicists, who realized the inadequacy of visual perception when they discarded the visual imagery of the solar system atom after 1923. In their scientific papers they often expressed their dismay over being forced to work without visual images. Bohr offered a stopgap solution in 1927, in a motif that has striking parallels to cubist multiple perspectives. According to Bohr's "principle of complementarity," the atomic entity has two sides—wave and particle—only one of which can be revealed by any experimental arrangement. Depending on how you look at it, that is what it is.[68]

Although Einstein continued to produce notable works after 1915, he never agreed with the quantum mechanics of Bohr and Heisenberg. Besides lack of classical imagery, its violation of classical causality remained abhorrent to him.[69] Yet some of Einstein's subsequent work became an integral part of quantum mechanics. In 1916, he formulated a new theory of how large numbers of atoms interact with light, which became the basis of the laser; and in 1924 he discovered a new way of "counting" atoms that came to be called "Bose-Einstein statistics." In 1935 he proposed an intriguing thought experiment, known as the Einstein-Podolsky-Rosen experiment, that caught the essence of the terribly counterintuitive properties quantum mechanics predicted for atomic matter.[70] The experiment still reverberates among physicists and philosophers.

After *Les Demoiselles d'Avignon,* Picasso embarked with Braque on further experiments with space, resulting in such masterpieces as *Portrait of Kahnweiler* and *Nude Woman.*[71] In 1910 there appeared such spin-offs of cubism as Duchamp's *Nude Descending a Staircase* and Kandinsky's *Improvisation,* the first totally nonfigurative painting, in addition to products of Gris's highly intellectual approach to cubism. In 1917 Mondrian produced paintings in which nature was reduced to horizontal and vertical lines, all in pursuit of a higher aesthetic that would connect somehow with the scientist's search for invariant laws of nature.[72] The following year Malevich produced *White on White*, which, even compared with Kandinsky's *Improvisation*, represented nothing. Complete deobjectivation had arrived. Abstraction in art preceded that in science.

Although Picasso never left cubism entirely behind, in 1915 he began to return to a more classical genre and never attempted the extreme abstractions of Kandinsky, Malevich and Mondrian. Picasso did plunge into some of the new styles spawned by cubism, such as surrealism, and embarked on dazzling experiments in painting, photography, sculpture, ceramics, stage design and graphic arts. In 1937 he produced the great painting *Guernica.* Picasso continued to seek freshness of expression, and most of all "fresh ambitions."[73] In their own ways Einstein and Picasso were intent on extending classical figurative science and art. Pierre Daix tells the story of how, in about 1915, the Swiss conductor Ernest Ansermet expressed astonishment at Picasso's ability to move between figurative and cubist styles. Picasso's immediate reply was, "But can't you see? The results are the same."[74]

A hallmark of classicism in art and science is a visual imagery abstracted from phenomena and objects we have experienced in the daily world. There is no such visual imagery in quantum mechanics or in highly abstract art. Artists and scientists had to seek it anew rather than extrapolate it from the everyday world. Just as it is pointless to stand in front of a Mondrian or Pollock, for instance, and ask what the painting is *of*, so it's pointless to ask what the electron under quantum mechanics looks like. Neither question has an answer, and neither Einstein nor Picasso could accept such a radical break with classical thinking. We have noted how Picasso and Braque drew back from this trend with such devices as stenciled letters and by inserting such clues as keys and trompe l'oeil nails, appropriate titles, as well as bits of newspapers, wallpaper and wood glued to canvas. In physics, the visual imagery imposed on atomic theories led to inconsistencies and confusions in interpretation. It turned out that the proper visual imagery is generated by the mathematics of quantum mechanics, and it consists entirely of schematic representations of events, not pictures of objects.[75] Cubist imagery is also, to a large extent, generated by mathematics. This transformation in the role of imagery is one of the main distinguishing features of art and science in the twentieth century.

Let us pursue this line of exploration and compare the bubble chamber photograph in Figure 8.4(a) with that of *Les Demoiselles d'Avignon* in Figure 8.4(b). An important and basic problem faced by scientists in

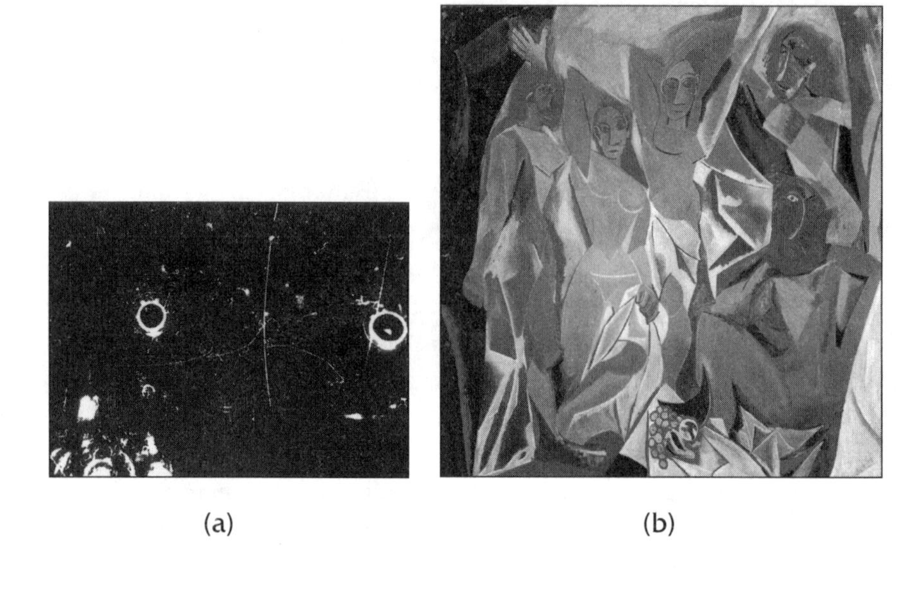

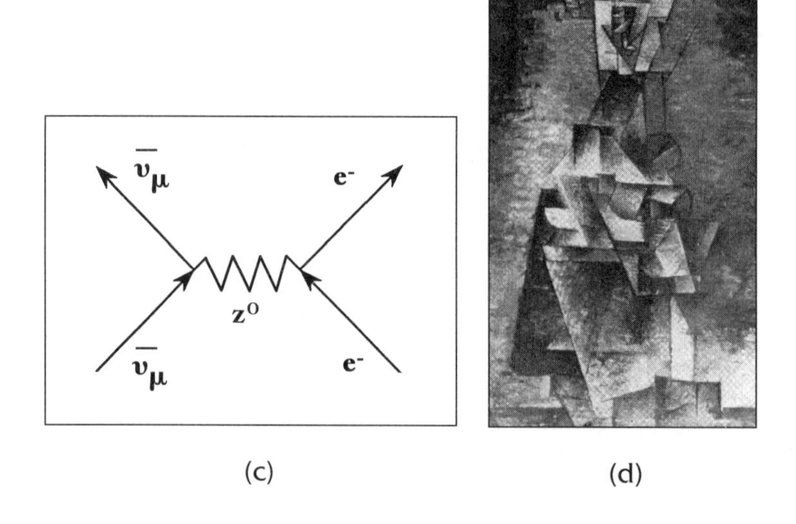

FIGURE 8.4 The "deep structure" in the bubble chamber photograph in Figure (a) is the Feynman diagram in Figure (c), where the muon anti-neutrino $\bar{\nu}_\mu$ interacts with an electron (e^-), by exchanging a (Z^0) particle. While the "deep structure" in *Les Demoiselles d'Avignon* in Figure (b) is Picasso's *Female Nude* in Figure (d).

1973 was to seek the deep structure in this photograph—the meaning that lay behind appearances. It depicts a collision between an electron (e^-) and a muon anti-neutrino.[76] But owing to the complexity of the experimental setup, the bubble chamber photograph is many layers removed from the primordial process of two elementary particles colliding with one another. Taking as a guide the Feynman diagram in Figure 8.1(b), physicists effectively arrived at the diagram in Figure 8.4(c).[77] According to this diagram the muon anti-neutrino ($\bar{\nu}_\mu$) interacts with an electron (e^-) by exchanging a particle called a Z^0-meson. The theory that explained the phenomenon in the bubble chamber photograph—the so-called electroweak theory—thus also anticipated the existence of the Z^0-meson, which was discovered soon after. This offers further evidence that Feynman diagrams are not merely mnemonics or diagrammatic aids to calculation but indicators of phenomena at a level beyond appearances.

The deep structure in *Les Demoiselles d'Avignon* turned out to reside in geometry. To illustrate this point let us look at one of Picasso's most abstract paintings, the magnificent *Female Nude,* Figure 8.4(d). Here the visible subject, the nude female, has almost completely disappeared, and what stands revealed is an object, like the muon anti-neutrino, whose essence is beyond vision. The painting, like a Feynman diagram, expresses a deeper visualization. *Female Nude* and other drawings from that summer of 1910 are the closest Picasso ever came to a completely abstractive art.

Like *Female Nude,* Feynman diagrams represent a world beyond the forms to which our perceptions restrict us. They are the most abstract scientific art existing today. The only barrier to further abstraction is that we must still represent a figure separate from its background. Thus we cannot represent or imagine something that is a wave and particle simultaneously. But subatomic particles are just that, and Feynman diagrams are presently our only means to view them in a way consistent with their properties.

But if we are going to say that the trend toward abstraction reflects parallel progress in art and science through the twentieth century, we must be very careful. Progress in art and progress in science are very different, and any comparison is a minefield. While scientific theories sometimes knock out their predecessors, Braque's and Picasso's cubism, of course, did not.[78] After cubism was recognized as a major art movement, no one threw out their Rembrandts and Titians. But after the advent of Heisenberg's quantum mechanics, in 1925, no one used the old Bohr theory anymore for serious calculations. Still, cubism represented a single-minded and unprecedented effort by a group of artists to reduce forms to geometry. In this restricted sense there was progress in art between 1907 and 1914.

After the second decade of the twentieth century neither Einstein nor Picasso produced any works comparable in the depth, breadth and virtuosity to what went before. Einstein produced no more momentous thought experiments: the radical transformation of visual imagery in the atomic domain rendered his signature visualization a less fruitful approach. As classical purists, Einstein and Picasso would not—probably could not—pursue the abstractions their great works had engendered. But they were no longer brash young men willing to immerse themselves totally. To have one's life consumed by an idea—that role fell to others: Bohr and Heisenberg were next in line in physics, Duchamp and Mondrian in art. By the 1920s, Einstein was principally preoccupied with discovering a unified view of nature that encompassed electromagnetic and gravitational forces. This effort became completely quixotic in the 1930s when other fundamental forces were discovered, including the attractive force between neutrons and protons, which is so extremely nonclassical that it could not be imaged at all before the discovery of Feynman diagrams. Einstein and Picasso did not participate in the bold new experiments toward a pure abstraction. Each of them lost touch with the advances they created.[79]

In 1912 Einstein embarked on his affair with Elsa Löwenthal ("Meine liebe Elsa"), and Picasso walked out on Fernande for Eva Gouel ("Ma Jolie"). Both women appealed to their mates' bourgeois side. Elsa offered relief from Mileva's moodiness, and Eva's calm demeanor contrasted with Fernande's stormy ways. They provided a sense of peace to men who were emerging from great intellectual struggles as well as an opportunity for them to experience almost-forgotten passions. Love letters, much like the ones he wrote to Mileva as a young man, Einstein now sent Elsa; Picasso's love letters to Eva were discreetly slipped into works such as *Guitar* (1912),[80] where he inserted "J'aime Eva," and *Woman with Guitar* (1911), on which Picasso painted "Ma Jolie."[81] Many of his paintings became highly sexually charged. In addition to needing romance, both Einstein and Picasso were beginning to suffer health problems that required rigorous diets, and Elsa and Eva were willing to care for them.

Einstein went on to marry Elsa, who did take care of him for many years. Eva was diagnosed with cancer in 1914 and died 14 December, 1915. She and Picasso had planned to marry, and he grieved over the loss. "My poor Eva is dead. . . . This has been a great sorrow for me," he wrote to Gertrude Stein soon after.[82] Picasso quickly resumed his usual complex love life, which became Montmartoise ratcheted up to the nth power by wealth, fame and the mystique of artistic genius.[83] A roll call of his most well-known mistresses and overlapping marriages is: Fernande Olivier, 1904–1912; Eva Gouel, 1912–1915; marriage to Olga Kokhlova in 1918–1934 (with the birth of his first child, Paulo, in 1921); Marie-Thérèse Walter, in 1927 through the early 1940s (with the birth of their child, Maya, in 1935); Dora Maar, 1936 to the early 1940s; Françoise Gilot, early 1940s to 1953 (with whom he had two children, Claude and Paloma). In 1954 Picasso met Jacqueline Roque, whom he subsequently married.[84]

By the mid-1920s, having achieved cult status, Einstein found his attractiveness to women raised to phenomenal heights. At five feet eight inches, broad shouldered, in fair to good shape, touted to be of legendary intelligence, and the recipient of Elsa's makeover, the middle-aged Einstein cut a striking figure. The first of his affairs that we know of during his marriage to Elsa was with a young secretary at the Kaiser Wilhelm Institute, in about 1924.[85] Reportedly Einstein broke off a passionate relationship with the excuse that "he had to seek in the stars what was denied to him on Earth,"[86] which sounds like what he told Marie Winteler some two decades earlier. Others followed. Women came in chauffeur-driven limousines to take him out for the evening. Sometimes he returned several days later. It was customary for Einstein's companion to bring a box of candy for Elsa, who was then dismissed. All of this greatly dismayed Elsa and her daughters, but it fueled the machine.[87]

Einstein's attitude toward women, like Picasso's, was typical of his era. Esther Salaman, who studied with Einstein in Berlin, wrote to him some years later about her own creative inadequacies. Einstein replied, "Very few women are creative. I should not have sent a daughter of mine to study physics. I'm glad my wife doesn't know any science; my first wife did."[88] When Salaman reminded Einstein about Curie, whom he had met several times, he responded, "Madame Curie never heard the birds sing."[89]

Their incredible fame made both Einstein and Picasso players on the stage of world events. Einstein was offered numerous speaking engagements and trips, as well as photo opportunities. He reveled in them. His correspondence, always large and varied, grew voluminous. In 1945 his name became linked with the enormous destructive force that emerged, in part, from another of his great discoveries of 1905, the mass-energy equivalence, $E = mc^2$, an equation that is practically a trademark of the twentieth century. This is fame beyond fame. Yet, he could write in 1936 to his sister, "As in my youth, I sit here endlessly and think and calculate, hoping to unearth deep secrets. The so-called Great World, i.e., men's bustle, has less attraction than ever, so that each day I find myself becoming more of a hermit."[90]

Picasso did not write many letters, he just saw people. Françoise Gilot recalled Picasso's constant complaints about the endless stream of visitors: "I asked him why he didn't shut out the world, and with it the interruptions. 'But I can't,' he said. 'What I create in painting is what comes from my interior world. But at the same time I need the contacts and exchanges I have with others. If I tell Sabartès I'm not available, and people come and I know they're there and I don't let them in, then I'm tormented by the idea that maybe there's something I ought to know and don't and I can't concentrate on my work.'"[91] The immensely wealthy and worldly Picasso's work habits in his seventh decade were a far cry from those of the intensely focused, almost penniless twenty-six-year-old artist, who refused to open the door of his dilapidated atelier even when close friends were banging on it. Einstein at twenty-six was little different, working with fanatical concentration in the cramped apartment at 49 Kramgasse, with wife and child bustling around him, as well as holding down an eight-hour-per-day, six-day-a-week job.[92] Neither man ever again achieved the total monastic dedication of those heady days.

LASTING IMAGES

We remember them best as young men. The group photograph in Figure 8.5, from the 1911 Solvay Conference, is revered among physicists. The giants who laid the foundations of twentieth-century physics are all present. It was the first summit meeting of physics. The style of the photograph would be repeated until the 1933 Solvay Conference, the last one before the war. As the years passed, those in the rear moved to the front and new faces filled gaps. The young Turks stand and the old guard sit. The 1911 photograph reveals much about the group through its intentional posing. Lorentz, chairman and conciliator, sits serenely at the head of the table, in the place of honor next to the conference's organizer, Ernest Solvay. Madame Curie and Poincaré are blasé about the whole ceremony, perhaps discussing something about radioactivity, on which she would win her second Nobel Prize later that year.

FIGURE 8.5 The famous Solvay Conference photograph. The conference took place 30 October to 3 November, 1911. During the period 1905–1911 Einstein progressed from an unknown patent clerk, third class (1902–1909), to associate professor at the University of Zurich (1909–1911), to professor at the German University in Prague and a participant at the first summit conference of physics. Many of the *dramatis personae* of this book are in this photograph. Seated from left to right: Nernst, Brillouin, Solvay, Lorentz, Warburg, Perrin, Wien, Mme. Curie, Poincaré. Standing: Goldschmidt, Planck, Rubens, Sommerfeld, Lindemann, de Broglie, Knudsen, Hasenöhrl, Hostelet, Herzen, Jeans, Rutherford, Kamerlingh Onnes, Einstein, Langevin. (Institut international de physique Solvay, courtesy AIP Emilio Segre Visual Archives)

Einstein stands next to his new friend Paul Langevin, and directly behind Poincaré—poised near the pinnacle of science.[93] Einstein's earliest major supporter, Max Planck, stands second from the left, looking shy and withdrawn. Seated on the far left is Einstein's other powerful friend from Berlin, Walther Nernst, in pince-nez and monocle, an exemplar of Prussian power.

The journey was almost over. The man who just eleven years before had been unemployed and apparently unemployable stood with the best and was their brightest star. He looks pensive and is perhaps already composing the letter he wrote to his friend Besso: "In general, the congress in Brussels resembled the lamentations on the ruins of Jerusalem. Nothing positive came out of it. . . . I did not find it very stimulating, because I heard nothing that I had not known before."[94] Few of the others would have said that.

Although Brod had caught Einstein well, nevertheless he was the catalyst of a scientific century. The popular press of Einstein's day understood this, and today it still does. His name is synonymous with genius and he has the highest recognition factor of the twentieth century. That barometer of popular culture, *Time* magazine, chose him as *the* person of the twentieth century. Whatever one may think of Einstein and Picasso on the personal level, they achieved their enormous successes under conditions that would have utterly defeated most people.

Picasso's photographic self-portrait in Figure 8.6 is my favorite because it is such a magnificent statement of an artist's confidence in his creative powers. It is a carefully arranged studio setup in rue Schoelcher, taken sometime between 1914 and 1916. The artist sits in his spacious studio amid works in progress that date back to 1907. *Les Demoiselles d'Avignon,* created in the exhilarating days of the Bateau Lavoir, is the great presence staring down from his right. On the wall at Picasso's left is the cubist construction *Guitar and Bottle.*[95] *The Smoker,*[96] nearby on his right and in line with Picasso and the *Demoiselles,* is on the very cusp of Picasso's return to natural forms. The time line runs through these two paintings: it signals the end of an era. The great experiment with cubism was coming to a close.

Picasso situates himself at the apex of a triangle formed by works of art that set the tenor of the twentieth century. He sits calmly contemplating the next era of his work. Gone are the dazed hallucinatory looks in self-portraits of thirteen years earlier. He has become the master of the Parisian art scene, the toast of the avant-garde, a beacon on the high road of civilization. A shaft of sunlight shines down on him, as if in recognition.

FIGURE 8.6 Pablo Picasso, *Self-Portrait with "The Smoker,"* Paris, rue Schoelcher Studio (1914–1916).

NOTES

CHAPTER 1

1. Arnheim (1962), p. 13.

2. Richardson (1980), p. 24.

3. Avant-garde has been rightly called a "slippery concept" (Cottington, 1998, p. 37). Since I am not writing a history of art that depends on detailed comparisons between eras, I am rescued from participating in this debate. I will use the term "avant-garde" in the following direct way in which, in fact, according to Cottington, it was commonly understood from about 1900. Avant-garde groups of artists and writers were those who rebelled against academic convention and bourgeois taste, which included its social, moral and aesthetic dimensions. As the cultural capital of Europe, aspiring artists and writers began to migrate to Paris in the 1880s. At the turn of the twentieth century these loose groups began to form communities, the principal one in Montmartre. At that time an identifiable and separate subcommunity of artists emerged with the definite goal of a countercultural transformation, and the term "avant-garde" began to be applied to them. For further discussion see Cottington (1998), esp. Chapter 2. See also Weiss (1994), p. xvi, and elsewhere, where he makes the point that certain present-day theories of the avant-garde bear little relation to archival evidence. Weiss uses "avant-garde" and "modernism" interchangeably. I have chosen to avoid the term "modernism" altogether. Certain contemporaneous theories of art attempt to furnish definitions of modernism that are separate from avant-gardism and that are *imposed* on historical events. These definitions are based on views formulated decades after the fact and are usually rooted in the formalist art theory of Clement Greenberg. Rather I want to demonstrate how Picasso's discoveries *emerge* from the artistic and scientific cultures as we are informed from archival evidence or from secondary sources as near to 1907 as is currently possible to obtain. Regarding cubism, Christopher Green has pungently argued that "something dubbed 'Modernism'" presents a "diversionary caricature" of historical developments (Green, 1987, p. 2). In his fascinating overview of intellectual developments during 1880–1918, Kern (1987) avoids altogether the terms

"avant-garde" and "modernism." In a spirited survey that goes beyond the years considered by Kern, Everdell (1999) defines modernism as high intellectual culture and also avoids use of "avant-garde," which, in his analysis, is the same as modernism. He has separate chapters on Einstein and Picasso but makes no attempt to connect them. In their interesting book, Vargish and Mook (1999) try to provide an analysis that has at least one point in common with mine, which is to establish connections between the work of Picasso and Einstein at the beginning of the twentieth century. They add the dimension of looking for connections with literature, too. The difference is that their goal is a new definition of modernism, which is that "modernism is a culture made up of specialised but comparable phenomena" (p. 2). For this purpose they propose what they call cultural diagnostics, which, in their book, are threefold: physics, art and narrative. They go on to define a "'value' which is an underlying but identifiable characteristic common to the three cultural diagnostics" (p. 6). One of their "values" turns out to be a "field theory" for the spread of knowledge as well as for the equal interdependency of all constituents as in, for example, parts of a canvas or subjects in a novel. In physics, as we all know, iron filings are affected by a magnet not necessarily by direct contact but by the magnet's magnetic field. They steer between taking the "field" notion as metaphoric and as real. Their discussion of Einstein's preference for field theories is followed by their assessment of cubism as a "visual continuum of the painting surface" (p. 115), and then in certain examples from literature they claim a trend toward multiple viewpoints in which the narrative becomes plural and collective (p. 123). Their arguments are based on a paucity of primary historical sources and are often bolstered by selected quotations from art critics such as Clement Greenberg that are ripe for interpretation into the space-time language of relativity. Neither Richardson's two-volume biography of Picasso (1991, 1996) nor Rubin's (1994) study of *Les Demoiselles d'Avignon* is even mentioned. All of this in addition to a dependence on post-structuralist literature with its accompanying jargon.

4. Strictly speaking we should include Sigmund Freud. Although Freud's exploration of the unconscious had repercussions in art, literature and science, to the best of my knowledge he had no effect at all on Einstein and Picasso in 1905 and 1907, respectively. At the century's end, Einstein and Picasso have survived critical scrutiny and their original works continue to be influential. Einstein has emerged as the twentieth-century figure with the highest recognition factor worldwide. See too Gardner (1993).

5. I am not claiming any cause-and-effect relation between what Einstein did in 1905 and Picasso in 1907 with the *Demoiselles*. This is unnecessary and, moreover, incorrect, as has been shown in some detail by the art historian Linda Dalrymple Henderson in her important book, *The Fourth Dimension and Non-Euclidean Geometry in Modern Art* (1983). The influence of mathematics, science and technology is indisputable, however, on how cubism developed after 1910. Further documents to this effect can be found in Henderson (1998).

6. See Daix (1995), pp. 671–675.

7. Brassaï (1964), p. 125.

8. Letter of Einstein to Jakob Laub on 19 May, 1909, in Einstein (1993), *The Collected Papers of Albert Einstein: Volume 5*, hereafter CPAE5, English translation, p. 121.

9. Einstein (1967), p. 5.

10. Parmelin (1969), p. 116.

11. Stein (1984), p. 19.

CHAPTER 2

1. Richardson (1991), p. 17.

2. Vallentin (1963), p. 5.

3. Richardson (1991), p. 27.

4. Picasso's family destroyed all of his drawings previous to the age of nine.

5. Quoted from Richardson (1991), p. 95.

6. This story was further embroidered by Picasso's confidant and secretary Jaime Sabartés. Actually don José went on painting well into the twentieth century, despite failing vision, which was probably the reason for his producing fewer paintings. See Richardson (1991), p. 51.

7. Richardson (1991), pp. 64–65.

8. Zervos (1932–1975), vol. 21, p. 49.

9. Zervos (1932–1975), vol. 21, p. 56.

10. Quoted from Richardson (1991), p. 90.

11. Sabartés (1949), p. 41.

12. Blunt and Pool (1962), p. 7.

13. Johnson (1980b), p. 99.

14. Shattuck (1955), pp. 32–33.

15. Daix (1987), p. 25.

16. One of Picasso's dealers at that time, Berthe Weill, could obtain only on the order of three to five francs for what became Blue Period masterpieces. This was a pittance. In desperate straits, Picasso offered Vollard twenty Blue Period paintings for 150 francs. Vollard refused. See Crespelle (1978), p. 215. For a reference point, sales personnel at a typical large Parisian store were earning about 3,000 francs per year.

17. Jacob (1927), p. 199.

18. Crespelle (1978), p. 162.

19. On 1 December, 1969, André Malraux, then minister of culture, prevailed upon the French government to declare the Bateau Lavoir a historical monument. Unfortunately, on 12 May, 1970, it burned down. Crespelle remarks that it was rather a miracle that it had not burned down earlier, with its haphazard collection of heating devices and careless installation of electricity (Crespelle, 1978, p. 82). According to Maurice Raynal, one of *la bande à Picasso,* no company would insure the Bateau Lavoir against fire (Raynal, 1922, p. 27).

20. Olivier (1933), p. 27.

21. Olivier (1933), p. 14.

22. Richardson (1991), p. 310.
23. Olivier (1933), p. 27.
24. Crespelle (1978), p. 84.
25. Crespelle (1978), p. 85. For example, writes Crespelle, "Juan Gris, who moved into the Bateau Lavoir in 1906, lived even worse. He had no furniture, and slept on a bed of newspapers. Indifferent to the least discomfort, Gris let his atelier degrade itself beyond all description. He lived there for seven years with a wife and child. Without a carriage they hung the child in swaddling clothes from their window."
26. Olivier (1933), p. 26.
27. In Picasso's Carnet 8, on p. 44R (p. 44 recto—see Léal, 1988, p. 241), is a shopping list in Picasso's handwriting: "Tobacco, coffee, eggs, butter, cheese." This supports Fernande's memoirs in which she wrote that Picasso locked her up like a recluse: "Picasso was very jealous and would never let his companion go out alone, so it was he who did the shopping" (Olivier, 1933, p. 49).
28. Crespelle (1978), p. 97.
29. Crespelle (1978), p. 30.
30. Carco (1927), p. 68.
31. Quoted from Shattuck (1955), p. 202.
32. See Richardson (1991), p. 333, where he uses this term to refer to *la bande à Picasso.*
33. Seckel (1994a), pp. 48, 49.
34. Salmon (1956), pp. 199–200. Raynal recounted how *la bande à Picasso* would watch through the window of the Bateau Lavoir as Degas and Renoir climbed up the Butte, debating whether to ask them in, as if they would accept the invitation. Quoted from Read (1997), p. 222.
35. Some fifty years later Roché wrote the novel *Jules et Jim.*
36. Carco (1927), p. 190.
37. Shattuck (1955), p. 196.
38. Carco (1927), p. 190. Francis Carco writes from firsthand knowledge of *la bande à Picasso* days.
39. Salmon (1945), p. 68.
40. Salmon (1922), p. 45.
41. Kahnweiler (1961), p. 65.
42. Raynal (1922), p. 40. Read (1995, p. 46) rightly emphasizes Richardson's calling Picasso's Rose Period "The Apollinaire Period."
43. Vallier (1954), p. 18.
44. Apollinaire (1905), p. 16.
45. See, for example, Miller (2000), especially Chapter 5.
46. Beaumont (1984), p. 56.
47. Richardson (1991), p. 203. Bergson was extremely well known. In 1900 he moved from Lycée Henri IV to the Collège de France where his lectures were so popular that only the largest auditoriums could be used. His nontechnical fluid style led to his books being widely read by the French public at large.
48. Bergson (1907), p. 53.

49. Quoted from Mitchell (1977–1978), p. 177.
50. See Antliff (1988), p. 341, for further discussion.
51. Blunt and Pool (1980), p. 6.
52. As Stuart Merrill expressed this in 1893, "By Word she speaks and thinks, by Music she sings and dreams" (quoted from Décaudin, 1981, p. 22).
53. Quoted from ibid., p. 256. It is of interest to mention here the fauves' admiration for Cézanne, who had written, "Nature, I wanted to copy it; I did not succeed. But I have been content with myself, when I discovered that the sun, for example, could not be reproduced, but it was necessary to represent it by other things" (p. 256).
54. See Richardson (1991), p. 41.
55. Breunig (1972), p. xxii.
56. Henderson (1988), p. 337.
57. See Kern (1983) for discussion of these currents.
58. Henderson (1988), p. 326.
59. Péladin (1904), p. 609.
60. See Richardson (1991), pp. 339–340.
61. Prieur (1904).
62. Réja (1904).
63. Unsigned (1905). Although not a professional scientist, LeBon was on friendly terms with such eminent members of the Académie des Sciences as Henri Poincaré. For biographical details on LeBon see Nye (1974).
64. See Mitchell (1977–1978), p. 179.
65. It would go through twelve printings and sell 44,000 copies (Nye, 1974, p. 179).
66. Weber (1903).
67. At the Lycée Henri IV during 1891–1893, Jarry encountered Bergson and benefited from his philosophy courses (see Beaumont, 1984, p. 30).
68. See Beaumont (1984), pp. 100–101.
69. Shattuck (1955), p. 164.
70. Jarry had bad luck with publication of this work, which was bewildering even to avant-garde publishers. He had to content himself with partial publication in the *Mercure de France* of May 1898. Ever resourceful, Jarry combined printed pages with manuscript pages and distributed these private versions to three friends. One carried the inscription, "This book will only be published in full when the author has acquired enough experience to savour all its beauties" (for details see Beaumont, 1984, pp. 180 and 330–331). Shattuck has interpreted such comments as these as follows: "at twenty-five Jarry suggested that he was writing over everyone's head, including his own; he had to 'experience' death in order to catch up with himself" (Shattuck, 1965, p. 13). It is entirely reasonable that these friends, figures of importance in the French avant-garde world of letters, discussed them in cafés, perhaps even at the Tuesday evening Vers et Prose salons at the Closerie des Lilas. No doubt Jarry, himself, held forth on *Faustroll.* Jarry's book was published in 1911, four years after his death.
71. Beaumont (1984), p. 200.

72. Jarry (1911), p. 192.
73. Jarry (1911), p. 193. Unless stated otherwise, all italics are in original.
74. For discussion of this point see Beaumont (1984), Chapter 11.
75. Quoted from Johnson (1980a), p. 111. Bougrereau is a pun on the name of the famous academic painter Adolphe Bouguereau with the slant of buggering and homosexuality.
76. For example, see Richardson (1991), Chapter 23, who argues that the two never met, while Leighten (1987 and 1988b) and Henderson (1983) claim they did.
77. For a short time Picasso possessed Jarry's own pistol. The situation was as follows. In a letter of 22 April, 1905, Jarry asked Apollinaire to take care of his pistol. The problem was that in the midst of a dinner at Raynal's, Jarry accused the sculptor Manolo of not being drunk enough and took a shot at him. In the ensuing confusion Jarry dropped his pistol and ran. Apollinaire, who was there, lent the pistol to a "friend"—as he recalled in a memoir of 1909. Six months later Jarry came to Montmartre to retrieve it. The "friend" was almost certainly Picasso. Quoted from Read (1995), p. 215.
78. Salmon (1945), p. 35.
79. From Richardson (1980), p. 24.
80. Olivier (1933), p. 127.
81. Read (1995), p. 45.
82. Daix (1988a), XII, p. 35.
83. See Reff (1971).
84. When asked in 1939 for advice on how to camouflage parachutists' uniforms, Picasso replied, "Dress them as Harlequins" (quoted from Reff, 1971, p. 31).
85. Salmon (1912), p. 41, also used this terminology.
86. See also the discussion in Read (1995), pp. 46, 47.
87. Daix (1988a), XII.27, p. 265.
88. Rubin (1994), p. 38; and Rubin (1984), p. 242.
89. Quoted from Richardson (1991), p. 301.
90. Daix (1987), p. 81.
91. Ibid.
92. Richardson (1991), p. 381.
93. Leo Stein (1947), p. 170.
94. Quoted in Richardson (1991), p. 400.
95. Richardson (1991), p. 378.
96. Richardson (1991), p. 408.
97. Stein (1947), p. 171.
98. Gertrude took a particular liking to Apollinaire, who was likewise trying experiments with syntax in order to seek new literary forms. Apollinaire did not reciprocate, very likely because of Gertrude's poor and never improving spoken French as well as her seemingly intentional ignorance of French literature. She was vain to the point of declaring that she was not only present at the birth of cubism but was virtually its muse, as she wrote in the rather turgid Chapter 5 of

her *Autobiography of Alice B. Toklas* (1933; actually her own autobiography). Braque and Salmon took her to task in 1935 for this revisionist history, in which Braque is hardly mentioned. They recalled her woeful spoken French as the principal reason for her never really understanding what young French painters, particularly the cubists, were striving for. As Salmon recalled in 1935, their conversations were often in the Bateau Lavoir style of making fun of everything in French slang. See McCully (1981), pp. 62–63. Picasso resented her overuse of the word "genius" and her claim to have had a unique understanding of his thinking. Reading her prose in translation, Picasso, whose instinct for literature was legendary, never saw the point. Richardson writes that the "artist remained baffled to the end" (1991, p. 407).

99. Soulié sold canvases and art supplies opposite to the Cirque Médrano. It was his style to lean the paintings to be sold against his storefront, indifferent to the weather and the homages of dogs. See Crespelle (1978), p. 227–228.

100. Kahnweiler, three years younger than Picasso, was from a wealthy Mannheim banking family. But his real passion was art. He succeeded in convincing his parents that if he did not succeed as a dealer in one year, then he would return to the family business. They agreed and generously provided him with 25,000 francs and he was off to Paris. Applying his commercial flair he scouted the art scene. Realizing that he was too late for the fauves, which Vollard was selling virtually by the hundreds, he looked around for what was new in the younger painters.

101. Fernande describes him in a rather unfair way as a "typical Jewish businessman, who knew how to take the risks which make a profit" (Olivier, 1933, p. 96).

102. For details of Picasso and his dealers see Crespelle (1978), pp. 240–242; Richardson (1991), esp. Chapter 22; and Fitzgerald (1995), who writes forcefully for the case that the modern art market was, in fact, created by Kahnweiler and Picasso.

103. For details of the art market during the first decade of the twentieth century see Fitzgerald (1995), esp. Chapter 1.

104. The scandal was so great that the president of France, Émile Loubet, refused to come to inaugurate the Salon d'Automne at the Grand Palais. See Daix (1995), pp. 328, 329.

105. Richardson (1991), p. 414.

106. Daix (1988c), p. 495. Daix points out that interest in primitivism reached all over Europe, for example, to the painters of Die Brücke in Germany.

107. Olivier (1933), p. 94.

108. Daix (1994), p. 60. See also Daix (1988c).

109. Gertrude's legend, that is. Her concept of a genius thinking up ideas in a flash recurs throughout her writings.

110. Apollinaire (1912a), p. 260.

111. Daix (1988a), XVI.15, p. 323.

112. Daix (1988a), XVI.28, p. 327.

113. Richardson (1991), p. 472.

114. Stein (1947), p. 172.

115. Gilot (1964), p. 77. The complete version of this oft-quoted Picasso comment as related by Gilot—who, to Picasso's chagrin, escaped this process, is "For me, there are only two kinds of women—goddesses and doormats."

CHAPTER 3

1. Winteler-Einstein (1924), p. xvi.

2. When Albert's parents told him of the arrival of a little sister with whom he could play, the little boy imagined a toy. So, when confronted with his new sister, Albert asked with great disappointment, "Yes, but where are the wheels" (Winteler-Einstein, 1924), p. xviii.

3. Winteler-Einstein (1924), p. xvi. Einstein's sister's name was actually Maria; Maja was a nickname.

4. Frank (1949), pp. 24–25.

5. Einstein to Sybille Bintoff, 21 May, 1954, quoted from Fölsing (1998), p. 19.

6. Frank (1949), p. 32. An earlier recollection of Einstein of this incident is in a draft of a letter to Frank, 1940, quoted in Einstein (1987), *The Collected Papers of Albert Einstein: Volume 1*, hereafter CPAE1, p. lxiii, note 56.

7. Einstein (1946), p. 5.

8. Ibid., p. 9.

9. Ibid.

10. Ibid.

11. Ibid., p. 15.

12. Renn (1993), p. 326.

13. Einstein (1946), p. 11.

14. Ibid., p. 9.

15. Winteler-Einstein (1924), p. xx.

16. CPAE1, pp. 370–371.

17. Ibid., p. 21.

18. Einstein (1946), p. 5.

19. Ibid.

20. Although conscription in Germany was at age twenty, males over the age of seventeen could not be released from citizenship, guaranteeing, thereby, a large number of potential draftees.

21. Frank (1949), p. 31. The Luitpold Gymnasium was dissolved in 1921. During the Second World War the old building was destroyed and all records were lost. Another gymnasium was built subsequently on the site and, ironically, named the Albert Einstein Gymnasium.

22. From Hans Albert Einstein, in the *Ladies Home Journal*, April 1951, quoted from Clark (1972), p. 41.

23. CPAE1, pp. 6–9.

24. CPAE1, English translation, p. 6.

25. Ibid., p. 5.

26. Letter of Otto Neustätter to Einstein, 12 March, 1929, in ibid., p. lxiv, note 62.

27. See ibid., pp. 10–12. The minimum age for entrance into the Swiss Polytechnic was eighteen. But this was waved in Einstein's case because a family friend convinced the Polytechnic's authorities that he was a child prodigy. As proof Einstein produced the informal letter from the mathematics teacher at the Luitpold Gymnasium praising his mathematical ability.

28. Pestalozzi's first publication on the power of visual thinking was in the same year that Kant published his *Critique of Pure Reason*, 1781, in which similar thoughts are expressed. See Miller (1986a), pp. 242–243, 273–275, as well as Holton (1973a), pp. 370–372.

29. Quoted from Arnheim (1969), p. 299.

30. Winteler-Einstein (1924), p. xxii.

31. Eventually Einstein would become related to the Wintelers. Einstein's sister Maja attended the Aarau teacher-training college from 1899 to 1902 and married Winteler's son Paul. Michele Besso, Einstein's lifelong friend, married Anna, the Winteler's eldest daughter. See Fölsing (1998), p. 43.

32. Einstein later recalled Jost's distrust of Germany's political climate as prophetic. Jost died in 1929. See CPAE1, p. 388.

33. Seelig (1954), p. 16.

34. See Miller (1986a), pp. 242–246; and Holton (1973a), Chapter 10.

35. Einstein (1946), p. 53. Einstein first discussed this experiment in 1916 with the Gestalt psychologist Max Wertheimer, who was a colleague at the University of Berlin. See Wertheimer (1959), p. 214. Wertheimer's scenario of Einstein's discovery of special relativity is specially constructed along the guidelines of Gestalt psychology, as Wertheimer wrote in the book's introduction. See Miller (1986a), Chapter 5.

36. CPAE1, English translation, pp. 12–13.

37. Letter of Marie Winteler to Einstein, 30 November, 1896, in CPAE1, English translation, p. 31.

38. Ibid., pp. 32–33.

39. Vallentin (1954), p. 40.

40. See CPAE1, p. 385.

41. Renn (1992), p. xviii.

42. CPAE1, English translation, pp. 135–136. In 1911 Marie married Albert Müller, a factory manager. Perhaps there is no coincidence in the husband's first name. The couple had two children and divorced in 1938. Twenty years later Marie died in a mental asylum. Auda Einstein, a descendent of Albert's, believed that Marie's unhappy affair with Einstein "confused her," subsequently ruining her life. Interview of Aude Einstein with Roger Highfield on 20 March, 1993; quoted from Highfield and Carter (1993), p. 32. For a discussion of Albert and Marie see Highfield and Carter (1993), pp. 24–32. What has to be factored in here is a tragedy that occurred in the Winteler family. In 1906, Julius, one of their sons, who was mentally unbalanced, shot and killed his mother, Pauline, as well as the husband of another of their daughters, and then killed himself.

43. For details see CPAE1, pp. 23–42.

44. Ibid., p. 24, note 6.

45. See ibid., pp. 43–44, for further details about the Swiss Polytechnic Institute, which, in 1911, received its present name, Eidgenössische Technische Hochschule (ETH).

46. Boltzmann (1897), p. 225.

47. Einstein (1946), p. 17. Apparently the classes that Einstein attended most regularly were the advanced laboratories, "fascinated by the direct contact with experience" (Einstein, 1946, p. 15). Evidence of his seriousness, and perhaps clumsiness as well, is a letter he wrote to a female acquaintance about a serious injury he suffered in an accident in the physics laboratory that required stitches in his right hand. Letter of Einstein to Julia Niggli, 28 July, 1899, in CPAE1, English translation, p. 128.

48. Einstein (1956), p. 11.

49. CPAE1, p. 61, note 10.

50. Seelig (1954), p. 40.

51. Ibid., p. 35.

52. See *Electrotechnische Zeitschrift* 18, 515–616 (1897) and *Electrical Review* 41, 526 (1897).

53. Seelig (1954), p. 35.

54. Ibid.

55. Ibid., p. 47.

56. From the unpublished reminiscences of Dr. Leon L. Watters, quoted from Clark (1972), p. 50. Watters was a wealthy pharmaceutical manufacturer who advised Einstein on finances.

57. Ibid.

58. Seelig (1954), p. 53.

59. Ibid., p. 45.

60. Ibid., p. 53.

61. Upon Mileva's death in 1948, her literary effects were transferred from Zurich to her eldest son, Hans Albert, who was a professor of civil engineering at the University of California, Berkeley. In 1986 a great-granddaughter, Evelyn Einstein, discovered them in a bank vault in Los Angeles. From the period 1897–1903 there are fifty-four letters; of that eleven are from Mileva. Since it appears that they replied to each other by return mail, most likely Einstein destroyed many of the ones in his possession. For other discussions of their relationship see Highfield and Carter (1993); Holton (1995), pp. 45–73; Renn and Schulmann (1992), xi–xxviii; Stachel (1996).

62. Renn and Schulmann (1992), p. 26.

63. Letter of Einstein to Mileva, on 13 September, 1900, ibid., p. 32.

64. Ibid., p. xi.

65. For development of this theme see ibid., pp. xxvii–xxviii.

66. See Einstein to Mileva, early August 1899, where he writes: "But in Zurich you are the mistress of our house, which isn't such a bad thing" (ibid., p. 9).

67. Ibid., p. 10.

68. Ibid., p. 11.
69. Ibid., pp. 29–30.
70. Letter of Einstein to Mileva, 20 August 1900, in ibid., p. 27.
71. Einstein (1946), p. 19.
72. Quoted from Brush (1986), p. 295.
73. There were successful theories spun along these lines. For example, H. A. Lorentz's 1892 formulation of his electromagnetic theory. As a calculational model, Lorentz assumed that macroscopic electrical conductors and magnets were comprised of charged "ions," as he then called electrons. He devised a procedure to average over the electrical effects of these billions of ions to obtain the measured electrical and magnetic phenomena produced by conductors and magnets. No trace of the ions appeared in the equations to be compared with measurement. See Miller (1998a), Chapters 1 and 9.
74. Miller (1986b), pp. xiv–xvi, and Heilbron (1982).
75. Mach (1960), p. 273. For discussions of Mach's influence on Einstein see Holton (1973c) and Miller (1998a), which contains an extensive bibliography to the literature.
76. Mach (1960), p. 589.
77. Ibid., p. 589.
78. For example, in 1897, the eminent German experimental physicist Walter Kaufmann missed out on the accolades for discovery of the electron because he was a follower of Mach's positivism. Consequently, Kaufmann could not bring himself to interpret his data as caused by submicroscopic particles. In 1901 Kaufmann thought otherwise and became the most prominent experimenter seeking to explain the electron's structure. See Miller (1998a), Chapter 1.
79. Einstein (1946), p. 21.
80. This measurement was first accomplished by the Danish astronomer Olaf Roemer in 1676. Roemer noticed that the intervals of time between the eclipses of one of Jupiter's satellites increased as the Earth receded from Jupiter and decreased as it approached Jupiter. Taking a large number of observations over the course of a year, he found a mean difference of 996 seconds between eclipses. Roemer attributed this to the finite velocity of light. Consequently, Roemer could argue that it required 996 seconds for light to traverse the diameter of the Earth's orbit, which was known to a good approximation. From this came the first measured value for the velocity of light, which is a fair estimation to today's.
81. Einstein (1936), p. 58.
82. Lorentz shared the Nobel Prize with his former student, Pieter Zeeman. The award was for using Lorentz's electromagnetic theory to explain the splitting of certain spectral lines in a strong magnetic field. Yet another triumph for the theory.
83. See Miller (1998a), p. 82.
84. In many previous versions of Maxwell's electromagnetic theory the ether had mechanical properties such as elasticity, which became cumbersome and counterproductive. See Schaffner (1972).
85. See Swenson (1972).

86. See Miller (1998a), Chapter 1, for details.

87. Lorentz and Poincaré often produced a quick calculation demonstrating that the difference between the local time and the "real" physical time for two clocks one kilometer apart on the moving Earth was much too small to be measured and so of no physical consequence. It is the order of a billionth of a second. For example, see Poincaré (1901), p. 535.

88. There were ether-based theories of optics in which the ether participated in the Earth's motion. In the paper in which Lorentz first published the contraction hypothesis, he gave his essentially aesthetic reason for why he chose a stagnant ether. Lorentz explained that as a result of comprehensively studying these theories he concluded that "being more complicated [they] were less worthy of attention" (Lorentz, 1892, p. 219). See also Miller (1998a), Chapter 1.

89. Renn and Schulmann (1992), p. 10.

90. For example, in the theories of von Helmholtz and Hertz it was difficult to disentangle the ether from the bodies moving through it.

91. Renn and Schulmann (1992), p. 14.

92. Ibid., p. 39.

93. Ibid., p. 41.

94. Letter of Einstein to Mileva, 17 December, 1901, in ibid., p. 69. See CPAE1, pp. 223–225, for details on Einstein's musings on an electrodynamics of moving bodies in his letters of 1899–1901.

95. See letter of Mileva to Helen Savic, 20 December, 1900, in CPAE1, pp. 272–273. Most likely Einstein's attempted dissertation concerned thermoelectricity, a topic of interest to Weber. Among his problems with Weber was Einstein's unrealistic hope to do a theoretical thesis based on experimental data gathered by others, or already existing in the literature. This was neither Weber's style nor to his liking because he expected his students to do their own experimental work. See Renn (1997), p. 9.

96. See, for example, Renn and Schulmann (1992), p. 19.

97. Ibid., p. 20; see also p. xvii for further analysis.

98. Ibid., p. 19.

99. Ibid.

100. For example, letter of Einstein to Mileva, 30 August or 6 September, 1900, in ibid., p. 29.

101. Letter of Einstein to Mileva of 9 August, 1900, in ibid., p. 24.

102. Ibid.

103. CPAE1, English translation, p. 143.

104. Letter of Einstein to Grossmann, 14 April, 1901, in ibid., English translation, p. 165. In the same vein, Einstein writes to Mileva on 27 March, 1901: "I'm absolutely convinced that Weber is to blame [for his employment problems]. . . . I'm convinced that under these circumstances it doesn't make any sense to write to any more professors, because they'll surely turn to Weber for information about me at a certain point, and he'll just give me another bad recommendation" (Renn and Schulmann, 1992, pp. 38–39).

105. Ibid., p. 42.

106. CPAE1, English translation, p. 162.
107. Ibid., English translation, p. 164.
108. Renn and Schulmann (1992), p. 55.
109. Letter of Einstein to Mileva on about 7 July, 1901, ibid., p. 57.
110. See Renn (1997), p. 11.
111. Letter of Einstein to Mileva on 15 April, 1901, Renn and Schulmann (1992), p. 44.
112. Starting in 1911 the Swiss Polytechnic granted its own doctorate degrees. Before that time its graduates could submit a thesis to the University of Zurich, without any further examinations (CPAE1, p. 61).
113. Renn and Schulmann (1992), p. 71. Einstein alludes to this experiment in a letter to Grossmann of about 6 September, 1901. See CPAE1, p. 316. For discussion see CPAE1, p. 224, note 9.
114. See CPAE1, p. 331, note 2 of Document 132.
115. Renn and Schulmann (1992), pp. 69–70.
116. CPAE1, English translation, p. 183.
117. Letter of Friedrich Adler to Victor Adler, 19 June, 1908; quoted from Fölsing (1998), p. 88.
118. Letter of Einstein to Heinrich Zangger, summer 1912; quoted from ibid., p. 79. Zangger was an important professor at the University of Zurich. His field was anatomy and developmental biology. Zangger first met Einstein in Bern owing to their common interest in the phenomenon of Brownian motion. He would serve as an important figure in Einstein's appointment to the Swiss Polytechnic in 1912. In the course of their professional ties, Zangger, along with Besso, became Einstein's confidant in personal and legal matters. In 1919 he was appointed guardian of Einstein's sons after Einstein's divorce from Mileva in 1918. See Einstein (1993), *The Collected Papers of Albert Einstein: Volume 5*, hereafter CPAE5, pp. 642–643.
119. Speziali (1972), p. 133.
120. Renn and Schulmann (1992), p. 54.
121. Ibid., p. 69.
122. Einstein (1901).
123. Letter of Einstein to Johannes Stark, 7 December, 1907, in CPAE5, English translation, p. 46.
124. Letter of Einstein to Grossmann, on 14 April, 1901, in CPAE1, pp. 290–291. See also note 8, p. 291.
125. Ibid., English translation, p. 165. See Fölsing (1998), Chapter 5, for details of what transpired during the period of Einstein's application for this position.
126. Einstein was reunited with Max Talmud in spring 1921 during a trip to New York City. Talmud had become a successful physician and had changed his name to Talmey. Einstein was delighted to see him again (Fölsing, 1998, p. 500).
127. CPAE1, p. 303, note 7.
128. Letter of Einstein to Mileva 17 February, 1902, in Renn and Schulmann (1992), p. 76. Einstein also remained grateful to Marcel Grossmann. He wrote to

Frau Grossmann on 26 September, 1936: "I might not have died [had he not received the Patent Office position], but I would have been intellectually stunted." Quoted from Fölsing (1998), p. 101.

129. Einstein (1965), not paginated.

130. Seelig (1954), p. 68.

131. Einstein (1965), not paginated.

132. Einstein (1956), p. 12.

133. Renn and Schulmann (1992), p. 78.

134. Private communication from Helen Dukas to Abraham Pais, in Pais (1982), p. 47.

135. Letter of Einstein to Carl Seelig, 5 May, 1952, quoted from Fölsing (1998), p. 106.

136. Letter of Einstein to Besso, on about 22 January, 1903, CPAE5, English translation, p. 7.

137. See Stachel (1996), p. 209, who describes an interview by one of Einstein's biographers of his son Hans Albert. Indeed the couple for whom the opinion of others meant nothing felt such shame that they never discussed Lieserl with anyone.

138. Letter of Einstein to Carl Seelig, 5 May, 1952, quoted from Fölsing (1998), p. 106.

139. Letter of Einstein to Carl Seelig, 5 May, 1952, quoted from ibid., pp. 114–115.

140. Solovine (1956), p. viii.

141. Ibid.

142. The translators were F. and L. Lindemann. The title page indicated that it was the "authorised German edition with explanatory notes" that were extensive and detailed; pp. 251–347.

143. Solovine (1956), p. x.

144. See ibid., p. ix.

145. Ibid., p. xiv.

146. Ibid., p. ix.

147. Ibid., p. x. Ellipses in original.

148. Ibid., p. xi.

149. Letter of Einstein to Conrad Habicht, written sometime during 30 June to 22 September, 1905, in CPAE5, English translation, p. 20.

150. Letter of Einstein to Maurice Solovine, 25 November, 1948, in Seelig (1954), p. 69.

151. Informative biographical material on Besso is in Speziali (1972), pp. xv–lxiii.

152. Letter of Einstein to Besso, on 6 March, 1952, in ibid., pp. 464–465.

153. Speziali (1972), p. xxii.

154. Letter of Einstein to Vero Besso, 21 March, 1955, in ibid., p. 538.

155. See letter of Einstein to Mileva, sometime during 30 August to 6 September, 1900, in Renn and Schulmann (1992), pp. 29–30, and note 8 on p. 88.

156. Letter of Einstein to Mileva, 4 April, 1901, in ibid., p. 41.

157. Letter of Einstein to Mileva, 27 March, 1901, in ibid., p. 39.
158. Letter of Einstein to Besso, 4 April, 1901, in ibid., p. 41.
159. Seelig (1954), p. 85.
160. Letter of Einstein in remembrance of Besso to his son Vero Besso, 21 March, 1955, in Speziali (1972), p. 538. Michele Besso had died on 15 March, 1955.
161. Besso was appointed a technical expert II class, with the salary of 4,800 Swiss francs. Actually Einstein had also applied for this position but was turned down since his experience was deemed inappropriate. On 6 September, 1904, however, Einstein was made permanent and his salary was raised to 3,900 Swiss francs, although he remained a technical expert III class. In March 1906 he would be advanced to technical expert II class. See CPAE1, p. 383, and CPAE5, p. 41, note 4.
162. Letter of Einstein to Vero Besso, 21 March, 1955, in Speziali (1972), p. 538.
163. Letter of Einstein to Heinrich Zangger, 21 December, 1926, in ibid., p. 544.
164. Ibid., p. xxxi. Haller had been director of the Patent Office from 1888 to 1921. He died in 1936 at age ninety-two.
165. CPAE5, English translation, p. 25.
166. Ibid. Einstein never lost contact with Habicht as we know from their collaboration starting in 1907 on a machine to measure minute quantities of electricity. Conrad Habicht's brother Paul also participated.
167. CPAE5, p. 12, note 3.
168. Ibid., p. 79.
169. Ibid., English translation, p. 7.
170. Letter of Einstein to Besso, on 17 March, 1903, in Speziali (1972), p. 14.
171. CPAE5, English translation, p. 17. In 1904 Habicht took employment as a teacher of mathematics and physics at a Protestant school in Schiers, canton of Graubünden.
172. Einstein (1904).
173. For discussion of these papers see Klein (1967), pp. 510–511.
174. See Renn (1997), p. 19.
175. For development of this point see Renn (1993), pp. 326–327.
176. Fluctuation phenomena can be explained as follows. The atoms in a gas pass through many configurations before they arrive at one of equilibrium. But this process is not linearly determined. At any time there can be erratic movements of the system as a whole away from the "normal" course of the system's unfolding. These erratic movements are called fluctuations and are usually dismissed as "oddities" that quickly die out, and so make no difference in the long run.
177. Einstein (1946), p. 45.
178. Ibid., p. 47.
179. CPAE5, pp. 27–28.
180. Quoted from Fölsing (1998), p. 222.
181. Seelig (1954), p. 68.

CHAPTER 4

1. Linda Henderson's 1983 study focused on developments in cubism and particularly on such theoretically inclined artists as Jean Metzinger and Marcel Duchamp. I will address her formidable research at the end of this chapter.

2. See commentary on Picasso's correspondence during this era in Rubin (1989), pp. 47–50. Daix notes that about twenty-five years later Picasso mentioned to Kahnweiler that there were "two periods of work" on the *Demoiselles*. They are denoted as "campaigns": the first is from March to about end of May or beginning of June, and the second from about mid-June through about mid- to end of July. See Daix (1988c), pp. 508–509.

3. For details see Rubin (1984), pp. 260–262, whom I am summarizing.

4. See Rubin (1984), pp. 260 and 336, notes 60, 61, and 62, for details.

5. Rubin (1984), pp. 260 and 336, note 67. See also Richardson (1996), pp. 24–25.

6. Alfred H. Barr occupied a central place in the New York City art scene. Besides being an eminent art historian, he was the first director of MOMA, created in 1929, as well as organizer for two major shows on cubism: "Cubism and Abstract Art" of 1936, and "Picasso, Forty Years of his Art" in 1939, the year when MOMA purchased Picasso's *Les Demoiselles d'Avignon*.

7. Barr (1975), p. 259.

8. See the discussion in Henderson (1983), pp. 71–72, from which I have quoted from Vauxcelles's article of 29 December, 1918, entitled, "Le Carnet des ateliers: Le Père du cubisme."

9. Vlaminck (1942). At the time Vlaminck, "long past the talents of his youth, enjoyed the flattering attentions of the German ambassador and was perhaps hoping to please the new masters" (Daix, 1994, p. 266).

10. Vlaminck (1942), quoted from Seckel (1994b), p. 264.

11. Interview with Picasso in *Paris-Journal*, 1 January, 1912.

12. Quoted from Rubin (1984), p. 336, note 64.

13. Richardson (1991), p. 26.

14. Olivier (1933), p. 20.

15. Usually designated as an interview with Marius de Zayas in 1923 in Spanish and translated in Barr (1975), p. 271. Actually de Zayas was not the interviewer, whose identity is unknown. De Zayas's supposed interview is actually only an edited version of the one published in *The Arts* in 1923. He edited out the passage that disagreed with his thesis that African art is the source of modernist abstraction, including Picasso's thinking in 1907. For details see Rubin (1984, pp. 260 and 336, notes 61, 63 and 64). This episode is one of the causes of Picasso's contradictory statements about the importance to him of African art.

16. Quoted from Leighten (1987), p. 53.

17. Leighten (1988), p. 275. See Barr (1975), p. 286, where Picasso claimed that the "letter is spurious."

18. See, for example, the letter of 7 July, 1906, from Jacob to Picasso in Chapter 2 as well as recollections Apollinaire and Raynal quoted throughout this book.

19. Leighten (1987), p. 51.

20. Zervos (1935).

21. Chipp (1968), p. 266. See Daix (1995), pp. 914–915. Picasso forbade anyone to take notes during an interview.

22. Another source regarding Picasso's creativity is Hélène Parmelin's book (1969). She met Picasso in Paris during the war and remained friendly with him for the rest of his life. Daix (1995, p. 670) writes of "Picasso speaking freely of his private problems" with Parmelin and her husband.

23. Rubin sets the dealer Wilhelm Uhde with Kahnweiler; see Rubin (1989), esp. pp. 46–47.

24. Kahnweiler (1920). The terms "analytic" and "synthetic cubism" emerge from Kahnweiler's book. They are loosely based on Kant's philosophy and folded into Kahnweiler's own conception that cubism had two distinct periods of development, which was not the case. For convenience, however, I will use the terms "analytic" and "synthetic cubism" as shorthand for the periods late 1907 to about 1912, and 1912 and beyond, respectively. I will return to this point of historiography in Chapter 5.

25. Kahnweiler neither liked nor understood Apollinaire, and dismissed his recollections as "literature" (see Daix, 1994, p. 374).

26. See Daix (1988d), p. 532.

27. Daix (1979), p. 215, no. 131.

28. Daix (1988b), p. 151.

29. Quoted from Rubin (1989), p. 47.

30. See Daix (1988d), p. 544. For a number of years Picasso was asked whether the *Demoiselles* was actually completed. Daix recalled such a conversation in 1970. He was taken aback by Picasso's irritation about this point owing to Picasso's legendary fear of anything to do with death. Completion [*l'achèvement*] carries with it the connotation of execution, or finality, "like a revolver shot in the head," as Picasso put it. According to Daix, whenever Picasso spoke about his work on the *Demoiselles* he included the entire year of 1907, which does not exclude the possibility of retouching after July. See Daix (1988d), p. 533, note 97. See, too, Daix (1979), p. 185, note 28, where Daix quotes Picasso's "almighty fury with Salmon" regarding this article. Taking Picasso's comments in their totality, we can infer that Picasso was most annoyed over Salmon's suggestion that he had taken a vacation amidst hard work. Picasso seems to have taken this as an insult that he might have been weak enough to need one.

31. See Rubin (1994), pp. 110–112, and Daix (1988c), pp. 509 and 532. There are caricatures of Salmon in Picasso's Carnet 8 from this period as well as a letter from Fernande to Gertrude dated 8 August, 1907, that mentions Salmon being present in the Bateau Lavoir at just that moment.

32. Salmon (1912), p. 46.

33. Olivier (1933), p. 130, and Crespelle (1978), p. 114. Fernande and Picasso began inviting people for dinner after 1907 when their finances began to improve. They were, however, at this time still short of serviettes.

34. Apollinaire's book is essentially a pastiche of earlier writings that go back to 1905. See, for example, Read (1995), pp. 101–105, the "Introduction" by Breunig

and Chevalier to the 1980 edition of Apollinaire (1913), and Steegmuller (1986), p. 129.

35. Rubin (1994), p. 112.

36. Ten years later Salmon wrote that Apollinaire and Jacob were also involved in the baptism (Salmon, 1922, p. 16). Then, in a letter of 7 December, 1960, to Euro Civis, Salmon again claimed sole credit for suggesting the title "the philosophical brothel," "which lasted half a day" (quoted from Seckel, 1994b, p. 225, note 3).

37. Burgess (1910), p. 408.

38. Picasso was irritated by the name because it could be interpreted as a facetious joke about the red light district in Avignon. See Seckel, 1994b, pp. 222–223. Picasso never signed the painting.

39. Daix (1988a), XV, 40.

40. Picasso is known to have contracted venereal disease in 1902 and underwent the uncomfortable traditional mercury treatment, four centuries old. See Rubin (1994), p. 57, and Richardson (1996), p. 18.

41. See, for example, Leja (1985) and Rubin (1994), pp. 56–57 and p. 130, note 166.

42. As Françoise Gilot wrote of him, "Pablo's definition of a perfect Sunday, according to Spanish standards, was 'mass in the morning, bullfight in the afternoon, whore-house in the evening'" Gilot and Lake (1990), p. 226.

43. Rubin (1984), p. 252, and Rubin (1994), p. 69.

44. See, for example, Richardson (1996), p. 26.

45. Flam (1984), p. 217. According to Rubin, in that era the term "primitivism" applied to all tribal art whether it be Iberian or African (Rubin, 1994, pp. 2–3).

46. See Richardson (1996), pp. 68–77, and Daix (1995), pp. 257–269.

47. In 1911 Picasso found out to his horror that Pieret had stolen them from the Louvre. For this episode see Rubin (1994), p. 129, and Richardson (1996), pp. 199–205.

48. Malraux (1994), p. 11.

49. Rubin (1984), p. 255.

50. Through exhaustive research, Rubin has shown that none of the African masks supposedly taken as candidates by Picasso for the *Demoiselles* existed in France at the time, or anywhere else in Europe for that matter. All of the extant African masks of that era were symmetrical in their facial features. Rubin additionally established that, as far as we know, all resemblances between the demoiselles and tribal masks are fortuitous (Rubin, 1984, p. 265).

51. See Johnson (1980a, 1980b).

52. Patricia Leighten has written on the effect of anarchy and world events on Picasso (Leighten, 1989). But the fact remains that while in Paris during the period of the *Demoiselles*, and for some time afterward, we know that Picasso was essentially nonpolitical and steered clear of Catalan anarchists for fear of the French police and deportation (Richardson, 1991, p. 172).

53. Apollinaire (1913), p. 77.

54. Apollinaire (1913), pp. 77–78; translation in Chipp (1968), pp. 231–232. I have slightly altered Chipp's translation.

55. Décaudin (1991), p. 147. Apollinaire's journals were published by M. Décaudin as *Journal intime, 1898–1918.* Quoted from Seckel (1994b), p. 227.

56. See, for example, Cousins and Seckel (1994), p. 148.

57. Uhde (1938), p. 42.

58. Cousins and Seckel (1994), p. 148.

59. As Fernande wrote: "In 1909 Picasso, who was richer by now, made up his mind to move—to leave his old *Bateau Lavoir,* where he had two studios; one just for working in, the other for his private life" (Olivier, 1933, p. 132).

60. Quoted from Seckel (1994b), pp. 239, 240.

61. Quoted from ibid., p. 240.

62. Quoted from Chipp (1968), p. 273.

63. Salmon (1912), p. 42.

64. Ibid.

65. See Chapter 1 for a definition of *passage.* For example, Picasso could have seen paintings by Cézanne at Vollard's gallery in 1901, at the 1906 Salon d'Automne, at an exhibition at the Bernheim-Jeune gallery to commemorate Cézanne's death on 22 October, 1906. Picasso had entrée to view certain of Cézanne's works at leisure: Matisse owned Cézanne's *Three Bathers* (bought from Vollard in 1901), Derain had a print of Cézanne's *Five Bathers* hung in his atelier and the Steins owned several Cézannes.

66. Quoted from Chipp (1968), p. 21.

67. As Cézanne wrote on 8 September, 1906, to his son Paul: "Here on the edge of the river, the motifs are very plentiful, the same subject seen from a different angle gives a subject for study of the highest interest and so varied that I think I could be occupied for months without changing my place, simply bending a little more to the right or left." Quoted from Chipp (1968), p. 22.

68. Apollinaire (1908), p. 51.

69. Parmelin (1969), p. 116.

70. Zervos (1935), quoted from Bernadac and Michael (1998), p. 36.

71. Picasso to Lieberman, quoted from Lieberman (1988), p. 105.

72. Warnod (1945), p. 56.

73. Olivier (1933), p. 55.

74. The example I have in mind is Cézanne's *The Kitchen Table.* See Miller (2000), pp. 414–416.

75. Shiff (1984), p. 127.

76. Olivier (1933), p. 53.

77. Ibid., p. 56.

78. Sabartés (1949), pp. 78–79.

79. Richardson (1996), p. 20.

80. Ibid., p. 18.

81. Quoted from ibid., p. 47.

82. As Picasso was supposed to have said to Gertrude on the matter of women: "If you love a woman you give her money. Well now it is when you want

to leave a woman you have to wait until you have enough money to give her" (Stein, 1933, p. 19).

83. Richardson (1991), p. 324.

84. Through the many connections between France and the Far East, opium was cheap and easy to obtain in Paris, and was rather chic to smoke from ornate glass pipes. Picasso's first source was through friends at the Closerie des Lilas. On 1 June, 1908, their neighbor and friend, the artist Karl-Heinz Wiegels, hung himself after having lost his senses as a result of taking ether, opium and hashish in succession. *La bande à Picasso* ceased smoking opium, although from time to time there would be hashish sessions. See Richardson (1996), p. 87.

85. Olivier (1933), p. 50.

86. Picasso feared becoming an addict like his friend Amedeo Modigliani, who virtually could not work without his steady diet of opium, hashish and alcohol, which eventually killed him.

87. See Crespelle (1978), especially p. 191.

88. Richardson (1996), p. 189.

89. See, for example, Warnod (1947), p. 109.

90. Crespelle (1978), pp. 191–192.

91. Ibid., p. 191.

92. The erudite Maurice Raynal moonlighted as a movie critic. In his movie reviews Raynal's sophisticated language from his art criticism gave way to enthusiastic earthy descriptions. For example, in a review of 1913 Raynal describes what it was like to watch a movie in the American Californy Vitograph Company near Place Pigalle, where anything went: "First of all, no intermissions. A continuous spectacle; one can see the same thing several times. . . . No lighted hall . . . ; delicious tarts come there to be able to practise those poses which relative obscurity favours; the little electric flashlights of the usherettes, who all are pretty, sometimes intercept hands upon legs and other places; one can smoke, one can drink; the orchestra plays 'deliciously out of tune, *on purpose*', at least we hope so, etc., etc." (quoted from Staller, 1989, p. 207; from Raynal, 1913, p. 6).

93. This is a point of contention among certain art historians who claim not much besides newspapers. See Rubin (1989), p. 54.

94. Raynal (1922), pp. 52–53.

95. Quoted from Read (1995), p. 49. Jacint's brother is Ramón, seated at Picasso's left in Figure 2.2 of Chapter 2.

96. Kahnweiler (1961), p. 65.

97. There are conflicting testimonies. Fernande recalled her surprise that Picasso was au courant on literary matters, since he "never reads" (Olivier, 1988, p. 231). But, then, as we will see in Chapter 6, Fernande seemed to have been also unaware of her lover's expertise in photography. In this vein, sometime later Gertrude Stein discussed with Picasso his attempts to write poetry. She told him that it is not possible to write without reading. Gertrude claimed to have criticized him in her circumlocutionary prose: "Pablo . . . you never read a book that was not written by a friend and then not then" (Stein, 1937, p. 37). But, then, this was decades later.

98. See the letter of 7 July, 1906, of Jacob to Picasso, which we discussed in Chapter 3.

99. By definition Euclidean geometry is the one in which the interior angles of a triangle add up to 180 degrees. This is equivalent to Euclid's fifth postulate, which states that for any given line, only one parallel line can be drawn through an exterior point. Euclidean geometry is formulated on flat surfaces in two or three dimensions. A straightforward example of a non-Euclidean geometry is the one obeyed by lines on a sphere, such as latitude and longitude lines. So, for example, the interior angles for a triangle drawn on a spherical surface add up to more than 180 degrees, and there are no parallel lines in the sense that this term is used in Euclidean geometry. In non-Euclidean geometry one naturally explores the curvature of the surface in question. Then there are *n*-dimensional geometries, which are constructed by simply adding spatial coordinates to ordinary three-dimensional analytic geometry. Analytic geometry discusses relations between points in space. If the dimensionality of the geometry is greater than three then, using methods developed in the mid-eighteenth century by the German mathematician Georg Riemann, one can demonstrate the connection with surfaces in non-Euclidean geometry. At the beginning of the twentieth century, however, for the most part non-Euclidean and *n*-dimensional geometries were discussed separately. Philosophical discussions on the nature of geometrical axioms ranged around the non-Euclidean geometries. In contrast, adding more spatial coordinates to an analytic geometry offered no philosophical challenges. The most popular *n*-dimensional geometry was four-dimensional. See Miller (2000), Chapter 6, for an introduction to the historical, mathematical and philosophical issues associated with non-Euclidean geometries and Henderson (1983) for further discussion of the development and connections between these two geometries.

100. According to Poincaré the mind is prestructured in the sense that we are born with certain organizing principles hard-wired in. These principles enable us to organize the potpourri of sensations that we glean from interacting with the world about us. Thus, we realize that relations between objects can be better understood by reasoning based on three-dimensional Euclidean geometry, rather than by any of the myriad of geometries intellectually possible. In this way we are also able to reason toward exact laws of nature by considerations based on experimental data, which are necessarily inexact. For a discussion of Poincaré's views on geometry see Miller (2000), especially Chapter 6.

101. Poincaré (1902), p. 45.

102. See Henderson (1983), p. 44.

103. See Salmon (1955), p. 187; Salmon (1956), p. 24; and Crespelle (1978), p. 120.

104. The earliest biographical account of Princet is in Henderson (1983), pp. 67–72.

105. Ibid., p. 72. Henderson cites other treatises that Princet may also have read. See Joffre (1903), pp. 152 and 153, for relevant figures.

106. Metzinger (1972), p 43.

107. For biographical details of Alice, see Daix (1995), pp. 256–257.
108. Crespelle (1978), p. 120.
109. Stein (1947), pp. 175–176.
110. Carco (1927), p. 34.
111. Crespelle (1978), p. 120.
112. Warnod (1975), p. 119.
113. Léal (1988), pp. 230 and 246.
114. Richardson (1991), p. 324.
115. Olivier (1933), pp. 133–134.
116. Quoted from Richardson (1991), p. 306. Richardson's dating of the marriage is incorrect. See his (1996), p. 76, and Daix (1995), p. 256.
117. Salmon (1910). This article by Salmon was discovered by Gamwell (1977), p. 163.
118. The only time that Salmon moved to somewhat qualify Princet's role was in response to Vauxcelles's satirical article of 1918 that we discussed previously in this chapter. For details see Henderson (1983), p. 68, note 58.
119. Salmon (1919), p. 485.
120. Ibid., p. 486.
121. Ibid., p. 488.
122. Ibid.
123. Ibid., p. 187.
124. Ibid., p. 43.
125. Ibid., p. 44.
126. Quoted in McCully (1981), p. 69; from Pla (1930).
127. See Henderson (1983), Chapter 1.
128. See ibid. and Gibbons (1981).
129. See the detailed bibliography in Henderson (1983).
130. Jarry (1899). Jarry also displayed some knowledge of non-Euclidean geometry in this essay. Another response to Wells's tale is by Jarry's friend Paul Valéry, a poet about to turn mathematician. Valéry wrote an obscure article on language functioning like a time machine. See Valéry (1899).
131. The ether Jarry describes is the mechanical one formulated by Lord Kelvin, whose popular articles Jarry had read. See Chapter 2.
132. See Johnson (1980a), p. 111.
133. Poincaré (1902), p. 92.
134. Rubin (1994), p. 14.
135. See ibid., pp. 14 and 64, as well as comments throughout Daix (1988d).
136. Léal (1988), p. 241, Figure 44R.
137. Daix (1988d), p. 493.
138. Originally there were five women, a sailor and a student. There is a sizable literature on these transformations and who Picasso may have meant these figures to be. See Rubin (1994) for discussion and references to the literature.
139. The earlier Carnet 3 has geometrized faces for the squatter that contain the *quart de Brie* nose as well. However, Daix (1988d) has found that Carnets 1–3 actually contain material dating from summer 1906 to winter 1907. Picasso did not always use the carnets in chronological order.

140. Salmon (1912), p. 46. Geometrical construction of figures is, of course, an old technique of which Dürer and Leonardo immediately spring to mind. At art school Picasso surely saw studies in which Dürer built up a face from parallelepipeds and triangles for a nose. But Picasso's Figure 4.4 differs dramatically from any Dürer I have seen and is definitely more comparable to Jouffret's figures. Amusingly, a critical article on cubism published in a 1912 issue of *L'Art et les artistes* carries the heading "Albert Dürer, Cubiste" and shows two of Dürer's geometrical constructions of heads and faces. See Weiss, 1994, p. 79. The article's thrust was to "demonstrate" precedents for cubism and so expose its originality as a sham.

141. See Jouffret (1903), pp. 70–71 and 153. See also Henderson (1983), pp. 57–58. The rather high degree of mathematical sophistication in Jouffret's text speaks well of Princet's abilities.

142. Salmon (1912), p. 42.

143. Ibid., p. 45.

144. Picasso's competitiveness emerges also in his insistence on *la bande à Picasso* referring to the *Demoiselles* in progress with the code name "the Egyptians." See Richardson (1996), p. 20.

145. Daix (1988d), p. 516.

146. Daix (1979), p. 208, no. 95.

147. Quoted from Johnson (1980a), p. 111.

148. Salmon (1912), p. 46.

149. Daix (1988d), p. 530.

150. Rubin (1994), p. 112.

151. Daix (1988d), p. 532.

152. Salmon (1912), p. 46.

153. Ibid., p. 44.

154. Daix (1988d), p. 522.

155. Léal (1988), pp. 230, 246.

156. Quoted from Golding (1988), p. 103; translated from Huyghe (1935), p. 80.

157. Léal (1988), p. 238.

158. Salmon (1912), pp. 49–50.

159. See Henderson (1983), pp. 353–365, for proof that there was no relation between Einstein's 1905 special relativity theory and developments by Picasso and Braque. (See Henderson, 1983, pp. 353–365.)

160. Baldassari (1997), pp. 45–61.

161. Ibid., p. 50.

162. See ibid., pp. 45, 48 and 49. In Chapter 6 we will have more to say about Picasso's extraordinary ability as a photographer and its relation to his painting.

163. Salmon (1912), p. 47.

164. Steinberg describes this demoiselle as being in a position that can be maintained comfortably only horizontally or by defying gravity. She is the shameless counterpart to the squatting whore. He writes: "She lies back, sexually unfurled, *une horizontale*, as the Parisians called their cocottes" (Steinberg, 1972, pp. 24–25).

165. See Braun (1997) for details on Marey and Muybridge.
166. Daix (1979), p. 223, no. 172.
167. Ibid., p. 195, no. 22.
168. Ibid., p. 212, no. 116.
169. Baldassari (1997, p. 62) goes on to note that *Composition à la tête de mort* is an "idiogrammatic *momento mori*, perhaps inspired" by Wiegels' suicide."
170. See Teuber (1997) for details.
171. Stein (1933), p. 80.
172. Teuber (1997), p. 262, quotes a passage from James's *Principles of Psychology* (1890, vol. 1, p. 90) that has an uncanny resemblance to Lhote's recollection of the problem posed by Princet: "My table-top is named square, after but one view of an infinite number. All the other views show two acute and two obtuse angles; but I call the latter perspective views, and the four right angles the true form of the table, and I erect the attribute squareness into the table's essence, for aesthetic reasons of my own. In like manner, the real circular form (of the opening of a glass) when seen perpendicularly from above is the true form, all others [e.g., ovals in perspective] are only signs [of the true form]."
173. Incidentally, ambiguities such as this one would eventually erode reliance on sense perceptions, as was advocated by Mach. Besides possible ambiguities in sense perceptions, our interpretation of them depends on our cognitive framework, or simply how much we know about the world about us as gleaned, for example, from scientific studies. In philosophy of science this became known as the "theory-ladeness" of data.
174. The principal results in this chapter conflict with certain of Linda Henderson's in her 1983 pioneering study on the effect of science and mathematics on developments in art at the beginning of the twentieth century. Like many art historians then and now, Henderson claims that the origins of the *Demoiselles,* and so of cubism too, "are to be found within art itself, primarily in African sculpture and the paintings of Cézanne" (p. 58). Yet historians of science some time ago showed that the roots of science are not within science per se. Then why should the roots of modern art be within art itself? While Henderson rightly states the difficulty in ascertaining the "degree to which Picasso was influenced by Princet and the cubist discussions going on around him," in my opinion she wrongly concludes that Princet's influence "does not seem to have been very great" (p. 59). What is offered in this book, for the first time, is the up-to-now overlooked importance of Princet for Picasso's discovery of *Les Demoiselles d'Avignon*. Henderson goes on to write that "in no way is a causal relationship being suggested between *n*-dimensional geometry and the development of the art of Picasso and Braque. Picasso's art was the product of his own artistic genius in its quest for alternatives to the classical figural tradition and to Renaissance perspective space" (p. 58). This phrasing, from her dissertation, reflects dominant art historical views in the 1970s from which she has since written otherwise (private communication). What seems to have made an even larger impact on some art historians is her 1988 study of the role played by the discovery of X rays

in the thinking of Picasso, Kupka and Duchamp, among others. Her ideas in this area are considered acceptable and cited by, for example, Richardson (1996), pp. 158–160. Why? After all, there are no cut-and-dry archival documentation to the effect that Picasso can be quoted as saying that "I was directly affected by this scientific discovery." It is unlikely, anyway, that Picasso would ever have made such a comment. Mutually confirming secondary sources should be equally acceptable for the role of mathematics and technology in Picasso's thinking toward the *Demoiselles.*

175. For example, after a lengthy analysis of the squatter, Rubin writes that "the distorted axe-blade features of the crouching demoiselle were arrived at intuitionally through the extended incremental metamorphosis of Picasso's drawings for her face" (Rubin, 1994, p. 116). Appeal to emotions or some unspecified intuition is unsatisfying.

176. Salmon (1912), pp. 47–48.

177. Kahnweiler shifts the dating of his recollection from March 1907 through the fall. I doubt whether Picasso's versions of the *Demoiselles* during his first campaign merited such an extreme comment from Derain. It must have been the completed version. Kahnweiler (1916), quoted from Seckel (1994b), p. 231.

178. Olivier (1933), p. 88.

179. Quoted from Richardson (1996), p. 45.

180. Salmon (1912), p. 51.

181. Quoted from Richardson (1996), p. 45.

182. Kahnweiler (1920), p. 252; Chipp (1968).

183. Burgess (1910), p. 408.

184. Ibid.

185. See Daix (1992), p. 311.

186. Picasso kept the painting in his various ateliers until the 1916 Salon d'Antin exhibition organized by Salmon, who also gave the painting its present-day name. The reviews were severe and included such summings-up as "overwhelming," "nightmare," and "There is yet another big daub by Picasso" (see Cousins and Seckel, 1994, pp. 165–168). See Cousins and Seckel (1994), especially pp. 164–205, for the fascinating story of the painting's journey to MOMA, which purchased it in April 1939 for $28,000.

CHAPTER 5

1. Vallier (1954), p. 14.

2. Ibid.

3. Ibid.

4. Some products from Braque's summer of 1907 are *Terrace of the Hotel Mistral* and *Landscape with Houses.* See Rubin (1977), pp. 159–165, for analysis of these paintings.

5. Vallier (1954), p. 14.
6. Richardson (1996), p. 38. See, too, Golding (1988), p. 61.
7. The sketch was published in Burgess (1910), p. 405, and has since been lost.
8. For example, the critic from *Le Rire* singled out Braque's painting and wrote in the issue of 11 April, 1908 (p. 3): "It's *Ubu Roi*, but in painting. I particularly recommend the painting *Hunger, Thirst, Sensuality* [which seemed to be the title at the exhibition], in which a woman—if one may call her such—is eating her right leg, drinking her blood, and with her left hand . . . No, I could never tell you where her left hand is wandering, no doubt in memory of Titian." Quoted from Cousins (1989), p. 351.
9. Read (1995), p. 93.
10. See Richardson (1996).
11. Quoted from Rubin (1977), p. 167.
12. Quoted from ibid., p. 169.
13. Cabanne (1960), p. 10.
14. Vallier (1954), p. 14.
15. Richardson (1996), p. 97.
16. Vallier (1954), p. 16.
17. Quoted from Rubin (1977), p. 198.
18. See, for example, Shiff (1991). A typical philosophical article is by Dunan (1888), which is an in-depth review of theories of perception that discusses the origins of visual and tactile space. Art critics used the term "tactility" to denote the artist's use of brush strokes or paint application. The critical issue, particularly with the impressionists, was their technique of applying paint liberally with a palette knife. Representing the foreground thus often resulted in a clash that resulted from trying to fold foreground, middle ground and background into one another. See Ward (1996), especially pp. 92–95. I thank Professors Christopher Green, Michael Leja and Richard Shiff for discussions on this aspect of Braque's thought.
19. Poincaré (1902), pp. 80–81. Poincaré's discussions of visual and tactile space were one among many, for example, Dunan's (1888), in which there is an extensive bibliography that covers English, French and German literature, and which grew by leaps and bounds by the first decade of the last century.
20. See Poincaré (1902), Chapter 4.
21. Ibid., p. 82.
22. Ibid.
23. Ibid.
24. Ultimately, for Poincaré, the number of dimensions of space is decided by the geometry that best describes our daily world, which he argued was three-dimensional Euclidean geometry. See ibid., Chapter 4, and Miller (2000), pp. 196–202.
25. Poincaré (1902), p. 80. Dunan (1888), for example, insists on the "absolute heterogeneity of the two forms of spaces, visual and tactile" (p. 619).
26. Poincaré (1902), p. 81.
27. Ibid.

28. See Kahnweiler (1961).
29. Crespelle (1978), p. 120.
30. See Cousins (1989), pp. 435–436, and Richardson (1996), p. 450, for details. Vauxcelles, who had coined the term "fauvism" for Matisse's followers, took up Matisse's term "cubes." In a review of Kahnweiler's exhibition of Braque in 1908, Vauxcelles wrote of Braque as "a daring young man [who] despises form, reduces everything, places and figures and houses, to geometrical schemas, to cubes" (Vauxcelles, 1908, quoted from Fry, 1966, p. 50). Actually Vauxcelles's term "cubism" applies only to Braque's *Maisons à l'Estaque* and detracts from Braque's handling of space, planes and light in the other paintings. See Rubin, 1977, p. 180.
31. Statement made by Picasso in *The Arts* of 1923, and quoted in Chipp (1968), p. 265.
32. Daix (1979), p. 240, no. 269.
33. See Henderson (1983), especially pp. 167–174.
34. For example: "This sentence is a lie." If it's false then it's true; if it's true then it's false.
35. From Picasso's interview in *The Arts* of 1923, quoted from Chipp (1968), p. 264.
36. Poincaré (1908), especially pp. 201–206.
37. Poincaré (1909).
38. Richardson (1996), p. 103.
39. Malraux (1994), p. 138.
40. Baldassari (1997), p. 7.
41. Olivier (1933), pp. 133–134. This was the session in which Princet "wept for [Alice] who left him" for Derain.
42. Baldassari (1994), p. 11.
43. Bernadac and Michael (1998), p. 31. From Zervos (1935).
44. Baldassari points out that Picasso's apprenticeship in photography occurred in Barcelona at the urging of his friend Joan Vidal Ventosa, who was a sculptor, restorer and eventually official photographer of Barcelona's municipal museums (Baldassari, 1997), pp. 14, 245. Ventosa took the photograph in Figure 2.2 of Chapter 2.
45. Daix (1988), V, 64.
46. Ibid., VI, 25.
47. Ibid., II, 20.
48. Ibid., V, 2.
49. Private Collection.
50. See Baldassari (1994), pp. 43 and 48, and Baldassari (1997), pp. 19–20.
51. Varnedoe (1996), p. 120. See also Daix (1995), pp. 704–705.
52. Moreover, if it were the result of printing two superposed negatives, then there would have been telltale lines. I thank Chris Dawes for technical discussions.
53. Baldassari (1997), pp. 19, 22, 246.
54. Ibid., p. 22.

55. Ibid., p. 22.
56. Bernadac and Michael (1998), p. 31. From Zervos (1935).
57. Daix (1995), p. 326.
58. Daix (1988), IX, 23.
59. Despite its almost photographic realism and attempt to convey familial solidity, Nazi censors condemned *The Soler Family* on the ground that it is an insult to common sense (Baldassari, 1997, p. 35).
60. Baldassari calls this "triangle photography" (Baldassari, 1994, p. 51).
61. Quoted from ibid.
62. Ibid., p. 29.
63. Ibid.
64. For example, see, Figures 6, 7 and 8 in Baldassari (1997).
65. This analysis is based on details in Baldassari (1997), p. 66, and Baldassari (1994), p. 103.
66. The description that follows is based on Baldassari's analysis in her (1994), pp. 229–243, and (1997), pp. 116–123.
67. Daix (1979), p. 295, No. 557.
68. Baldassari (1997), p. 116.
69. Mélon (1986), p. 83.
70. Pictorialism, in turn, is a meeting point of two previous photographic movements: academicism and naturalism. One of its chief practitioners, Oscar Rejlander, explains academicism thus: "The time will come when a work will be judged by its merits and not by the method of its production" (quoted from Mélon, 1986, p. 62). It advocated illusions such as those obtained from printing two negatives side by side and "brushing" over the joining line. Naturalism, on the other hand, as advocated by its chief proponent, Peter Henry Emerson, is a form of photography that sought to produce photographs as the images are supposed to occur on the retina with all accompanying distortions. For this purpose Emerson proposed a theory of focusing in which the camera's lens is put slightly out of focus. See, for example, Newhall (1964), pp. 97–100, and Mélon (1986), pp. 84–85. In neither case does the camera record "reality."
71. An interesting example of the scientist's quest to see everything through scientific photography is the "optograms" that legally minded doctors advocated in about 1868. These were photographs of the internal retina of murder victims in the hope of obtaining a portrait of the murderer. The assumption is that for a short period of time the eye acted like a camera that retained its final image. See Dibi-Huberman (1986), p. 74.
72. See ibid., pp. 71–75.
73. These drawings were so far advanced that Palau's catalogue raisonné of Picasso's works placed them in the year 1911. See Richardson (1996), p. 452, note 8.
74. Quoted from Baldassari (1997), p. 74; from a letter of Fernande to Alice B. Toklas of 15 June, 1909.
75. See, for example, Coke (1964), pp. 180–187. Coke points out that Picasso specifically made use of the perspective distortion of feet and legs in such paintings as *The Fisherman* (1917) and *By the Sea* (1920).

76. Picasso to Stein, 24 June, 1909, quoted from Baldassari (1994), p. 177.

77. As Tucker points out, Cézanne's *passage* had been discovered some time earlier by photographers and was known as "halation." This is a blurring effect due to how a high-contrast scene appears on photographic paper. As a result of the development process, roofs of houses will not be sharply bordered but will blend into the sky giving the impression of the background being tilted upwards. At Horta, Picasso rediscovered halation. See Tucker (1982), p. 293.

78. Richardson (1996), p. 97.

79. Daix (1979), p. 242, no. 278.

80. Stein (1933), p. 32. Picasso sent Gertrude this photograph and she purchased the painting in autumn of 1909.

81. Daix (1979), p. 242, no. 279.

82. Gertrude went on in her *Autobiography of Alice B. Toklas* (which is really Gertrude's autobiography) to argue that cubism was purely a Spanish conception (Stein, 1933, p. 91). For this statement, as well as other historical revisionisms, she was thoroughly lambasted by Braque and others present in the early days of cubism. See Braque (1935) and Salmon (1935).

83. Daix (1979), p. 244, no. 287.

84. Ibid., p. 246, no. 301.

85. Ibid., p. 243, no. 284.

86. Ibid., p. 245, no. 293.

87. Ibid., p. 242, no. 280.

88. Ibid., p. 242, no. 279.

89. Ibid., p. 246, no. 299.

90. Ibid., p. 245, no. 292.

91. The two heads on the right are in ibid., p. 243, no. 285, while the one on the left is in ibid., p. 243, no. 284. According to Kahnweiler, the collector and financier Paul Level cut the canvas in half (ibid., p. 243).

92. Ibid., p. 247, no. 302.

93. Ibid., p. 244, no. 290.

94. Daix (1994), pp. 246–247.

95. Baldassari (1994), p. 192.

96. Ibid., pp. 192 and 194.

97. Daix (1979), p. 243, no. 282.

98. Ibid., p. 251, no. 328. This photograph corrects Daix's dating from "winter 1909–10(?)" to summer 1909.

99. Ibid., p. 253, no. 337.

100. Ibid., p. 253.

101. See Golding (1988), especially pp. 85–88, for discussion.

102. Vallier (1954), p. 16.

103. Ibid.

104. Henderson (1988), p. 334. See, too, Daix (1979), p. 256, no. 351, and p. 256, no. 352.

105. Golding (1988), p. 83.

106. Daix (1979), p. 257, no. 358.

107. Vallier (1954), p. 16.

108. Daix (1979), p. 278, no. 466.
109. Golding (1988), p. 111.
110. Kahnweiler (1920), quoted from Chipp (1968), pp. 256–257.
111. Ibid., p. 256.
112. See Miller (2000), p. 106.
113. Apollinaire (1912b).
114. Apollinaire (1913), p. 62.
115. Ibid.
116. Ibid., p. 61.
117. The import of this spacing was first noticed by Bohn (1980).
118. Ibid., p. 167.
119. Ibid., especially pp. 167–168.
120. Francis Steegmuller, Apollinaire's biographer, reports a conversation with Picasso regarding Apollinaire in which Picasso spoke with deep feeling about "a great distinction made between three aspects of Apollinaire: the poet, the man, and the journalist art critic." Picasso realized early on that "Apollinaire's powers of visual comprehension were limited" in a way so different from his acute verbal powers. See Steegmuller (1986), p. 132.
121. Apollinaire (1913), p. 68.
122. For example, see Chipp (1968), p. 227.
123. Charensol (1924), p. 5.
124. Breunig (1993), p. 11.
125. Daix (1994), p. 131.
126. See, too, Chipp's comment that Apollinaire's book has been "under-rated for its insights into contemporary painting" (1968, p. 220). Daix writes that Picasso "hadn't agreed with Kahnweiler's opinions, which were vehemently negative. There was a great deal which would seriously annoy a rigid aesthetic theoretician like Kahnweiler" (Daix, 1994, p. 131). After heroic war service for which he was awarded the Croix de Guerre, and serious wounds, Apollinaire died of the Spanish influenza on 9 November, 1918. Picasso never forgot him. On his deathbed, Picasso's physician heard him "often speak of Apollinaire" (quoted from Read, 1995, p. 302).
127. Kahnweiler's strategy to establish an exclusivity for his artists was to discourage their exhibiting in Paris and market them outside of France, particularly in Germany. He encouraged other young dealers to do likewise, such as Uhde. Besides establishing an international reputation for them, Kahnweiler generated a great deal of acclaim for cubism in Germany. At the outbreak of hostilities in 1914, however, this strategy tragically backfired. Cubism was declared to be German art and so unpatriotic in France. Hand in hand with this chauvinism went anti-Semitism because, after all, were not cubism's principal dealers German Jews, such as Kahnweiler? This anticubist sentiment was among the events that led to Picasso's shift away from this genre in about 1915. For further discussion see Richardson (1996), Chapters 20 and 26.
128. *Gil Blas*, 25 March, 1909, quoted from Golding (1988), p. 151.
129. See ibid., pp. 117–118.

130. See ibid., pp. 118 and 120, and especially Robbins (1988), especially p. 9, and Cottington (1998), pp. 155–158, for the modern historical view.

131. The attack on Metzinger, in particular, was carried forward some years later by Douglas Cooper who wrote of him as a painter of little talent. See Robbins (1988), p. 10.

132. Cottington (1998), p. 155.

133. Metzinger (1910), p. 60.

134. Ibid.

135. Metzinger (1911), pp. 66–67. This was picked up by the art critic Michel Puy who wrote in a 1911 review of the cubist show at the Salon des Indépendants: "Cubism seems to be a system with a scientific foundation . . . [Cubists] are hungry for objective truth" (Puy, 1911, p. 65).

136. From an interview with Herschel Chipp, in (1968), p. 223. This recollection appears in Metzinger's autobiography, as well. See Metzinger (1972), p. 43. In an unpublished section of his memoirs, Metzinger mentions that, before arriving in Paris in 1905, he had read a treatise on crystallography in four dimensions. See Henderson (1983), p. 70, note 62.

137. The most in-depth discussion of Metzinger's role in scientific aspects of cubism is in Henderson (1983), especially pp. 65–73.

138. Richardson (1996), p. 211.

139. See Henderson (1983), pp. 66–71.

140. See Richardson (1996), Chapter 14, for amusing discussions of the complex clashes among the enormous egos involved, in no small measure due to Metzinger's reputation as *Chef d'École.*

141. Vauxcelles (1911).

142. Delaunay was critical about the lack of color in Braque's and Picasso's paintings at this time. "They paint with cobwebs, these fellows," he remarked. Quoted from Golding (1988), p. 157. Delaunay had known Picasso and may well have visited the Bateau Lavoir.

143. See ibid., pp. 159–161, for a description of this exhibit.

144. Gleizes and Metzinger (1912), p. 49.

145. Ibid., p. 75. Perhaps as a bit of scientific one-upmanship to the futurists, and to be more subtle generally, X rays are not mentioned in *Du Cubisme.* Rather, Gleizes and Metzinger write of the more scientifically sophisticated Fraunhofer spectral lines, which, like X rays are also invisible and so provide further evidence against positivist science. Gleizes and Metzinger (1912), p. 10. And so, too, against neoimpressionist color theory, which the futurists take as a source for their movement. See Henderson (1988), p. 335.

146. See Henderson (1983), especially pp. 81–85; while Antliff (1988) argues for the influence of Bergson on *Du Cubisme.*

147. Gleizes and Metzinger (1912), p. 75. To his credit, Metzinger tried to experiment with non-Euclidean geometries. From his studies with Princet, and perhaps readings on his own as well, he understood the importance to Euclidean geometry of the indeformability of figures in movement. Metzinger's *Cubist Landscape* (1911) indicates that he explored the notion of how objects become

deformed as they move about on curved space. See Henderson (1983), p. 96, for further discussion.

148. Gleizes and Metzinger (1912), p. 68.

149. For further analysis see Golding (1988) and Robbins (1988).

150. See Henderson (1983), pp. 66 and 74.

151. See, for example, Richardson (1996), p. 179.

152. The view of Gris is developed in Green (1992), Chapter 2. Green eloquently argues that Gris was not seeking strictly logical solutions to problems posed by Picasso and Braque.

153. Quoted from Golding (1988), p. 102. Contrary to Golding, while it is reasonable that Gris "became a serious student of the works of Poincaré" during the war, it is unlikely that he read any Einstein at this time—perhaps after Einstein's visit to Paris in March 1922 when he became all the rage.

154. After two years of seriously declining health, Gris died in great agony from uraemia, on 13 May, 1927, at age forty. On the complexity of the relationships of Gris with Picasso and Braque, see Richardson (1996). This is nicely summed up by Gertrude Stein: "Later when Juan died and Gertrude Stein was heart broken Picasso came to the house and spent all day there. I do not know what was said but I do know that at one time Gertrude Stein said to him bitterly, you have no right to mourn, and he said, you have no right to say that to me. You never realised his meaning because you did not have it, she said angrily. You know very well I did, he replied" (p. 212).

155. In Michel-Eugène Chevreul's theory of simultaneous contrasts, alterations in the appearance of a color is due to neighboring colors. Ogden Rood explored theories of color as well as differences in their appearance depending on how they are perceived. To illustrate the concepts involved, Rood's books contain disks for relating colors that bear some similarity to certain of Delaunay's paintings based on color disks. For further discussion see Vitz (1984).

156. Delaunay (1957), p. 146.

157. Ibid., p. 178.

158. Metzinger (1972), pp. 62–63.

159. Apollinaire (1913), p. 69.

160. Spate (1979), p. 3.

161. See Henderson (1998).

162. Ibid.

163. See Miller (2000), Chapter 10.

164. Derain (1994), p. 199, note 3. Editor Dagen's note is to a letter of Derain to Vlaminck of 23 August, 1909, in which Derain makes some technical comments on the construction of airplanes. Derain was an inveterate builder of model airplanes replete with engines.

165. Raynal (1912), p. 94.

166. Ibid., pp. 94–95.

167. See Henderson (1998).

168. Stein (1933), p. 11.

169. Quoted from Kern (1983), p. 303.

170. See Rich and Janos (1994), pp. 19–21.

INTERMEZZO

1. Presently there is no detailed biography of Poincaré. An informative biographical sketch can be found in Darboux (1913). See, too, Miller (1992) and Miller (2000).

2. These papers concerned Poincaré's results on configurations of equilibrium for a system containing three bodies. See Stewart (1990), especially Chapter 4.

3. See Miller (2000), pp. 340–350.

4. Toulouse (1910), p. 200, and for discussion see Miller (1992) and Miller (2000), especially pp. 343–350. Poincaré gave Toulouse's book his endorsement.

5. Poincaré (1908), p. 59.

6. Poincaré (1905a), p. 186.

CHAPTER 6

1. Haller made far and away the highest salary of anyone in the patent office, 8,000 Swiss francs per year. The next highest salary was 6,600 Swiss francs paid to Hermann Oberlin, adjunct technician. These two top salaries stayed fixed over three years during 1906–1909. See Einstein (1965), unpaginated. Einstein's salary was above that of tradesmen. For example, a merchant earned about 2,500 Swiss francs, a salesperson about 2,000 Swiss francs and a laborer about 1,500 Swiss francs. I thank Dr. Adolf Meichle for this information.

2. Einstein (1956), p. 12.

3. Winteler-Einstein (1924), p. xxii.

4. Seelig (1954), pp. 123–124. Tanner began his doctoral thesis under Einstein at Zurich and then left when Einstein went to Prague in 1911. He completed his Ph.D. at University of Basel. See CPAE5, p. 334, note 1.

5. See letter of Einstein to Habicht, written sometime between 15 and 25 May, 1905, CPAE5, p. 31. We will discuss this letter in a moment. As for the ten book reviews, see Klein and Needell (1977) and CPAE5, p. 618.

6. CPAE5, p. 40.

7. Whitrow (1967), p. 18.

8. Ibid.

9. Seelig (1954), p. 120.

10. Yet Einstein's course material at Zurich was well organized, complete with extensive references to the literature. See Einstein (1993a), *The Collected Papers of Albert Einstein: Volume 3*, hereafter CPAE3, pp. 3–10.

11. Whitrow (1967), p. 19.

12. Hans Albert Einstein attended his father's alma mater, the Swiss Polytechnic, where he received a Ph.D. in civil engineering in 1936. He emigrated to the United States in 1938 and was a professor of hydraulic engineering at University of California, Berkeley, from 1947 to 1971. Hans Albert remembered his father as being extremely authoritarian. This was in contrast to Einstein's de-

mands for his own personal freedom when he was a young man. With great remorse, Hans Albert recalled that "probably the only project he ever gave up on was me. He tried to give me advice, but soon discovered that I was too stubborn and that he was wasting his time." (From the *New York Times,* 27 July, 1973, quoted from Pais, 1982, p. 453.) See Fölsing (1998) for discussion of their relationship. Hans Albert died in 1973.

13. Whitrow (1967), p. 22.
14. Poincaré (1902), p. 242.
15. Ibid., p. 182.
16. See Chapter 3 for detailed discussion.
17. Poincaré (1902), p. 157.
18. See Chapter 3, footnote on p. 63.
19. See Chapter 3, note 87.
20. See Holton (1973c) for discussion.
21. For example, that he became interested in the experiment "only after 1905" (Shankland, 1963, pp. 47–48), and that he was unaware that it had influenced him directly "during the seven years that relativity occupied my life" (p. 55).
22. For example, consider the ceremonial gathering at the California Institute of Technology, 15 January, 1931, attended by, among others, Einstein and Michelson himself, by then frail at the age of seventy-nine. Einstein was in the presence of a man he greatly admired who, in 1907, had been America's first Nobel Prize winner. Einstein rose to the occasion and said that "it was you who led the physicists into new paths, and through your marvellous experimental work paved the way for the development of the Theory of Relativity. . . . You uncovered an insidious defect in the ether of light . . . out of which the Special Theory of Relativity developed." Quoted from Holton (1973d), p. 319.
23. Letter of Einstein to Robert Shankland, 19 December, 1952, quoted from Holton (1973c), p. 285. Holton goes through this episode in detail. See, too, Miller (1998a) for further proof. More recent evidence is in Einstein's letter to Mileva of about 28 September, 1899, where he tells her about a paper he had just read by Wien, which outlines the Michelson-Morley experiment. Most important is that before leaving the Swiss Polytechnic Einstein read Lorentz's 1895 treatise in which the Michelson-Morley experiment is discussed along with Lorentz's contraction hypothesis.
24. The archival material and historical research included in the volumes of *Collected Papers of Albert Einstein* have borne out conjectures by myself and others as to what philosophical and physics literature Einstein was aware of before June 1905. In the meanwhile, the editorial staff of the Einstein Collected Papers project have brought to light other documents that have further illuminated Einstein's professional and personal life, for example, the love letters.
25. Poincaré (1902), p. 158.
26. Ibid., p. 176.
27. Wien (1900). For details and a bibliography to the literature see Miller (1998a), Chapters 1, 7 and 12.
28. The radiation, or light, that the electron emits as a result of accelerating acts back on it and exerts a force that tries to slow it down. In this way the elec-

tron develops a heaviness or inertia. The mathematical result obtained from Lorentz's theory can be partitioned into a mass multiplied into an acceleration, which is Newton's second law of motion. The electron's mass, in this case, depends on its charge and radius. This derivation of Newton's second law from electromagnetic theory, however, is far from exact and requires several highly restrictive conditions on the electron's acceleration. See Miller (1998a), Section 1.10.

29. Lorentz published the extension of his electromagnetic theory to a theory of the electron in March 1904. By extending his contraction hypothesis to cover Kaufmann's experiments, as well as two other failed ether-drift experiments of the same order of accuracy as that of Michelson and Morley, the contraction hypothesis was deemed no longer ad hoc. Poincaré was elated. See Poincaré (1905a), Chapter 8, as well as Poincaré (1905b) and (1906). At the time he wrote the relativity paper Einstein was unaware of Lorentz's 1904 paper but could have seen earlier reports by Lorentz elsewhere in the literature. See Miller (1998a), Section 1.15. Incidentally the Dutch journal in which Lorentz published his electron theory was so difficult to obtain that physicists in such major universities as in Berlin were also unable to get a copy immediately.

30. Poincaré (1902), p. 188. The photoelectric effect had puzzled physicists since its discovery in 1887 by Heinrich Hertz. According to electromagnetic theory, extremely intense light should be capable of knocking electrons out of metals. But this turned out not to be the case. Only light with frequency above a certain threshold accomplished this, no matter what its intensity.

31. Poincaré (1902), pp. 187–188. Brownian motion was thought to possibly violate energy conservation since the particles' energetic movements seemed to be without limit.

32. Whitrow (1967), p. 21.

33. Seelig (1954), p. 17.

34. Ibid.

35. Hoffmann and Dukas (1972), p. 252.

36. The quotations to follow are from Dukas and Hoffmann (1979), pp. 76–77.

37. Seelig (1954), p. 18. Einstein made this comment in March 1928.

38. Ibid., p. 86. The comment was made to Jakob Laub, about whom more in a moment. By 1939 Einstein's opinion of Wagner was stronger: "To me his musical personality is indescribably offensive so that for the most part I can listen to him only with disgust" (Dukas and Hoffmann, 1979, p. 77).

39. Seelig (1954), p. 82.

40. Einstein (1934b), p. 61.

41. Einstein (1936), p. 59.

42. Solovine (1956), p. x. This was inspirational to the psychologist Jean Piaget whose work on time "was promoted by a number of questions kindly suggested by Albert Einstein" around 1930 (Piaget, 1971, p. vii).

43. Wertheimer (1959), p. 213.

44. Einstein (1970), p. 1. This is an edition of Einstein's Stafford Lectures, delivered in May 1921 at Princeton University.

45. Einstein (1970), p. 1.
46. Poincaré (1908), p. 48. See Miller (2000), Chapter 9, for analysis.
47. Einstein (1946), p. 7.
48. Ibid.
49. Einstein's argumentation bears similarities to descriptions of thinking by Hermann von Helmholtz and Ludwig Boltzmann in books that Einstein read while a student at the Swiss Polytechnic. But Einstein goes beyond them. See Miller (1986a), pp. 48–51. No doubt, at the turn of the twentieth century, the rise of Freudian psychoanalysis with its accompanying "discovery" of the unconscious led several prominent scientists to publish introspections.
50. Philosophically Einstein likened concepts to Kant's organizing principles. Although in opposition to Kant, Einstein's system of concepts was not absolutely fixed a priori. For example, according to Kant there can be no geometry other than Euclidean geometry. The formulation of non-Euclidean geometries in the late 1820s was, therefore, a serious blow to Kant's philosophical system. But for Einstein, of lasting importance was Kant's insistence on the need for organizing principles. As we will see in a moment, Einstein's relativity theory is based on two principles—which act like organizing principles—one of which asserts that light travels in a straight line at a velocity that never varies. Einstein found, however, that in order to generalize the 1905 relativity theory to include gravity he had to relax this principle to permit light to travel in curved paths. In *La Science et l'Hypothèse* Einstein had studied Poincaré's cognitive theory of the origins of geometry in which organizing principles play a key role. See Chapter 4, note 100; Poincaré (1902), Chapter 4; and Miller (2000), Chapter 6.
51. Einstein (1946), p. 7.
52. Ibid., p. 9.
53. Letter of Einstein to Jacques Hadamard, 17 June, 1944, in Hadamard (1954), pp. 142–143.
54. Einstein (1923), p. 484.
55. Letter of Einstein to Max Laue, 17 March, 1952, in Miller (1998a), p. 126. Einstein did not publish these results until 1909.
56. CPAE5, English translation, p. 20.
57. Einstein (1905d) is dated 30 April, 1905, and submitted to the University of Bern on 20 July, 1905. Einstein dedicated it to Marcel Grossmann.
58. Einstein (1905a).
59. Ibid., p. 145.
60. Ibid., p. 132.
61. On 9 November, 1922, Einstein was awarded the 1921 Nobel Prize for discovering the law of the photoelectric effect, which he was able to explain with his concept of light quanta. Among other uses, Einstein's law is the basis of automatic opening doors. We can only conjecture why Einstein did not win the Nobel Prize for relativity. First there was no one on the committee capable of assessing the full content of relativity theory, which included the general theory. Then there were criticisms such as Bergson's. As we discussed in Chapter 2, Bergson attributed no physical reality to scientific time. He believed that actually

interrogating observers in different reference systems would show they all registered the same time, contrary to what special relativity theory predicted. Consequently, in his 1922 book Bergson placed himself in the physically impossible situation of essentially having each foot on the platforms of different moving observers. Bergson's criticisms can be responded to in an elementary manner. So influential was Bergson, however, that his criticisms were mentioned in the presentation speech by S. Arrhenius, chairman of the Nobel Committee for Physics of the Royal Swedish Academy of Science's presentation of Einstein's Nobel Prize in 1921. See Einstein (1967), p. 479. For discussion see Miller (1998a), pp. 242–248, 273–274. Bergson and Einstein became friends and enjoyed discussing a wide variety of issues. But of Bergson's philosophy of relativity, Einstein is recalled as saying, "Gott verzeih ihm [God forgive him]" (Pais, 1982, p. 510). For further discussion see Pais (1982), pp. 510–511.

62. See Planck (1910), p. 758. This situation persisted until 1927 when light quanta were systematically incorporated into the new quantum mechanics with a probabilistic interpretation. By then Einstein had abandoned light quanta in favor of theories that could explain matter in terms only of continuous quantities. See Miller (1986a), Chapters 4 and 6, and Miller (2000), pp. 124–127.

63. Einstein (1905d). This is one of Einstein's most fundamental papers and was his most cited paper during 1961–1975. See Pais (1982), pp. 88–92, for discussion.

64. Einstein (1905b).

65. In both papers Einstein used his method for calculating fluctuations to deduce a way to relate Avogadro's number to quantities that could be measured from macroscopic systems. In the doctoral thesis it was the rate at which sugar granules dissolve in solution and in the *Annalen* paper it was the erratic dance of pollen granules under a microscope, that is, Brownian motion. This was entirely unexpected because to measure Avogadro's number all you needed was "a stopwatch and a microscope" (Pais, 1982, p. 97). Einstein's solution to the problem of Brownian motion is that it is the result of fluctuations of the system about equilibrium and the process conserves energy. In 1908 the French physicist Jean Perrin carried out the necessary experiments and measured a value for Avogadro's number that squared with those from other phenomena, some of which were also suggested by Einstein, such as the scattering of light that results in the blueness of the sky. In addition there was Planck's determination of Avogadro's number from his work on cavity radiation. Taking into account Perrin's data, and that Avogadro's number can be calculated from such diverse phenomena, from 1908 no serious scientist could argue against the reality of atoms. For details see Pais (1982), Chapter 5, and Brush (1986), especially Volume 1.

66. Einstein (1907a), p. 372.

67. See Miller (1998a), Chapter 2, for a detailed discussion of this point.

68. Max Planck first used the term *Relativtheorie* (relative theory) in 1906 in order to distinguish the Lorentz-Einstein electron theory from others (see Planck, 1906, p. 756). At first Einstein's results in the relativity paper were interpreted as generalizing Lorentz's electron theory—more on this in a moment. In

the discussion session to Planck's paper, the German experimentalist Alfred Bucherer referred to Einstein's theory as the "Relativity theory (*Relativitätstheorie*)," p. 760. In 1907, Einstein referred to his theory as "*Relativitätstheorie* (relativity theory)" (Einstein, 1907a, p. 373).

69. The kind of title Einstein gave his paper customarily signaled a discussion of the properties of bulk magnetic or dielectric matter. But Einstein analyzed neither of these topics in detail, only in principle.

70. Einstein (1905c), p. 371.

71. In 1905 the Curatorium was composed of F. Kohlrausch, M. Planck, G. Quicke, W. C. Röntgen and E. Warburg "with the participation of the German Physical Society and especially M. Planck" (from the title page of Volume 17).

72. See Klein and Needell (1977).

73. Besides a few typos, there was only one error in reasoning that resulted in an incorrect prediction for the electron's mass. This error was quickly noted by Kaufmann and corrected. See Miller (1998a), pp. 310–311. When corrected, Einstein's result for the electron's mass was considered to be important because it was carried out with no approximations, and was independent of electromagnetic theory. See note 28 of this chapter.

74. Einstein (1946), p. 53.

75. This came about through his own research in his last three papers in the *Annalen*, as well as in his Patent Office work where he scrutinized patents that claimed to be for a *perpetuum mobile*, or perpetual motion machine.

76. By this I mean the following: The fundamental quantities of thermodynamics are pressure, volume and temperature. As these quantities stand in the various gas laws there is nothing indicating the systems' constitution: gases can be composed of anything, atoms, liquid, etc.

77. Einstein (1919), p. 54.

78. Letter of Einstein to Walter Dällenbach, on 31 May, 1915, in Einstein (1998), *The Collected Papers of Albert Einstein: Volume 8*, hereafter CPAE8, English translation, p. 102.

79. Throughout much of *La Science et l'Hypothèse*, Poincaré referred to the principle of relativity in Newton's mechanics as the principle of relative motion.

80. Poincaré (1902), p. 129.

81. Ibid., p. 25.

82. The assumption is made that the moving Earth is an inertial reference system. But in actuality it is not because the Earth rotates on its axis as well as moving in an orbit about the Sun. Yet to a sufficient degree of accuracy we can, in many instances, neglect these complex motions. For discussion see Miller (2000), pp. 25–26, 80–86.

83. Poincaré (1902), p. 182.

84. Ibid.

85. Ibid., p. 180.

86. Ibid.

87. See Poincaré (1902), p. 185, and (1900), pp. 484–488.

88. As Klein (1962), p. 476, wrote of this situation: "Planck's concept of energy quanta went practically unrecognised in the literature of physics for over

four years. His radiation formula was accepted as describing the experimental facts in a simple and adequate way, but the theory which he had proposed as a basis for this formula drew no attention until [Einstein's paper of] 1905."

89. Einstein (1905c), p. 370.

90. Certain important design problems concerned the so-called unipolar machines. See Miller (1986b), Essay 3, and Miller (1998a), Chapters 3 and 9. In the relativity paper, after proposing his own principle of relativity, Einstein audaciously declared them to be "meaningless" (Einstein, 1905c, p. 384). Although they may be meaningless in principle, in practice details have yet to be worked out to everyone's satisfaction.

91. This manuscript was brought to light by Holton in his (1973a), pp. 363–364. Reproduced and further discussed in Miller (1998a), p. 137. Einstein's 1905 relativity theory is referred to as the special theory of relativity because it restricts measurements to be made only from inertial reference systems, whereas in the generalized theory of relativity that Einstein completed in 1915, measurements can be made by observers in accelerated reference systems as well.

92. Einstein (1905c), p. 370.

93. Furthermore, the source of this force was unclear and confusing even to Lorentz because it seemed to violate the principle of relativity. Yet such problems in electromagnetic induction were not considered to be of basic importance. For discussion, which includes issues in electrical dynamo design, see Miller (1986b), Essay 3, and Miller (1998a), Chapters 3 and 9.

94. Miller (1998a), p. 137.

95. In Einstein (1905c), p. 370, *en passant* Einstein mentions ether-drift experiments of lowest-order accuracy, that is, "first-order accuracy."

96. Einstein (1905c), p. 371.

97. Poincaré (1902), p. 111.

98. Einstein (1905c), p. 371.

99. In the opening sentence of the relativity paper Einstein used the plural asymmetries. What others could he have had in mind other than the one concerning redundancy of explanation? In 1909 Einstein wrote that "it is totally unnatural to single out" reference systems fixed in the ether from the collection of inertial reference systems to which they are, in fact, connected mathematically. A satisfactory resolution is reached "only if one drops the ether hypothesis." Einstein (1909), p. 819.

100. Einstein (1946), p. 53.

101. Einstein's unpublished 1919 recollection, as quoted from Miller (1998a), p. 137.

102. Einstein (1946), p. 53.

103. Poincaré (1902), p. 111.

104. Ibid., p. 112. In the German-language edition, the word "intuition" is translated as *Anschauung*, p. 92.

105. Ibid.

106. Lucien Chavan was born in Lausanne, Switzerland, and so was fluent in French. Chavan recalled of his first meeting with Einstein that "Einstein speaks correct French, with a slightly foreign accent" (Seelig, 1954, p. 71). Solovine is

from Romania and so was also fluent. Einstein, however, was more comfortable with German-language prose material.

107. Among them is Wilhelm Wien's in which he sounded the call for research toward an electromagnetic world-picture. See Wien (1900).

108. Einstein (1906), p. 627. Additionally, in Section 8 of the relativity paper Einstein explores certain characteristics of a light pulse, in particular its energy and frequency. Poincaré was the first person to examine this problem using the local time. I showed that Einstein's intent in this section was to relate the relativity and light quantum papers. See Miller (1998a), Chapter 11. Further study of Poincaré's 1900 paper has convinced me that another of Einstein's intents was to clean up this problem situation as he had seen Poincaré develop it within Lorentz's electromagnetic theory to lowest-order accuracy. See Miller (1996), pp. 95–96.

109. The cities are Berlin, Munich, Stuttgart, Karlsruhe and Ludwigshafen. Most of Switzerland was on Berlin time.

110. See Kern (1983), pp. 12–13.

111. See Howse (1997), pp. 120–121.

112. See Bartky (1989), p. 32.

113. See Howse (1997), p. 118.

114. See, for example, Whittaker (1987), pp. 227–231. Precise knowledge of transmission delays is extremely important because four seconds of time is equal to one minute of longitude. For a description of the astronomical methods that were combined with telegraphy for determining delays in transmission times, see Hayden (1905), p. 10.

115. Howse (1997), pp. 120–125. Dowd was principal of Temple Grove Ladies' Seminary in Saratoga Springs, New York.

116. Until 1884 the United States used two prime meridians: Greenwich for sea charts, and Greenwich and Washington, D.C., for land maps. See Howse (1997), p. 130.

117. See Bartky (1989), pp. 34–39.

118. Poincaré wrote of "two simultaneous psychological facts linked so closely together that analysis cannot separate without mutilating them" (Poincaré, 1898, p. 49). Einstein wrote, "We shall not here discuss the inexactitude which lurks in the concept of simultaneity of two events at (approximately) the same place, which must be removed through introducing an abstract concept" (Einstein, 1905c, p. 371).

119. See Howse (1997), Chapter 5.

120. Right from the beginning of the 1884 conference the two French delegates, A. Lefaivre, minister plenipotentiary and consul general, and Mr. Janssen of the Institute and director of the Physical Observatory of Paris, resisted Greenwich as the prime meridian in favor of a neutral one not cutting a major continent such as Europe or America. Furthermore, they continued, in their view the conference's main order of business was to examine principles upon which a prime meridian would be based and not to choose one. The consensus, however, was just the opposite, yet this resolution provoked a great deal of dis-

cussion. The issue was decided by arguments from Sandford Fleming, the British delegate representing Canada, and William Thomson, the scientific delegate of Great Britain. Thomson, elevated to Lord Kelvin, was one of Britain's greatest scientists, responsible, among other things, for solving the practical problems of signal transmission through the transatlantic cable. Fleming and Thomson pointed out the impossibility of a strictly neutral prime meridian and that what was at issue was not politics but practicality. It became quite clear that the French government's not so hidden agenda was to have the prime meridian running through the Paris observatory. But Fleming and Thomson brought to everyone's attention the incontrovertible fact that 72 percent of the world's shipping already used the prime meridian through Greenwich, with Paris a distant second. The remaining 28 percent were divided up among ten different prime meridians. What France wanted in return was for the United States and Great Britain to adopt the metric system. To which the American scientific delegate, Cleveland Abbe, replied that the metric system was a French system and so not neutral. And, anyway, he continued, the metric system was already in use by scientists in the United States and Britain. In the vote for a prime meridian through Greenwich, twenty-two nations voted yes, San Domingo voted no and two abstained, Brazil and France. On 27 October, 1896, a bill was introduced into the Chamber of Deputies that proposed Greenwich Mean Time (GMT) as the standard time in France. It was passed on 24 February, 1898, in an amended form where instead of GMT it was expressed as PMT (Paris Mean Time), diminished by 9 minutes 21 seconds, which is just GMT. Because of opposition by the Navy and the Ministries of Public Instruction this bill languished for twelve years before being legalized. For discussion of the 1884 Prime Meridian Conference see Howse (1997), Chapter 5.

121. See Wilford (1982), Chapter 8.

122. Poincaré (1898), p. 52. Poincaré immediately eliminated a standard time based on the Earth's rotation because of variations due to the tidal drag of the Earth's oceans. Poincaré was well acquainted with this aspect of geodesy and navigation because his research on tidal theory was the best available and, in fact, had earned him entry into the navigation section of l'Académie. See Darboux (1913), p. lxvii.

123. See Chapter 3, note 80.

124. Poincaré (1898), pp. 53–54. For example, the time delay for the reply from Berlin is about one-thousandth of a second, omitting further delays due to relays.

125. Maxwell's theory predicted electromagnetic waves, which were discovered in 1888 by Heinrich Hertz. In 1894 Guglielmo Marconi invented a device for reception and transmission. Throughout Hertz's work, he was in correspondence with Poincaré whose assistance was invaluable. See Darboux (1913), pp. xl–xli. Practitioners considered Poincaré's courses at l'École de Télégraphie to be essential. See Darboux (1913), p. xli.

126. Poincaré (1898), p. 54.

127. Ibid., p. 53.

128. Ibid, p. 54.
129. In 1905 Einstein showed that this was not the case.
130. Poincaré (1900), p. 483.
131. The shift in meaning from the local time as the time of each community to the time of each reference system is not far, and may have been Lorentz's reason for naming his new time coordinate. I make this conjectural statement because, like Poincaré, Lorentz certainly kept up-to-date on such pressing technical matters as time standardization.
132. An exact but mathematically messy version of the local time is in Lorentz's 1904 paper and then in its familiar form in Poincaré's 1905 and 1906 papers. When he wrote his relativity paper Einstein was unaware of Lorentz's 1904 paper, as well as of Poincaré's 1905 paper. See Miller (1998a), pp. 81–86.
133. Kern (1983), p. 13.
134. It was the sinking of the *Titanic* on 14 April, 1912, that made clear the need for a single set of worldwide wireless telegraphy safety regulations. See Kern (1983), pp. 65–67.
135. Hayden (1905), p. 11.
136. Poincaré in *La Science et l'hypothèse* and Einstein in the relativity paper used the term "event."
137. Einstein (1905c), p. 372. One of these thought experiments concerned a master clock that sequences the arrival of light rays from clocks at relative rest and situated throughout space. But this method depends on the distances between clocks. By 1912, while passing each other, ships at sea often checked their times by wireless using transmissions called "time rushes" (Kern, 1983, p. 66).
138. Einstein (1905c), p. 371.
139. Einstein (1907b), p. 413. "The difficulty discussed" was the failure of all ether-drift experiments, including Michelson and Morley's. I use the term "reminiscence" here because at this point Einstein's relativity theory was considered by many physicists to have been disconfirmed. He thought otherwise and turned out to be correct.
140. Poincaré (1900), p. 483.
141. For this calculation see Miller (1998a), pp. 177–179.
142. See Chapter 3, note 87.
143. Quoted from Miller (1998a), p. 177.
144. Einstein (1970), p. 46.
145. CPAE5, English translation, p. 20.
146. Einstein (1905c), p. 393.
147. Letter of Einstein to Besso, on 6 March, 1952, in Speziali (1972), pp. 464–465. After writing this chapter I became aware of similar approaches to how Einstein's discovery of the relativity of simultaneity is connected with standard time determinations and longitude measurements. There is a throwaway line in Everdell (1999), p. 9. Galison's (2000) omits Poincaré's all-important paper of 1900.
148. Kaufmann (1905), p. 954. Kaufmann was the first one to note and correct Einstein's erroneous result for the moving electron's mass. He also appreciated

the exactness of Einstein's derivation compared to the severe approximations required by Lorentz. See note 73 in this chapter.

149. Kaufmann (1905), p. 954.

150. Kaufmann (1906), p. 495.

151. Letter of Einstein to Solovine on 27 April, 1906, in CPAE5, English translation, p. 25.

152. Planck (1906).

153. CPAE5, English translation, pp. 20–21.

154. In nuclear reactions mass is not conserved because the total mass of the final products is less than that of the initial ones. Rather energy is conserved and so the "missing mass" emerges as energy. Despite the fact that the missing mass is almost vanishingly small, it translates into a huge amount of energy because it is multiplied by the velocity of light squared. Historically, it was known in 1905 that mass was not conserved in chemical reactions involving radioactive substances. Einstein's mass-energy equivalence brought order to this situation. See, for example, Miller (1998a), pp. 333–335.

155. Letter of Einstein to Solovine on 27 April, 1906, in CPAE5, English translation, p. 25.

156. CPAE5, pp. 41–42.

157. Ibid., p. 42, note 10.

158. Seelig (1954), pp. 92–93.

159. Quoted from Fölsing (1998), pp. 211–212.

160. Letter of Laub to Carl Seelig, 11 September, 1959. Quoted from Fölsing (1998), p. 202.

161. Pais (1982), p. 505.

CHAPTER 7

1. Letter published in Miller (1998a), pp. 318–319. For discussion of this episode see Miller (1998a), Chapters 7 and 12.

2. Gardner (1997), pp. 140–141, 149.

3. Einstein (1907b).

4. Ibid., p. 439.

5. See, e.g., Gruber (1981).

6. This scenario can be developed for the 1895 thought experiment. However, it would not be as striking or informative because the discovery to which it partially led—principle of relativity and the relativity of time—was, itself, only a part of Einstein's discovery of special relativity. Other thought experiments were required, such as the one with magnet and conductor in relative motion.

7. Quoted from Miller (1999), p. 93.

8. Einstein (1922), p. 47.

9. Einstein (1919), quoted from Miller (1999), pp. 93–94.

10. Einstein (1919), quoted from Miller (1998a), p. 137.

11. Einstein (1905c), p. 406.

12. See Miller (1999), pp. 94–96, for details, and pp. 101–102 for Einstein's published verification in his 1907 review paper (1907b).

13. Einstein first used the term "equivalence principle" in 1912 (see Einstein, 1912). The 1907 result applied only to constant linear acceleration and took eight years of grueling work to generalize it to include any sort of acceleration.

14. For example, recall the force that holds you to the wall of, say, a rotating cylinder in an amusement park ride, when the cylinder's floor has dropped down. From the point of view of Newtonian mechanics this is not a real force because it does not depend on the masses and positions of other bodies. Through the equivalence principle such forces can be related to gravitational fields. This very brief statement belies how difficult it turned out for Einstein.

15. Einstein deduced these results in 1907 as follows: The equations relating times between two reference systems contain their relative velocity. For the situation Einstein considered in 1907—constant linear acceleration—it is straightforward to eliminate this relative velocity in terms of acceleration. Through the equivalence principle, acceleration is replaced by a gravitational field. In order that Lorentz's equations retain their form in an accelerated reference system, that is, satisfy the principle of relativity for accelerating reference systems, the velocity of light turns out to depend on the reference system's acceleration and, through the equivalence principle, the gravitational field.

16. High-precision experiments to demonstrate this equivalence had been carried out in 1891 by the Hungarian experimentalist Roland von Eötvös, but Einstein later claimed to be unaware of them. See Einstein (1934a), p. 80.

17. Poincaré (1902), p. 121.

18. See Poincaré (1905b, 1906). I have analyzed Poincaré's papers of 1905 and 1906 in my (1973) and (1998a).

19. See Miller (1986b), Essay 1.

20. Einstein (1946), p. 15. Minkowski's results are discussed in Galison (1979); Miller (1998a), Chapter 7; and Walter (1999).

21. Born (1958), p. 218.

22. He accomplished this by expressing the four-dimensional geometry in terms of mathematical functions that describe surfaces in a four-dimensional or non-Euclidean space. Physicists found the non-Euclidean formulation difficult to understand, Einstein included. See Walter (1999), pp. 94–105.

23. Poincaré (1913), pp. 108–109.

24. Minkowski (1908), p. 104.

25. Letter of Einstein to Arnold Heim, 14 July, 1952. Quoted from CPAE1, p. 44.

26. See CPAE1, p. 44, note 11.

27. Einstein (1905c), p. 375.

28. Poincaré (1902), p. 78.

29. Ibid., p. 111, and Einstein (1905c), p. 371. Poincaré distinguished between mathematical and physical spaces. See Miller (2000), pp. 196–202.

30. See Einstein (1916) and Miller (1998a), p. 241.

31. During 1912–1913 Einstein and Grossmann published two papers on gravitation. Grossmann died in 1936.

32. See Miller (1998a), Chapters 7 and 12; Corry (1999); and Walter (1999). In Minkowski's view Einstein had thought up a deeper way of expressing Lorentz's theory of the electron.

33. From a manuscript of a course given by Hilbert during 1916–1917, quoted from Corry (1999), p. 178.

34. Quoted from Fölsing (1998), p. 235. By summer of 1908 Laub and Einstein were in the midst of a lively correspondence on how to apply Einstein's relativity theory to dielectrics and magnets, which turned out to be an especially difficult problem, yet to be resolved to everyone's satisfaction. During April 1908 Einstein and Laub spent three weeks together in Bern and wrote two papers on the subject that later turned out to be incorrect.

35. CPAE5, English translation, p. 43.

36. CPAE5, p. 48, note 2.

37. Seelig (1952), p. 103.

38. The sole student was Max Stern whose principal interest was insurance mathematics, not science. See Fölsing (1998), pp. 237–238.

39. CPAE5, p. 96, note 5.

40. Letter of Einstein to Jakob Laub, CPAE5, English translation, p. 120.

41. Ibid.

42. This was a lecture that Einstein presented to the local Bern physical society on 11 February, 1909, entitled "Electrodynamics and the Principle of Relativity." CPAE5, p. 190, note 6.

43. Letter of Laub to Einstein, 16 May, 1909, in ibid., English translation, p. 117.

44. Letter of Einstein to Jakob Laub, ibid., English translation, p. 120. Einstein would have been even more amused at the comedy of academia had he known about what was happening behind the scenes. See Fölsing (1998), pp. 249–251.

45. Quoted from Fölsing (1998), p. 253.

46. See Einstein (1909).

47. CPAE5, English translation, p. 140.

48. See Chapter 6, note 62.

49. See Pais (1982), pp. 100–103.

50. See Miller (1998a), pp. 345–350.

51. In particular, for the advance of the planet Mercury's perihelion, which came out incorrect from Poincaré's gravitational theory. See Poincaré (1908), p. 261.

52. The deciding factor in unraveling the viewpoints of Einstein and Lorentz-Poincaré was clarifying the distinction between Lorentz's contraction hypothesis and how the contraction of moving bodies is described in Einstein's relativity theory. See Miller (1998a), pp. 245–253.

53. Letter from Einstein to Zangger on 15 November, 1909, in CPAE5, English translation, pp. 221–222.

54. Seelig (1954), p. 163. Marie Curie also wrote a glowing letter on Einstein's behalf. See Seelig (1954), p. 162.

55. See Miller (1998a), p. 240. For details of this assertion, which some philosophers of science still consider as tendentious despite objective historical evidence to the contrary, see Miller (1996).

56. Einstein (1921), p. 236.

57. CPAE5, English translation, p. 222.

58. Speziali (1972), p. 50.

59. Letter of E. G. Strauss to A. Pais, October 1979, quoted from Pais (1982), p. 239.

60. CPAE5, English translation, p. 222.

61. Letter of Einstein to Laub on 19 May, 1909, in CPAE5, English translation, p. 121.

62. Einstein (1957), p. 8.

63. CPAE5, English translation, p. 227.

64. See Miller (1998a), pp. 240–242. Lorentz and Poincaré knew this but insisted otherwise and they were wrong in doing so (see Miller, 1996).

65. CPAE5, English translation, p. 120.

66. Letter of Mileva to Helen Savic, around October 1909, and quoted from Stachel (1996).

67. Letter of Einstein to Anna Schmid, August 1899, in CPAE1, English translation, p. 128.

68. The young man's hormones really must have been in high gear because he also invited a female friend from Aarau, Juliet Niggli, to join him in Mettmenstetten. Some years later she recalled Einstein's invitation and her astonishment. Niggli confronted Einstein and he laughed it away saying that he meant nothing untoward since, after all, his mother and sister were there. Despite this episode, when Einstein was perhaps testing the waters, they were confidants. On about 6 August, 1899, from Mettmenstetten, Einstein responded to a letter of Niggli's in which she expressed concern over her involvement with an older man who had no intention of marrying her. Einstein replied with advice from a man of the world. His message was that men are a race apart, whose moods and feelings vacillate from day to day. So much so that not much should be expected from them; "I know this sort of animal personally—as I am one of them myself" (CPAE1, English translation, pp. 129–130). To some extent this captures Einstein's attitude toward women.

69. Letter of Einstein to Anna Meyer Schmid, in CPAE5, English translation, p. 115.

70. Ibid.

71. Ibid.

72. Ibid.

73. Ibid., p. 199, note 4. Mileva's letter to Georg Meyer, Anna's husband, is dated 23 May, 1909.

74. Letter of Einstein to Georg Meyer, 2 June, 1909, in ibid., English translation, p. 127.

75. Ibid., p. 140.

76. For example, Einstein felt it necessary to apologize to his mother: "The bad mood you noticed in me had nothing to do with you." Letter of Einstein to Pauline Einstein, 28 April, 1910, in ibid., English translation, p. 152.

77. Einstein (1998), *The Collected Papers of Albert Einstein: Volume 8*, hereafter CPAE8, English translation, p. 613.

78. Letter of Einstein to Erika Schaerer-Meyer, 27 July, 1951, quoted from CPAE5, p. 199, note 4.

79. In 1932 Eduard was diagnosed as having severe schizophrenia and interned in the Burghölzli psychiatric hospital where he died in 1965.

80. Frank (1949), p. 131.

81. From a reminiscence of David Reichinstein from 1934. Quoted from Highfield and Carter (1993), p. 130.

82. Ibid. Zangger was not new at this. He had achieved international prominence in 1906 when he urged the continuation of rescue efforts of workers trapped in a collapsed mine in Courrièrs. More than three hundred were saved, many requiring to be revived. See CPAE5, p. 642.

83. Seelig (1954), p. 119. Hans Tanner began his Ph.D. under Einstein in Zurich. He attended all of Einstein's classes during Einstein's tenure at the university, 1909–1911.

84. Quoted from Highfield and Carter (1993), p. 132.

85. Ibid.

86. So impressed, in fact, that Nernst set about organizing an international conference on the quantum hypothesis. He managed to persuade the Belgian industrialist and amateur scientist Ernest Solvay to provide the funds. See CPAE5, pp. xxi–xxviii.

87. From Planck's published lectures at Columbia University in 1909. See Fölsing (1998), p. 271.

88. CPAE5, p. xxxvi.

89. Ibid.

90. Letter of Einstein to Alfred and Clara Stern on 17 March, 1912, in ibid., English translation, p. 275.

91. See ibid., p. xxxvi, for references to relevant documents.

92. Letter of Einstein to Alfred and Clara Stern, on 2 February, 1912. Ibid., English translation, p. 255.

93. Ibid., p. xxxvii.

94. Ibid., p. 300. Einstein began visiting Elsa in Berlin in the latter part of April 1912. At her request, Elsa's letters to Albert were destroyed.

95. Ibid., p. 343.

96. Ibid., p. 355.

97. Ibid., p. 360.

98. Ibid., p. 366.

99. CPAE8, English translation, pp. 565–566.

100. Ibid., p. 565.

101. For example, see Pais (1994).

102. Seelig (1954), pp. 230–231. Charlie Chaplin gave an apt description of Elsa: "She was a square-framed woman with abundant vitality; she frankly enjoyed being the wife of the great man and made no attempt to hide the fact; her enthusiasm was endearing" (quoted from Pais, 1982, p. 301).

103. Quoted from the television production, *Einstein,* NOVA productions, March 1979.

CHAPTER 8

1. Gardner (1985), p. 8.
2. Ibid., p. 195.
3. This part of Copernicus's argument is based on Neoplatonism, a philosophical train of thought that can been traced back to the fifth-century A.D. Greek philosopher Proclus.
4. Einstein (1905a), p. 367.
5. Einstein (1905c), p. 370.
6. Although in the early days of perspective there was, of course, emphasis on the use of geometrical methods in setting out a painting, but not for the sake of seeking characteristics of nature beyond appearances.
7. Although Lorentz always believed that Einstein's special relativity and his electron theory were in every way equivalent, which is incorrect, he went on to inspire Einstein toward general relativity theory and made important contributions of his own. The reason is that general relativity offered a geometry of space-time on which light rays travel. For Lorentz this was a proper version of the ether. See Miller (1998a), pp. 255–257, and Miller (1986a), especially pp. 55–58.
8. See also Miller (1996).
9. Einstein accomplished this in his relativity paper and Poincaré in his paper on Lorentz's electron theory, i.e., Einstein (1905c) and Poincaré (1905b). Poincaré's paper was published 5 June, 1905, and Einstein's was received at the *Annalen* 30 June, 1905. There is no reason to believe that Einstein saw Poincaré's paper before sending off his own. Even if he had, it would have been of no help regarding conceptual matters because Poincaré never discussed simultaneity either in his short 1905 paper or in the longer one with the same title that he published in 1906, and was submitted 23 July, 1905.
10. Like Lorentz, Poincaré also considered special relativity theory to be in every way equivalent to Lorentz's electron theory. In fact, Poincaré never once cited Einstein's relativity theory in print, or discussed it in any explicit way. The nearest he came to discussing special relativity is in a lecture he gave at University College London (then the University of London), 17 July, 1912, where he summarized his position with regard to relativity theory: "Today some physicists want to adopt a new convention [for simultaneity and assert that] everything happens as if time were a fourth dimension of space. . . . It is not that they are constrained to do; they consider this new convention more convenient, that is all. And those who are not of this opinion can legitimately retain the old one in order not to disturb their habits. I believe, just between us, that this is what they shall do for a long time to come" (Poincaré, 1913, pp. 108–109).

11. In his 1906 paper, Poincaré began work on a theory of gravity that ran along the lines of the electromagnetic world-picture. Its principal prediction was an advance of the planet Mercury's perihelion, which did not agree with astronomical data. This was among the causes of Poincaré never elevating the principle of relativity to an axiom.

12. Quoted from Chipp (1968), p. 273.

13. Richardson (1991), pp. 48–49.

14. This episode cast a shadow over the rest of Picasso's life, affecting his relationship with women and leaving him with a terror of illness. See Richardson (1991), pp. 49–50.

15. CPAE1, English translation, pp. 32–33.

16. Frank (1949), p. 152.

17. Ibid.

18. Gilot and Lake (1964), p. 77.

19. See Richardson (1991), pp. 203–204.

20. See, for example, Richardson (1991), pp. 116–118, and Gilot (1964), pp. 168–171.

21. See Gardner (1997).

22. For Freudian-based studies of artists and scientists see Gombrich (1954) and Storr (1991). Although Picasso never made any direct statements on Freudian theory, Einstein did. Around 1927, to someone who suggested that he undergo psychoanalysis, Einstein drafted the following reply (which he never sent): "I regret that I cannot accede to your request, because I should like very much to remain in the darkness of not having been psychoanalysed" (Hoffmann and Dukas, 1979, p. 35).

23. See Zervos (1932), and for commentary Lipton (1976), pp. 279–282, and Golding (1994), pp. 214–215.

24. Lipton (1976), p. 288. For details see Lipton (1976), especially pp. 279–326, which has discussion of other subsequent psychological analyses of Picasso in the 1930s. Jung's psychoanalytic view had a great impact on the surrealists who were deeply concerned with the interplay between myth and visual imagery. See Lipton (1976), pp. 289–307; Golding (1994), pp. 214–215; and Green (1987), pp. 281 and 296.

25. See, for example, Miller (1986a), Chapters 5, 6 and 7; Miller (1992); Miller (1999); and Miller (2000).

26. See Miller (2000), especially Chapter 9.

27. See, for example, Simon et al. (1987). For an overview of discovery programs in science, mathematics and music see Boden (1990). For an interesting and provocative treatment of Faraday's discoveries in electricity and magnetism based heavily on laboratory data and that strives for the proper historical scenario, see Gooding (1988).

28. My remarks on discovery programs for scientific theories refer to work by Herbert Simon and collaborators, which is summarized in Simon et al. (1987). See also Miller (2000), Chapter 9.

29. See Miller (2000), pp. 340–360, Miller (1992) and Miller (1999).

30. Toulouse (1910), p. 146.

31. Poincaré (1908), p. 62.
32. Einstein (1946), p. 7.
33. Poincaré (1908), p. 54.
34. See, for example, Mandler (1994) and Simon et al. (1987).
35. Smith and Blankenship (1991). The architecture of long-term memory is a complex network in which information is stored as symbols and images. In a problem situation certain information is retrieved and then processed in short-term memory. Needless to say, accessing occurs in a massively parallel manner, else, for example, we could not recognize and respond to threatening situations.
36. Salmon (1912), p. 42.
37. Miller (2000), pp. 335–338.
38. Wertheimer's education in physics provided metaphors for his theory of creativity. So, for example, there is a trend in physical systems to reach configurations of maximal symmetry, which are also states of minimum energy and so of high stability.
39. Some of these principles are good continuation, proximity and symmetry. See Miller (2000), pp. 298–300.
40. Quoted from Rubin (1984), p. 225.
41. See Goldenberg, Mazursky and Solomon (1999). Conversely, complete freedom to let one's mind wander can inhibit creativity.
42. Poincaré (1908), p. 59.
43. Ibid., p. 58.
44. Picasso finished *Guernica* in about three months, from May to July 1937. He accomplished many of the key compositional studies in one day, on 1 May, 1937. See Arnheim (1962).
45. See Miller (2000), pp. 344–350.
46. Seelig (1954), p. 82.
47. Einstein (1946), p. 7.
48. By the "deep structure" of an entity, I mean its properties essential to understanding it and that lie beyond appearances.
49. By "see" in quotes I mean not visual perception, but seeing with the mind through understanding an entity or phenomenon's "deep structure." To "see" something is to understand it through a mixture of perception and cognition.
50. Einstein (1946), p. 15.
51. Ibid., p. 17.
52. See Gardner (1993), Chapter 10.
53. Einstein (1946), p. 21.
54. Ibid.
55. Ibid., p. 23.
56. Ibid., pp. 31 and 33.
57. Classical causality is an essential part of Newtonian mechanics and so, too, of Einstein's special and general theories of relativity. It states that knowledge of where an object is and how fast it is traveling is enough to predict with absolute accuracy the course of its future movements. This is not the case in quantum me-

chanics in which how a system develops in space and time depends on probabilities. For example, according to quantum mechanics, the instant in time when a radioactive atom will emit energy and de-excite cannot be predicted with certainty; only a probable time for de-excitation can be provided. Such examples led to Einstein's making his often quoted remark in the 1930s: "God does not throw dice." See Hoffmann and Dukas (1972), pp. 193–194, concerning Einstein's remark, and Miller (2000) for a discussion of causality in classical and quantum physics, as well as references to the abundant literature on this subject.

58. See Miller (2000), p. 50.

59. Born (1923), p. 537.

60. Data from the interaction between light and atoms could not be systematically interpreted on the basis of the atom being a minuscule solar system.

61. For example, Schrödinger, in one of the articles in which he presented his new version of atomic physics, called wave mechanics, wrote about what urged him to formulate it (Schrödinger, 1926, p. 128): "[I] felt discouraged, not to say repelled, by the methods of [Heisenberg's theory], and by the lack of visualisability." Heisenberg pulled no punches in a letter to his colleague Wolfgang Pauli on 8 June, 1926 (Pauli, 1979, p. 328): "The more I reflect on the physical portions of Schrödinger's theory the more disgusting I find it. . . . What Schrödinger writes on the visualisability of his theory . . . I consider trash."

62. For a detailed development see Miller (2000), Chapter 10; Miller (1986a), Chapter 4; and Schweber (1994).

63. This is the "simplest" of very many Feynman diagrams describing how two electrons interact.

64. I first wrote about this point in 1985, but was unable to locate the exact painting that Bohr had hung in his study. Recently Professor Mitchell Stephens was able to locate the correct one and I thank him for generously sharing this unpublished information with me. The painting depicts a female equestrian whose figurative rendering is mostly sliced up into facets and other geometrical constructions in a manner that is not at all as severe as in *Le Goûter*.

65. Gleizes and Metzinger (1912), p. 68.

66. Anderson (1967), p. 321.

67. See ibid., p. 322.

68. While complementarity sufficed as a possible explanation for measurement, it says nothing about a visual representation of atomic processes themselves. Bohr concluded this was not possible. Some physicists, particularly Heisenberg, were not pleased with this consequence of complementarity. See Miller (2000), Chapter 10.

69. Straightening out the issue of causality in quantum mechanics was a knotty one on which Bohr and Heisenberg struggled during late 1926 into spring of 1927. In 1927, as part of his complementarity principle, Bohr argued that in quantum mechanics, causality can be associated with the conservation laws of energy and momentum. But the probabilistic aspect of quantum mechanics remains because it is an intrinsic part of nature.

70. See Miller (2000), pp. 127–128.

71. The cordée between Braque and Picasso was effectively over by the end of 1913. In August of 1914, at the train station in Avignon, Picasso saw Braque and Derain off to war. Some years later Picasso told Kahnweiler that he "never saw [Braque and Derain] again" (quoted from Richardson, 1996, p. 345). He of course did but, as Kahnweiler recalled, Picasso "meant that it was never the same." Braque was mentioned in dispatches for bravery and was awarded the Légion d'Honneur and Croix de Guerre. The trauma of a devastating head wound that left him almost blind, and the subsequent treatment that included trepanning, drastically transformed him into an introvert. "According to Dora Maar," writes Richardson, "Braque always meant more to Picasso than any other man except the Catalan cronies of his youth. By the same token he was one of the few people capable of wounding him. Rejection was Braque's weapon" against Picasso's at times cruel sense of humor. See Richardson (1996), p. 195. By the end of his life in 1963, Braque had ventured into pure abstraction. As Braque said to Richardson toward the end of his career, in a vein that no doubt was affected by his near-death experience in the trenches, "You see, I have made a great discovery. I no longer believe in anything. Objects don't exist for me except in so far as a rapport exists between them or between them and myself. When one attains this harmony one reaches a sort of intellectual non-existence—what I can only describe as a sense of peace, which makes everything possible and right. Life then becomes a perpetual revelation. That is true poetry" (quoted from Golding, 1997, p. 10). Braque conveyed this mystical feeling in many of his late works. See, for example, Golding (1997).

72. On the physiological basis for this sort of art, see Miller (2000) and Zeki (1999), especially Chapter 12.

73. Daix (1994), p. 209.

74. Ibid., p. 155.

75. I have developed this view in a number of places. See, for example, Miller (2000), Chapter 10.

76. This bubble chamber photograph, from 1972, constitutes data toward verification of an important model in elementary particle physics that unifies the weak and electromagnetic interactions, the so-called electroweak theory. See Miller (2000), pp. 406–409.

77. For details, see Miller and Bulloch (1994).

78. But special relativity did not "knock out" Newtonian mechanics. What became clear during the first decade of the twentieth century is that theories have limits of validity. For example, Newtonian mechanics is perfectly satisfactory for calculating the motion of objects that are not moving close to the velocity of light, for which one uses special relativity. Going in another direction, Bohr would not have been able to formulate his atomic theory without Newton's theory as a starting point. For further discussion and references on this point see Miller (2000), especially pp. 65–68.

79. In later years Einstein became aware of Picasso's cubism. We know this from a letter that he wrote to the art historian Paul M. Laporte on 4 May, 1946. Laporte had sent to Einstein a manuscript entitled, "Cubism and Relativity."

Based on popularizations of relativity theory, Laporte tried to link developments in the analytic cubism of about 1911–1912 to physics at that time. Laporte's papers on this topic appeared in 1948 and 1949 (Laporte, 1948, 1949). The art historian Linda Henderson discussed their shortcomings in some detail and concluded (Henderson, 1983, p. 358): "The mistake of art historians dealing with Cubism and Relativity has been to read back into Cubist literature of 1911 and 1912 the development in physics of a non-Euclidean space-time continuum that was not completed until 1915 or 1916." In 1966 Laporte published a sequel article that contains Einstein's letter. In 1988 the editors of *Leonardo* reprinted Laporte's 1966 paper with an introduction by the psychologist of art Rudolf Arnheim and a supposedly improved translation of Einstein's letter (Laporte, 1988). The gist of Einstein's obliquely worded letter is essentially this: Works of art and science must be evaluated differently because art is culturally dependent while science is universal. The former is subjective and the latter objective. Einstein, however, goes on to confound the issue a bit by insisting that only one measurement platform (reference system) is actually necessary in relativity theory to understand a physical situation, while "this is quite different in the case of Picasso's painting, as I do not have to elaborate any further." Einstein's point is essentially that the name "relativity theory" is a misnomer because on a deeper level it reveals how the laws of nature remain the same in all reference systems. This is well and good, but how a phenomenon is understood measurement-wise depends on what reference system is used, that is, how you look at. This is a similarity between cubism and relativity. But, as we have seen a hint of already, great care has to be taken in how this relationship is explored, for instance the difference between simultaneities of spatial or timelike genre.

80. Daix (1979), no. 385, p. 282.

81. Ibid., no. 430, p. 272. Although *Woman with Guitar* is generally ascribed to 1911, the words *Ma Jolie* were probably put on some time later. See Richardson (1996), p. 222. Picasso took the phrase from a popular song often played at the Cirque Médrano, "O Manon ma jolie, mon coeur te dit bonjour."

82. Letter of Picasso to Gertrude Stein, 14 January, 1915, quoted from Daix (1994), p. 147. Almost at the same level of grief as the deaths of Eva and his father, in May 1914, was that of Picasso's dog Frika, in May 1913. Frika had been Picasso's loyal companion right from the beginning at the Bateau Lavoir. Just as on the occasions of the deaths of Eva and don Ruiz, when Frika was put to sleep Picasso's art took on an air of mourning. Even fifty years later, mention of Frika brought tears to Picasso's eyes (Richardson, 1996, p. 278).

83. See Daix (1994) and Gilot and Lake (1964), and for informative biographical sketches of Picasso's wives and mistresses see Daix (1995).

84. Picasso never saw Fernande again in person after they split up in 1912. She led a difficult life and would have died virtually penniless except for a bit of genteel blackmail. In 1957 she informed Picasso of her intention to publish a second volume of her memoirs. Picasso paid her a million old francs not to publish them in his lifetime; Marcelle Braque, Georges's wife, acted as the go-between. As it turned out, Picasso need not have been concerned. In 1956 she appeared on tele-

vision and spoke about her days in Montmartre. Picasso poked fun at what he considered a disgusting performance by an old and toothless woman. See Richardson (1996), pp. 232–233. Fernande died 29 January, 1966, at age eighty-five. Kokhlova died in 1955; Walter committed suicide in 1974 and Jacqueline in 1986; and Maar suffered serious mental problems after her breakup with Picasso and died in 1997.

85. Highfield and Carter (1993), p. 206.

86. According to Pais (1982), p. 320. Whatever we know about these liaisons is summarized in Highfield and Carter (1993).

87. Einstein made several asides regarding his less-than-enthusiastic view of marriage. To someone who asked him whether his joy in pipe smoking had anything to do with unclogging and refilling his pipe, he replied, "My aim lies in smoking, but as a result things tend to get clogged up, I'm afraid. Life, too, is like smoking, especially marriage" (quoted from Pais, 1982, p. 302). But, all in all, his failed marriages saddened him, as he wrote to Besso's son Vero, soon after his father's death: "What I admired most in Michele, as a man, is the fact of his having been able to live for many years with a woman, not only in peace, but also in a continuous harmony, an enterprise in which I have failed miserably twice" (Speziali, 1972, p. 538). Elsa died of circulatory and kidney problems on 20 December, 1936. Her last days were painful and Einstein was extremely caring. So much so that she is quoted as saying that "he went around miserable and depressed. I never thought he was so attached to me. That, too, helps" (Vallentin, 1954, p. 227).

88. Quoted from Highfield and Carter (1993), p. 158.

89. Ibid., p. 159.

90. Quoted from Hoffmann and Dukas (1979), p. 17.

91. Gilot and Lake (1964), p. 117.

92. The image of the struggling supergenius, producing sublime works in less-than-ideal conditions, had a major predecessor, Mozart. Mozart *chose* to work under these conditions in order to preserve his independence. Unlike Einstein and Picasso, Mozart could have lived otherwise, but his tastes for high living were costly and he squandered much of his money away. See Hildesheimer (1983), pp. 19–21.

93. At the time of the Solvay Conference, Curie and Langevin were at the storm's eye of a scandal over their love affair that had been revealed in the Paris press. Einstein's opinion of the matter was that if they are in love then who cares, and besides everyone knew that Langevin wanted to get divorced. See letter of Einstein to Zangger, 7 November, 1911, in CPAE5, English translation, p. 219. Langevin and Einstein became lifelong friends. Incidentally, Langevin's father was born in the Bateau Lavoir (Crespelle, 1978, p. 76).

94. CPAE5, English translation, p. 240.

95. Daix (1979), p. 311, no. 633.

96. Ibid., p. 332, no. 760.

Bibliography

Anderson, Mogens. 1967. "An Impression." In *Niels Bohr: His Life and Work as Seen by His Friends and Colleagues.* New York: Interscience Publishers. 321–324.

Antliff, Robert Mark. 1988. "Bergson and Cubism: A Reassessment." *Art Journal* (Winter): 341–349.

Apollinaire, Guillaume. 1905. "Young Artists: Picasso the Painter." *La Plume* (15 May). Quoted from *Apollinaire on Art: Essays and Reviews 1902–1918.* Edited by Leroy C. Breunig. Translated by Susan Suleiman. London: Thames and Hudson, 1972.

———. 1908. "Georges Braque." Preface to the *Catalogue de l'exposition Braque*, Kahnweiler Gallery. Quoted from *Apollinaire on Art: Essays and Reviews 1902–1918.* Edited by Leroy C. Breunig. Translated by Susan Suleiman. London: Thames and Hudson, 1972. 50–52.

———. 1912a. "Art and Curiosity: The Beginnings of Cubism." *Le Temps* (14 October). Quoted from *Apollinaire on Art: Essays and Reviews 1902–1918.* Edited by Leroy C. Breunig. Translated by Susan Suleiman. London: Thames and Hudson, 1972. 259–261.

———. 1912b. "La Peinture nouvelle, notes d'art." *Les Soirées de Paris* (April). Quoted from *Apollinaire on Art: Essays and Reviews 1902–1918.* Edited by Leroy C. Breunig. Translated by Susan Suleiman. London: Thames and Hudson, 1972. 222–225.

———. 1913. *Les Peintres cubistes: Meditations esthétiques.* Paris: Figuière. Reprinted with an introduction and annotation by L. C. Breunig and J.-Cl. Chevalier. Paris: Hermann, 1980. All page references are to the 1993 edition.

Arnheim, Rudolf. 1962. *The Genesis of a Painting: Picasso's Guernica.* Berkeley: University of California Press.

———. 1969. *Visual Thinking.* Berkeley: University of California Press.

Baldassari, Anne. 1994. *Picasso photographie: 1901–1916.* Paris: Éditions de la Réunion de musées nationaux.

———. 1997. *Picasso and Photography: The Dark Mirror.* Translated by Deke Dusinberre. Houston: The Museum of Fine Arts.

Barr, Alfred H., Jr. 1975 [1946]. *Picasso: Fifty Years of His Art.* London: Martin Secker & Warburg Ltd. [New York: Museum of Modern Art]. All references are to the 1975 edition.

Bartky, Ian R. 1989. "The Adaption of Standard Time." *Technology and Culture* 30: 26–56.

Beaumont, Keith. 1984. *Alfred Jarry: A Critical and Biographical Study.* Leicester: Leicester University Press.

Bergson, Henri. 1907. *L'Évolution créatrice.* Paris: Flammarion.

Bernadac, Marie-Laure, and Androula Michael. 1998. *Picasso: Propos sur l'art.* Paris: Gallimard.

Blunt, Anthony, and Phoebe Pool. 1962. *Picasso: The Formative Years, A Study of His Sources.* London: Studio Books.

Boden, Margaret A. 1990. *Creative Mind: Myths and Mechanisms.* London: Weidenfeld & Nicolson.

Bohn, Willard. 1980. "In Pursuit of the Fourth Dimension: Guillaume Apollinaire and Max Weber." *Arts* 54 (June): 166–169.

Boltzmann, Ludwig. 1897. *Vorlesungen über die Principe der Mechanik.* Edited by and translated in part in B. McGuiness, *Ludwig Boltzmann: Theoretical Physics and Philosophical Problems* (Boston: Reide, 1974). All quotations are from the McGuiness book.

Born, Max. 1923. "Quantentheorie und Störungsrechnung." *Die Naturwissenschaften* 27: 537–550.

———. 1958. *Physik im Wandel meiner Zeit.* Berlin: Braunschweig.

Braque, Georges. 1935. "Testimony Against Gertrude Stein." Transition 23(1) (supplement). The Hague: 13–14. Reprinted in Marilyn McCully, *A Picasso Anthology,* p. 64. London: Thames and Hudson, 1981.

Brassaï. 1964. *Conversations avec Picasso.* Paris: Gallimard.

Braun, Marta. 1997. "The Expanded Present: Photographing Movement." In *Beauty of Another Order: Photography in Science.* Edited by Ann Thomas. New Haven: Yale University Press. 150–185.

Breunig, LeRoy C. 1993. Introduction to Guillaume Apollinaire, *Les Peintres cubistes: Meditations esthétiques.* Paris: Figuière, 1913. Reprinted Paris: Hermann, 1980. xvii–xxx.

Breunig, LeRoy C., ed. 1972. Introduction to *Apollinaire on Art: Essays and Reviews 1902–1918.* Translated by Susan Suleiman. London: Thames and Hudson. xvii–xxx.

Brush, Stephen G. 1986. *The Kind of Motion We Call Heat: A History of the Kinetic Theory of Gases in the 19th Century.* 2 vols. New York: North-Holland.

Burgess, Gelett. 1910. "The Wild Men of Paris." *Architectural Record* 27(5) (May): 401–414.

Cabanne, Pierre. 1960. "Braque se retourne sur son passé." *Arts* 783 (July).

Carco, Francis. 1927. *De Montmartre au Quartier Latin*. Paris: Albin-Michel.

Charensol, Georges. 1924. "Chez Juan Gris." *Paris Journal* 25 (April): 5.

Chipp, Herschel, with P. Zelz and Joshua C. Taylor, eds. 1968. *Theories of Modern Art: A Source Book for Artists and Critics*. Berkeley: University of California Press.

Clark, Ronald, W. 1972. *Einstein: The Life and Times*. New York: Avon.

Coke, Van Deren. 1964. *The Painter and the Photograph*. Albuquerque: University of New Mexico Press.

Corry, Leo. 1999. "Hilbert and Physics (1900–1915)." In *The Symbolic Universe: Geometry and Physics, 1890–1930*. Edited by Jeremy Gray. Oxford: Oxford University Press. 145–188.

Cottington, David. 1998. *Cubism in the Shadow of War: The Avant-Garde and Politics in Paris 1905–1914*. New Haven: Yale University Press.

Cousins, Judith, with the collaboration of Pierre Daix. 1989. "Documentary Chronology." In *Picasso and Braque: Pioneering Cubism*. Edited by W. Rubin. New York: The Museum of Modern Art. 335–452.

Cousins, Judith, and Hélène Seckel. 1994. "Chronology of *Les Demoiselles d'Avignon*." In *Les Demoiselles d'Avignon*. Edited by William Rubin, Hélène Seckel, and Judith Cousins. New York: The Museum of Modern Art. 145–212.

CPAE1. See Einstein, 1987.

CPAE3. See Einstein, 1993a.

CPAE5. See Einstein, 1993b.

CPAE8. See Einstein, 1998.

Crespelle, Jean-Paul. 1978. *La Vie quotidienne à Montmartre au temps de Picasso: 1900–1910*. Paris: Hachette.

Daix, Pierre. 1966. *Picasso: 1900–1906*. With Georges Boudaille. Neuchâtel: Ides et Calendes.

———. 1979. *Picasso: The Cubist Years 1907–1916*. With Joan Rosselet. Translated by Dorothy S. Blair. Boston: New York Graphic Society. Originally published as *Le Cubisme de Picasso: Catalogue raisonné de l'oeuvre*. Neuchâtel: Ides et Calendes.

———. 1987. "Comment Picasso rompit-il avec son dessin classique?" *Revue des sciences morales et politiques* 1: 75–89.

———. 1988a [1966]. *Picasso: 1900–1906, Catalogue raisonné de l'oeuvre peint, 1900, 1901, 1906: Pierre Daix, 1902 à 1905: Georges Boudaille, Catalogue établi avec la collaboration de Joan Rosselet*. Neuchâtel: Editions Ides et Calendes.

———. 1988b. "Les Trois périodes de travail de Picasso sur Les Trois Femmes (Automne 1907–Automne 1908), Les Rapports avec Braque et les débuts du Cubism." *Gazette des Beaux Arts* (Jan.–Feb.): 141–154.

———. 1988c. "Dread, Desire and the Demoiselles." *Art News*: 133–137.

———. 1988d. "L'Historique des *Demoiselles d'Avignon* révisé à l'aide des carnets de Picasso." In *Picasso: Les Demoiselles d'Avignon—Carnet de dessins.* Edited by Hélène Seckel. 2 vols. Paris: Réunion des Musées Nationaux, Editions Adam Biro, 489–545.

———. 1992. "The Chronology of Proto-Cubism: New Data on the Opening of the Picasso/Braque Dialogue." In *Picasso and Braque: A Symposium.* Edited by L. Zelevansky. New York: The Museum of Modern Art. 306–321.

———. 1994. *Picasso: Life and Art.* Translated by Olivia Emmet. London: Thames and Hudson. Originally published as *Picasso créateur.* Paris: Editions du Seuil, 1987.

———. 1995. *Dictionnaire Picasso.* Paris: Éditions Robert Laffont.

Darboux, Gaston. 1913. "Élogie historique d'Henri Poincaré." In *Oeuvres d'Henri Poincaré.* Vol. 2, vii–lxxii. 11 vols. Paris: Gauthier-Villars.

Décaudin, Michel. 1981. *La Crise des valeurs symbolistes: Vingt ans de poésie Française, 1895–1914.* Paris: Slatkine.

———. 1991. *Journal intime, 1898–1918.* Paris: Limon.

Delaunay, Robert. 1957. *Du Cubisme à l'art abstrait.* Paris: SEVPEN.

Derain, André. 1994 [1955]. *André Derain: Lettres à Vlaminck.* Edited by Philippe Dagen. Paris: Flammarion.

Dibi-Huberman, Georges. 1986. "Photography—Scientific and Pseudo-Scientific." In *A History of Photography: Social and Cultural Perspectives.* Edited by Jean-Claude Lamagny and André Rouillé. Cambridge: Cambridge University Press. 71–76.

Dunan, Charles. 1888. "L'Espace visuel et l'espace tactile." *Revue philosophique de la France et de l'étranger* 25: 134–169, 354–386, 591–619.

Einstein, Albert. 1901. "Folgerungen aus den Kapillaritätserscheinungen." *Annalen der Physik* 4: 513–523.

———. 1904. "Allgemeine molekulare Theorie der Wärme." *Annalen der Physik* 14: 354–362.

———. 1905a. "Über einen die Erzeugung und Verwandlung des Lichtes betreffenden heuristischen Standpunkt." *Annalen der Physik* 17: 132–148.

———. 1905b. "Die von der molekularkinetischen Theorie der Wärme geforderte Bewegung von in ruhenden Flüssigkeiten suspendierten Teilchen." *Annalen der Physik* 17: 549–560.

———. 1905c. "Zur Elektrodynamik bewegter Körper." *Annalen der Physik* 17: 891–921. All quotations are from the English translation in Arthur I. Miller, *Albert Einstein's Special Theory of Relativity: Emergence (1905) and Early Interpretation (1905–1911).* New York: Springer-Verlag. 370–393.

———. 1905d. "Eine neue Bestimmung der Moleküldimensionen." Doctoral dissertation, University of Zurich.

———. 1906. "Prinzip von der Erhaltung der Schwerpunktsbewegung und die Trägheit der Energie." *Annalen der Physik* 20: 627–633.

———. 1907a. "Über die vom Relativitätsprinzip geforderte Trägheit der Energie." *Annalen der Physik* 23: 371–384.

———. 1907b. "Über das Relativitätsprinzip und die aus demselben gezogenen Folgerungen." *Jahrbuch der Radioaktivität und Elektronik* 4: 411–462.

———. 1909. "Über die Entwicklung unserer Anschauungen über das Wesen und die Konstitution der Strahlung." *Physikalische Zeitschrift* 10: 817–825.

———. 1912. "Prinzipielles zur allgemeinen Relativitätstheorie." *Annalen der Physik* 55: 241–244.

———. 1916. "Grundlagen der allgemeinen Relativitätstheorie." *Annalen der Physik* 49: 769–822.

———. 1919. "What Is the Theory of Relativity," written for the London *Times*, 28 November. Reprinted in Albert Einstein, *Essays in Science.* New York: Philosophical Library, 1934. 53–60.

———. 1920. "Relativity and the Ether." Lecture presented 27 October at Leiden University. Reprinted in Albert Einstein, *Essays in Science.* New York: Philosophical Library, 1934. 98–111.

———. 1921. "Geometry and Experience." Lecture presented 27 January. In A. Einstein, *Ideas and Opinions.* New York: Bonanza. 232–246.

———. 1922. Kyoto Lecture. 14 December. Translation in *Physics Today* (August 1982): 45–47.

———. 1923. "Fundamental Ideas and Problems of the Theory of Relativity." In *Nobel Lectures: Physics, 1901–1921.* Amsterdam: Elsevier, 1967. 482–490. Presentation address by S. Arrhenius, 479–481. This is referred to as Einstein (1923), because Einstein was in Japan at the time of the Nobel ceremonies and submitted this text as his "acceptance" lecture, which was delivered to the Nordic Assembly of Naturalists, Gothenburg, 11 July, 1923.

———. 1934a. "Notes on the Origin of the General Theory of Relativity." In A. Einstein, *Essays in Science.* New York: Philosophical Library. 78–84.

———. 1934b. "The Problem of Space, Ether and the Field in Physics." In A. Einstein, *Essays in Science.* New York: Philosophical Library. 61–77.

———. 1936. "Physics and Reality." *Franklin Institute Journal* 221: 73–77. Reprinted in A. Einstein, *Out of My Later Years.* Totowa, N.J.: Littlefield, Adams & Co., 1967. 58–94.

———. 1946. "Autobiographical Notes." In *Albert Einstein: Philosopher-Scientist.* Edited by P. A. Schilpp. La Salle, Ill.: Open Court, 1949. 2–94. This selection will be referred to as Einstein (1946) because Einstein completed the "Autobiographical Notes" in that year.

———. 1956. "Autobiographische Skizze." In *Helle Zeit-Dunkele Zeit.* Edited by Carl Seelig. Branschweig: Friedr. Vieweg Sohn.

———. 1957. "H. A. Lorentz, His Creative Genius and His Personality." In *H. A. Lorentz: Impressions of His Life and Work.* Edited by G. L. de Haas-Lorentz. Amsterdam: North-Holland Publishing Company. 5–9.

———. 1965. *Errinerungen an Albert Einstein.* Pamphlet issued by the Patent Office in Bern, about 1965, unpaginated.

———. 1967. *Out of My Later Years.* Totowa, N.J.: Littlefield, Adams & Co.

———. 1970. *The Meaning of Relativity.* Translated by E. P. Adams. Enlarged edition. Princeton: Princeton University Press [New York: Methuen, 1922].

———. 1987. *Collected Papers of Albert Einstein: Volume 1.* Edited by John Stachel. English translation by Anna Beck with Peter Havas, consultant. Princeton: Princeton University Press. Referred to as CPAE1.

———. 1993a. *Collected Papers of Albert Einstein: Volume 3.* Edited by Martin J. Klein, A. J. Kox, Jürgen Renn, and Robert Schulman. English translation by Anna Beck with Don Howard, consultant. Princeton: Princeton University Press, 1993. Referred to as CPAE3.

———. 1993b. *Collected Papers of Albert Einstein: Volume 5.* Edited by Martin J. Klein, A. J. Kox, Jürgen Renn, and Robert Schulman. English translation by Anna Beck with Don Howard, consultant. Princeton: Princeton University Press, 1993. Referred to as CPAE5.

———. 1998. *Collected Papers of Albert Einstein: Volume 8.* Edited by Robert Schulman, A. J. Kox, and A. M. Hentschel. English translation by Anna M. Hentschel with Klaus Hentschel, consultant. Princeton: Princeton University Press. Referred to as CPAE8.

Everdell, William R. 1999. *The First Moderns: Profiles in the Origins of Twentieth-Century Thought.* Chicago: University of Chicago Press.

Faraday, Michael. 1965. *Experimental Researches in Electricity.* 3 vols. New York: Dover Publications.

Fitzgerald, Michael G. 1995. *Making Modernism: Picasso and the Creation of the Market for Twentieth Century Art.* Berkeley: University of California Press.

Flam, Jack D. 1984. "Matisse and the Fauves." In *Primitivism in Twentieth-Century Art.* Edited by William Rubin. New York: The Museum of Modern Art. 211–239.

Fölsing, Albrecht. 1998. *Albert Einstein: A Biography.* Translated by Ewald Osers. London: Penguin Books. Originally published as *Albert Einstein: Eine Biographie.* Frankfurt: Suhrkamp Verlag, 1993. All references are to the English-language edition.

Frank, Philipp. 1949. *Einstein: Sein Leben und seine Zeit.* Munich: Paul List Verlag.

Frisch, J. 1899. *La Pratique de la photographie instantanée par les appareils à main.* Paris.

Fry, Edward, ed. 1966. *Cubism.* London: Thames & Hudson.

Galison, Peter. 1979. "Minkowski's Space-Time: From Visual Thought to the Absolute World." *Historical Studies in the Physical Sciences* 10: 85–121.

———. 2000. "Einstein's Clocks: The Place of Time." *Critical Inquiry* 26: 355–389.

Gamwell, Lynn. 1977. *Cubist Criticism.* Ann Arbor, Mich.: UMI Research Press.

Gardner, Howard. 1985. *Frames of Mind: The Theory of Multiple Intelligences.* New York: Basic Books.

———. 1993. *Creating Minds: An Anatomy of Creativity Seen Through the Lives of Freud, Einstein, Picasso, Stravinsky, Eliot, Graham, and Gandhi.* New York: Basic Books.

———. 1997. *Extraordinary Minds: Portraits of Exceptional Individuals and an Examination of Our Extraordinariness.* London: Weidenfeld & Nicolson.

Gibbons, T. 1981. "Cubism and the Fourth Dimension in the Context of the Late 19th Century and Early 20th Century Revival of Occult Idealism." *Journal of the Warbourg and Courtauld Institutes* 44: 130–147.

Gilot, Françoise, and Carlton Lake. 1964. *Life with Picasso.* London: Virago Press.

Gleizes, Albert, and Jean Metzinger. 1980. *Du Cubisme.* Paris: Éditions Présence [Paris: Figuière, 1912].

Goldenberg, Jacob, David Mazursky, and Sorin Solomon. 1999. "Creative Sparks." *Science* 285: 1495–1496.

Golding, John. 1988. *Cubism: A History and Analysis, 1907–1914.* 3d ed. rev. London: Faber and Faber.

———. 1994. *Visions of the Modern.* London: Thames and Hudson.

———. 1997. *Braque: The Late Works.* London: Royal Academy of Arts.

Golding, John, and R. Penrose, eds. 1973. *Picasso: 1881–1973.* New York: Paul Elek, Ltd.

., Ernest H. 1954. "Psychoanalysis and the History of Art." *The International Journal of Psycho-analysis* 35: 1–11.

Green, Christopher. 1987. *Cubism and Its Enemies: Modern Movements and Reaction in French Art, 1916–1928.* New Haven: Yale University Press.

———. 1992. *Juan Gris.* London: Whitechapel.

Gruber, Howard. 1981. "On the Relation Between 'Aha Experiences' and the Construction of Ideas." *History of Science* 19: 1–19.

Hadamard, Jacques. 1954. *The Psychology of Invention in the Mathematical Field.* New York: Dover.

Hayden, Edward Everett. 1905. "Appendix IV: The Present Status of the Use of Standard Time." Washington, D.C.: U.S. Naval Observatory. This text is a preliminary version meant for a report at the meeting of the International Railway Congress, Washington, D.C., May 1905.

Heilbron, John. 1982. "*Fin-de-Siècle* Physics." In *Science, Technology and Society in the Time of Alfred Nobel.* Edited by C. F. Bernhard, E. Crawford, and P. Sörbom, 51–73. New York: Pergamon Press.

Henderson, Linda Dalrymple. 1983. *The Fourth Dimension and Non-Euclidean Geometry in Modern Art.* Princeton: Princeton University Press.

———. 1988. "X-Rays and the Quest for Invisible Reality in the Art of Kupka, Duchamp, and the Cubists." *Art Journal* 47 (Winter): 323–340.

———. 1998. *Duchamp in Context: Science and Technology in the Large Glass and Related Works.* Princeton: Princeton University Press.

Highfield, Roger, and Paul Carter. 1993. *The Private Lives of Albert Einstein.* London: Faber and Faber.

Hildesheimer, Wolfgang. 1983. *Mozart.* New York: Vintage Books.

Hoffmann, Banesh, and Helen Dukas. 1972. *Albert Einstein Creator and Rebel.* New York: Viking Press.

Hoffmann, Banesh, and Helen Dukas, eds. 1979. *Albert Einstein: The Human Side.* Princeton: Princeton University Press.

Holton, Gerald. 1973a. "On Trying to Understand Scientific Genius." In *Thematic Origins of Scientific Thought: Kepler to Einstein.* Cambridge, Mass.: Harvard University Press. 353–380.

———. 1973b. "Influences on Einstein's Early Work." In *Thematic Origins of Scientific Thought: Kepler to Einstein.* Cambridge, Mass.: Harvard University Press. 197–217.

———. 1973c. "Mach, Einstein and the Search for Reality." In *Thematic Origins of Scientific Thought: Kepler to Einstein.* Cambridge, Mass.: Harvard University Press. 353–380.

———. 1973d. "Einstein, Michelson, and the 'Crucial' Experiment." In *Thematic Origins of Scientific Thought: Kepler to Einstein.* Cambridge, Mass.: Harvard University Press. 261–352.

———. 1995. *Einstein, History, and Other Passions.* New York: AIP Press.

Howse, Derek. 1997. *Greenwich Time and the Longitude.* London: Philip Wilson Publishers Limited. Originally published as *Greenwich Time and the Discovery of the Longitude.* Oxford: Oxford University Press, 1980.

Huyghe, René, ed. 1935. *Histoire de l'art contemporain: La Peinture.* Paris: Félix Alcan.

Jacob, Max. 1927. "Souvenirs sur Picasso contés par Max Jacob." *Cahiers d'Art* (Paris) 6: 199–203.

James, William. 1890. *The Principles of Psychology.* 2 vols. New York: Henry Holt & Co.

Jarry, Alfred. 1899. "Commentaire pour servir à la construction pratique de la machine à explorer le temps." *Mercure de France* 29: 387–396. Reprinted in *Selected Work of Alfred Jarry.* Edited and translated by Roger Shattuck and Simon Watson Taylor. London: Eyre Methuen, 1965. 114–121.

———. 1911. *Gestes et opinions du docteur Faustroll, pataphysicien.* Paris: Fasquelle. Reprinted in *Selected Works of Alfred Jarry.* Edited and translated by Roger Shattuck and Simon Watson Taylor. London: Eyre Methuen, 1965. 173–256.

Jouffret, Esprit. 1903. *Traité élémentaire de géometrie à quatre dimensions.* Paris: Gauthier-Villars.

Johnson, Ron. 1980a. "Picasso's 'Demoiselles d'Avignon' and the Theatre of the Absurd." *Arts* (October): 102–113.

———. 1980b. "The Demoiselles and Dionysion Destruction." *Arts* (October): 94–101.

Kahnweiler, Daniel-Henry. 1916. "Der Kubismus." *Der Weissen Blatte* 3: 209–222.

———. 1920. *Der Weg zum Kubismus.* Munich: Delphin. Translated in part in *Theories of Modern Art.* Edited by Herschel Chipp, with P. Zelz and Joshua C. Taylor. Berkeley: University of California Press, 1968. 248–259.

———. 1961. *Mes galaries et mes peintres: Entretiens avex Francis Crémieux.* Paris: Gallimard.

Kaufmann, Walter. 1905. "Über die Konstitution des Elektrons." *Sitzungsberichte der Königlich Preussischen Akademie der Wissenschaften* 45: 949–956.

———. 1906. "Über die Konstitution des Elektrons." *Annalen der Physik* 20: 487–553.

Kern, Stephen. 1987. *The Culture of Space and Time: 1880–1918.* Cambridge, Mass.: Harvard University Press.

Klein, Martin. 1962. "Max Planck and the Beginnings of Quantum Theory." *Archive for History of Exact Sciences* 1: 459–479.

———. 1967. "Thermodynamics in Einstein's Thought." *Science* 157: 509–516.

Klein, Martin, and Alan Needell. 1977. "Some Unnoticed Publications by Einstein." *ISIS* 68: 601–604.

Kramers, Hendrik, and H. Holst. 1923. *The Atom and the Bohr Theory of Its Structure.* Translated from the first Danish edition by R. B. and R. T. Lindsay. London: Gyldendal.

Laporte, Paul M. 1948. "The Space-Time Concept in the Work of Picasso." *Magazine of Art* 41: 26–32.

———. 1949. "Cubism and Science." *Journal of Aesthetics and Art Criticism* 7: 243–256.

———. 1988. "Cubism and Relativity with a Letter of Albert Einstein, with an Introduction by Rudolf Arnheim." *Leonardo* 21: 313–315. Reprinted from *Art Journal* 25 (Spring 1966): 246–248.

Léal, Brigitte. 1988. "Carnets." In *Picasso: Les Demoiselles d'Avignon—Carnet de dessins.* Edited by Hélène Seckel. 2 vols. Paris: Réunion des Musées Nationaux, Editions Adam Biro.

Leighten, Patricia. 1987. "The Dreams and Lies of Picasso." *Arts* (October): 50–55.

———. 1988a. "Editor's Comment: Revising Cubism." *Art Bulletin* (Winter): 269–276.

———. 1988b. "'La Propagande par le rire': Satire and Subversion in Apollinaire, Jarry and Picasso's Collages." *Gazette des Beaux Arts* (October): 163–172.

———. 1989. *Re-Ordering the Universe: Picasso and Anarchism, 1897–1914.* Princeton: Princeton University Press.

Leja, Michael. 1985. "'Le Vieux Marcheur' and 'Les Deux Risques': Venereal Disease and Maternity, 1899–1907." *Art History* 8: 66–81.

Lhote, André. 1935. "Naissance de cubisme." In *Histoire de l'art contemporain: La Peinture.* Edited by René Huyghe. Paris: Félix Alcan. 80.

Lieberman, Alexander. 1988. *The Artist in His Studio.* Rev. edition. New York: Random House.

Lipton, Eunice. 1976. *Picasso Criticism, 1900–1939: The Making of an Artist Hero.* London: Garland Publishing, Inc.

Lorentz, H. A. 1904. "Electromagnetic Phenomena in a System Moving with any Velocity Less Than That of Light." *Koninklijke Akademie van Wetenschappen te Amsterdam. Section of Sciences. Proceedings* 6: 809–831.

Mach, Ernst. 1960. *The Science of Mechanics: A Critical and Historical Account of Its Development.* Translated in 1893 by T. J. McComack from the second German edition of 1889, revised in 1942 to include additions and alterations up to the ninth German edition. La Salle, Ill.: Open Court. Originally published as *Die Mechanik in ihrer Entwicklung historisch-kritisch dargestellt.* Leipzig: F. A. Brockhaus, 1883. All page references are to the 1960 English-language edition.

Malraux, André. 1994. *Picasso's Mask.* Translated by June Guicharnaud with Jacques Guicharnaud. New York: Da Capo Press. Originally published as *La Tête d'obsidienne.* Paris: Gallimard, 1974. All references are to the English-language edition.

Mandler, George. 1994. "Hyperamnesia, Incubation, and Mind Popping: On Remembering Without Really Trying." In *Attention and Performance XV.* Edited by C. Umiltà and M. Moscovitch. Princeton: Princeton University Press.

McCully, Marilyn. 1981. *A Picasso Anthology.* London: Thames and Hudson.

Mélon, Marc. 1986. "Beyond Reality: Art and Photography." In *A History of Photography: Social and Cultural Perspectives.* Edited by Jean-Claude Lamagny and André Rouillé. Cambridge: Cambridge University Press. 82–101.

Metzinger, Jean. 1910. "Note sur la peinture." *Pan* (October–November): 649–651. Reprinted in Edward Fry, ed., *Cubism.* London: Thames & Hudson, 1966. 59–60.

———. 1911. "Cubisme et tradition." *Paris-Journal.* 16 August. Reprinted in Edward Fry, ed., *Cubism.* London: Thames & Hudson, 1966. 66–67.

———. 1972. *Le Cubisme était né: Souvenirs.* Paris: Éditions Présence.

Miller, Arthur I. 1973. "A Study of Henri Poincaré's 'Sur la Dynamique de l'Electron'." *Archive for History of Exact Sciences* 10: 207–328. Reprinted in A. I. Miller, *Frontiers of Physics: 1900–1911.* Boston: Birkhäuser, 1984. 29–150.

———. 1986a. *Imagery in Scientific Thought: Creating 20th-Century Physics.* Cambridge, Mass.: MIT Press; Boston: Birkhäuser, 1984.

———. 1986b. *Frontiers of Physics, 1900–1911: Selected Essays.* Boston: Birkhäuser.

———. 1992. "Scientific Creativity: A Comparative Study of Henri Poincaré and Albert Einstein." *Creativity Research Journal* 5: 385–418.

———. 1996. "Why Did Poincaré Not Formulate Special Relativity in 1905." In *Henri Poincaré: Science and Philosophy.* Edited by Jean-Louis Greffe, Gerhard Heinzmann, and Kuno Lorenz. Berlin: Akademie Verlag. 69–100.

———. 1998a. *Albert Einstein's Special Theory of Relativity: Emergence (1905) and Early Interpretation (1905–1911).* New York: Springer-Verlag. First edition published in Reading, Mass.: Addison-Wesley, 1981. All page references are to the 1998 edition.

———. 1998b. "The Gift of Creativity." *Roeper Reviews* 21: 51–54.

———. 1999. "Einstein's First Steps Toward General Relativity: *Gedanken* Experiments and Axiomatics." *Physics in Perspective* 1: 85–104.

———. 2000. *Insights of Genius: Imagery and Creativity in Science and Art.* Cambridge, Mass.: MIT Press; New York: Springer, 1996.

Miller, Arthur I., and Frederick W. Bullock. 1994. "Neutral Currents and the History of Scientific Ideas." *Studies in the History and Philosophy of Science* 25: 895–931.

Minkowski, Hermann. 1908. "Raum und Zeit." *Physikalische Zeitschrift* 20: 104–111. Lecture delivered to the eightieth Naturforscherversammlung at Cologne, 21 September 1908.

Mitchell, Timothy. 1977–1978. "Bergson, Le Bon, and Hermetic Cubism." *Journal of Aesthetics and Art Criticism* 36: 175–183.

Newhall, Beaumont. 1964. *The History of Photography from 1839 to the Present Day.* New York: The Museum of Modern Art.

Nye, Mary Jo. 1974. "Gustave LeBon's Black Light: A Study in Physics and Philosophy in France at the Turn of the Century." *Historical Studies in the Physical Sciences* 4: 163–195.

Olivier, Fernande. 1933. *Picasso et ses amis.* Paris: Librairie Stock. Translated by Jane Miller as *Picasso and His Friends.* London: Heinemann, 1964. All page references are to the English-language edition.

———. 1988. *Souvenirs intimes: Écrits pour Picasso.* Paris: Calmann-Lévy.

Pais, Abraham. 1982. *Subtle Is the Lord: The Science and the Life of Albert Einstein.* Oxford: Oxford University Press.

———. 1994. "Einstein and the Press." *Physics Today* (August): 30–36.

Parmelin, Hélène. 1969. *Picasso Says.* Translated by Christine Trollope. London: Allen and Unwin. Originally published as *Picasso dit.* Paris: Gonthier, 1966.

Pauli, Wolfgang. 1979. *Wissenschaftlicher Briefwechsel mit Bohr, Einstein, Heisenberg, U.A.: Volume I, 1919–1929.* Edited by A. Hermann, K. von Meyenn, and V. F. Weisskopf. Berlin: Springer-Verlag.

Péladan, Joséphin. 1904. "Le Radium et l'hyperphysique." *Mercure de France* 50: 608–637.

Piaget, Jean. 1971. *The Child's Conception of Time.* Translated by A. J. Pomerans. New York: Ballantine Books. Originally published as J. Piaget, *Le Développement de la notion de temps chez l'enfant.* Paris: Presses Universitaires de France, 1927.

Planck, Max. 1906. "Die Kaufmannschen Messungen der Ablenkbarkeit der ß-Strahlen in ihrer Bedeutung für die Dynamik der Elektronen." *Physikalische Zeitschrift* 7: 418–432.

———. 1910. "Zur Theorie der Wärmestrahlung." *Annalen der Physik* 31: 758–767.

Poincaré, Henri. 1898. "La Mesure de temps." *Revue de métaphysique de morale* 6: 371–384. Reprinted in Henri Poincaré, *La Valeur de la science*. Paris: Flammarion, 1904. 41–54. All references are to Flammarion's edition of 1970.

———. 1900. "La Théorie de Lorentz et le principe de réaction." In *Recueil de travaux offerts par les auteurs à H. A. Lorentz*. The Hague: Nijhoff. 464–488.

———. 1901. *Electricité et optique*. Paris: Gauthier-Villars.

———. 1902. *La Science et l'hypothèse*. Paris: Flammarion. All references are to the 1968 edition. German translation by F. and L. Lindemann as *Wissenschaft und Hypothese* (Leipzig: Teubner, 1904).

———. 1905a. *La Valeur de la science*. Paris: Flammarion. All references are to the 1970 edition.

———. 1905b. "Sur la dynamique de l'électron." *Comptes Rendus de l'Académie des Sciences* 140: 1504–1508.

———. 1906. "Sur la dynamique de l'électron." *Rend. del. Circ. Mat. Di Palermo* 21: 129–175.

———. 1908. *La Science et méthode*. Paris: Flammarion.

———. 1909. "La Logique de l'infini." *Revue de Métaphysique et Morale* 17: 461–482.

———. 1913. *Dernières pensées*. Paris: Flammarion.

Prieur, Albert. 1904. *Mercure de France* 50: 498–505.

Puy, Michel. 1911. "Les Indépendants." *Les Marges* (July): 27–30. Reprinted in *Cubism*. Edited by Edward Fry. London: Thames & Hudson, 1966. 65–66.

Raynal, Maurice. 1912. "Conception et vision." *Gil Blas*. Paris. Translated in Edward Fry, ed., *Cubism*. London: Thames & Hudson, 1966. 94–96.

———. 1913. "Chronique cinématographique." *Soirées de Paris* (December): 6.

———. 1922. *Picasso*. Paris: Crés.

Read, Peter. 1995. *Picasso et Apollinaire: Les Métamorphoses de la mémoire, 1905/1973*. Paris: Éditions Jean Michel Place.

———. 1997. "'*Au Rendez-vous des poètes*': Picasso, French Poetry, and Theatre, 1900–1906." In *Picasso: The Early Years, 1892–1906*. Edited by Marilyn McCully. New Haven: Yale University Press. 211–223.

Reff, Theodore. 1971. "Harlequins, Saltimbanques, Clowns and Fools." *Artforum* (October): 30–41.

Réja, Marcel. 1904. "H.-G. Wells et le merveilleux scientifique." *Mercure de France* 52: 40–62.

Renn, Jürgen. 1993. "Einstein as a Disciple of Galileo: A Comparative Study of Concept Development in Physics." *Science in Context* 6: 311–341.

———. 1997. "Einstein's Controversy with Drude and the Origin of Statistical Mechanics: A New Glimpse from the Love Letters." Preprint 55, *Max-Planck-Institut für Wissenschaftgeschichte.*

Renn, Jürgen, and Robert Schulmann, eds. 1992. *Albert Einstein—Mileva Marić: The Love Letters.* Translated from the German by Shawn Smith. Princeton: Princeton University Press. The letters in German are in CPAE1.

Rich, Ben R., and Leo Janos. 1994. *Skunk Works.* New York: Little, Brown & Company.

Richardson, John R. 1980. "Your Show of Shows." *The New York Review of Books* xxvii: 16–24.

Richardson, John R., with the collaboration of Marilyn McCully. 1991. *A Life of Picasso, Volume I, 1881–1906.* New York: Random House.

Richardson, John R., with the collaboration of Marilyn McCully. 1996. *A Life of Picasso, Volume II, 1907–1917: The Painter of Modern Life.* New York: Random House.

Robbins, Daniel. 1988. "Abbreviated Historiography of Cubism." *Art Bulletin* (Winter): 277–283.

Rubin, William. 1977. "Cézannism and the Beginnings of Cubism." In *Cézanne: The Late Work.* Edited by William Rubin. New York: The Museum of Modern Art.

———. 1984. "Picasso." In *Primitivism in Twentieth-Century Art.* Edited by William Rubin. New York: The Museum of Modern Art. 241–340.

———. 1989. "Picasso and Braque: An Introduction." In *Picasso and Braque: Pioneering Cubism.* Edited by William Rubin. New York: The Museum of Modern Art. 15–61.

———. 1994. "The Genesis of *Les Demoiselles d'Avignon.*" In *Les Demoiselles d'Avignon.* Edited by William Rubin, Hélène Seckel, and Judith Cousins. New York: The Museum of Modern Art. 13–144.

Sabartés, Jaime. 1949. *Picasso: An Intimate Portrait.* Translated from Spanish by Angel Flores. London: W. H. Allen, 1949. Originally published in France as *Picasso: Portraits et souvenirs.* Paris: Louis Carré, 1946.

Salmon, André. 1910. "Courrier des ateliers." *Paris Journal* (10 May): 4.

———. 1912. *La Jeune peinture Française.* Paris: Albert Messein.

———. 1919. "Les Origines et Intentions du Cubisme." *Demain* (Paris) 68 (26 April): 485–489.

———. 1922. *Propos d'Atelier.* Paris: Crés.

———. 1935. "Testimony against Gertrude Stein." *Transition* 23(1)(supplement). The Hague: 14–15. Reprinted in Marilyn McCully, *A Picasso Anthology,* 62–63. London: Thames and Hudson, 1981.

———. 1945. *L'Air de la butte.* Paris: Les éditions de la nouvelle France.

———. 1955. *Souvenirs Sans Fin: Première Époque (1903–1908).* Paris: Gallimard.

———. 1956. *Souvenirs Sans Fin: Deuxième Époque (1908–1920.)* Paris: Gallimard.

Schaffner, Kenneth. 1972. *Nineteenth-Century Aether Theories.* New York: Pergamon.

Schrödinger, Erwin. 1926. "Über das Verhältnis der Heisenberg-Born-Jordanschen Quantenmechanik zu der meinen." *Annalen der Physik* 70: 734–756. Translated in part in G. Ludwig, *Wave Mechanics.* New York: Pergamon, 1968. 127–150.

Schweber, Sylan S. 1994. *QED and the Men Who Made It: Dyson, Feynman and Tomonaga.* Princeton: Princeton University Press.

Seckel, Hélène, ed. 1988a. *Picasso: Les Demoiselles d'Avignon—Carnet de dessins.* Vol. 1. Paris: Réunion des Musées Nationaux, Editions Adam Biro.

———. 1988b. *Picasso: Les Demoiselles d'Avignon—Carnet de dessins.* Vol. 2. Paris: Réunion des Musées Nationaux, Editions Adam Biro.

———. 1994a. *Max Jacob et Picasso.* Quimper: Musée des Beaux Arts.

———. 1994b. "Anthology of Early Commentary on *Les Demoiselles d'Avignon.*" In *Les Demoiselles d'Avignon.* Edited by William Rubin, Hélène Seckel, and Judith Cousins. New York: The Museum of Modern Art. 213–256.

Seelig, Carl. 1952. *Albert Einstein und die Schweiz.* Zurich: Europa-Verlag.

———. 1954. *Albert Einstein: Eine dokumentarische Biographie.* Zurich: Europa Verlag.

———. 1956. *Helle Zeit–Dunkele Zeit.* Zurich: Europa Verlag.

Shankland, Robert. 1963. "Conversations with Albert Einstein." *American Journal of Physics* 31: 47–57.

Shattuck, Roger. 1955. *The Banquet Years: The Origins of the Avant-Garde in France, 1885 to World War I.* London: Faber and Faber.

Shattuck, Roger, and Simon Watson Taylor, eds. and transls. 1965. *Selected Works of Alfred Jarry.* London: Eyre Methuen.

Shiff, Richard. 1984. *Cézanne and the End of Impressionism: A Study of the Theory, Technique, and Critical Evaluation of Modern Art.* Chicago: University of Chicago Press.

———. 1991. "Cézanne's Physicality." In *The Language of Art History.* Edited by Salim Kemal and Ivan Gaskell. Cambridge: Cambridge University Press. 129–180.

Simon, Herbert A., P. Langley, G. L. Bradshaw, and J. M. Zytkow. 1987. *Scientific Discovery: Computational Explorations of the Creative Process.* Cambridge, Mass.: MIT Press.

Smith, S. M., and S. E. Blankenship. 1991. "Incubation and the Persistence of Fixation in Problem Solving." *American Journal of Psychology* 104: 61–87.

Solovine, Maurice, ed. and trans. 1956. *Albert Einstein: Lettres à Maurice Solovine.* Paris: Gauthier-Villars.

Spate, Virginia. 1979. *Orphism: The Evolution of Non-Figurative Painting in Paris.* Oxford: Clarendon Press.

Speziali, Pierre. 1972. *Albert Einstein, Michele Besso: Correspondance, 1903–1955.* Paris: Hermann.

Stachel, John. 1996. "Albert Einstein and Mileva Marić: A Collaboration That Failed to Develop." In *Creative Couples in the Sciences.* Edited by Helena M. Pycior, Nancy G. Black, and Pnina G. Abir-Am. New Brunswick: Rutgers University Press. 207–219.

Staller, Natasha. 1989. "Méliès' 'Fantastic' Cinema and the Origins of Cubism." *Art History* 12: 202–232.

Steegmuller, Francis. 1986. *Apollinaire: Poet Among Painters.* New York: Penguin.

Stein, Gertrude. 1933. *The Autobiography of Alice B. Toklas.* New York: Harcourt Brace. Reprinted, New York: Vintage Books, 1990.

———. 1937. *Everybody's Autobiography.* New York: Random House.

———. 1984 [1938]. *Picasso.* New York: Dover [London, Batsford].

Stein, Leo. 1947. *Appreciation: Painting, Poetry and Prose.* New York: Crown Publishers.

Steinberg, Leo. 1972. "The Philosophical Brothel." *Art News* 71 (September): 20–29 and (October): 38–47. Reprinted in *Art News* 44 (Spring 1988): 7–74.

Stewart, Ian. 1990. *Does God Play Dice? The Mathematics of Chaos.* London: Penguin.

Storr, Anthony. 1991 [1972]. *The Dynamics of Creation.* London: Penguin Books [London: Martin Secker & Warburg Ltd].

Swenson, Loyd S. 1972. *The Ethereal Aether: A History of the Michelson-Morley-Miller-Aether Drift Experiments, 1880–1930.* Austin: University of Texas Press.

Teuber, Marianne L. 1997. "Gertrude Stein, William James, and Pablo Picasso's Cubism." In *A Pictorial History of Psychology.* Edited by W. G. Bringmann. Chicago: Quintessence. 256–264.

Toulouse, Édouard. 1910. *Henri Poincaré.* Paris: Flammarion.

Tucker, Paul Hayes. 1982. "Picasso, Photography and Development of Cubism." *Art Bulletin* 69: 288–299. Reply by Edward Fry (1983) 65: 145–146.

Uhde, Wilhelm. 1938. *Von Bismark bis Picasso: Erinnerungen und Bekenntnisse.* Zurich: Oprecht.

Unsigned. 1905. *Mercure de France* 54: 623–625.

Valéry, Paul. 1899. "Méthodes." *Mercure de France* 30: 481–488.

Vallentin, Antonina. 1954. *Einstein: A Biography.* Translated by Moura Budberg. London: Weidenfeld and Nicolson.

———. 1963. *Picasso.* Garden City, N.Y.: Doubleday. Originally published as *Pablo Picasso.* Paris: Albin Michel, 1957.

Vallier, Dora. 1954. "Braque, la peinture et nous: Propos de l'artiste recueillis." *Cahiers d'art* (October): 13–24.

Vargish, Thomas, and Delo E. Mook. 1999. *Inside Modernism: Relativity Theory, Cubism, Narrative.* New Haven: Yale University Press.

Varnedoe, Kirk. 1996. "Picasso's Self-Portraits." In *Picasso and Portraiture: Representation and Transformation.* Edited by William Rubin. New York: The Museum of Modern Art. 110–179.

Vauxcelles, Louis. 1908. "Exposition Braque. Chez Kahnweiler, 28 rue Vignon." *Gil Blas* (14 November).

———. 1911. "La Salon d'Automne." *L'Intransigeant* (10 October).

Vitz, Paul C., and Arnold B. Glimcher. 1984. *Modern Art and Modern Science: The Parallel Analysis of Vision.* New York: Praeger.

Vlaminck, Maurice. 1942. "Opinions libres . . . sur la peinture." *Comoedia* (6 June): 1, 6.

Walter, Scott. 1999. "The Non-Euclidean Style of Minkowskian Relativity." In *The Symbolic Universe: Geometry and Physics, 1890–1930.* Edited by Jeremy Gray. Oxford: Oxford University Press. 91–127.

Ward, Martha. 1996. *Pissarro, Neo-Impressionism, and the Spaces of the Avant-Garde.* Chicago: University of Chicago Press.

Warnod, André. 1945. "En peinture tout n'est que signe." *Arts* (June 29). Reprinted partially in *Picasso: Propos sur l'art.* Edited by M.-L. Bernadac and A. Michael. Paris: Gallimard. 53–56.

———. 1947. *Ceux de la butte.* Paris: n.p.

Warnod, Jeanine. 1975. *Le Bateau-Lavoir: 1892–1914.* Paris: Les Presses de la Connaissance.

Weber, Louis. 1903. *Mercure de France* 46: 769–771.

Weiss, Jeffrey. 1994. *The Popular Culture of Modern Art: Picasso, Duchamp, and Avant-Gardism.* New Haven: Yale University Press.

Wertheimer, Max. 1959. *Productive Thinking.* New York: Harper.

Whitrow, George, ed. 1967. *Einstein: The Man and His Achievement.* New York: Dover Publications.

Whittaker, Edmund. 1987. *A History of the Theories of Aether and Electricity: Volume I. The Classical Theories.* New York: Tomash.

Wien, Wilhelm. 1900. "Über die Möglichkeit einer elektromagnetischen Begründung der Mechanik." In *Recueil de travaux offerts par les auteurs à H. A. Lorentz.* The Hague: Nijhoff. 501–513.

Wilford, John Noble. 1982. *The Mapmakers.* New York: Vintage Books.

Winteler-Einstein, Maja. 1924. "Albert Einstein: Biographical Sketch (Excerpt)." In CPAE1. Translated by Anna Beck. xv–xxii. All references are to the English translation.

Zeki, Semir. 1999. *Inner Vision: An Exploration of Art and the Brain.* Oxford: Oxford University Press.

Zervos, Christian. 1932. "Picasso étudie par le Dr. Jung." *Cahiers d'Art* 12: 352–354.

———. 1932–1975. *Catalogue général illustré de l'oeuvre de Picasso.* 33 vols. Paris: Éditions Cahiers d'art.

———. 1935. "Pablo Picasso, Conversation, 1935." In *Theories of Modern Art.* Edited by Herschel Chipp, with P. Zelz and Joshua C. Taylor. Berkeley: University of California Press, 1968. 267–272. Originally published as "Conversation avec Picasso," *Cahiers d'Art* 10: 7–10. Reprinted in French in *Picasso: Propos sur l'art.* Edited by Marie-Laure Bernadac and Androula Michael. Paris: Gallimard, 1998. 31–37.

Photo Credits

Figure 2.1 Musée Picasso, Paris. Réunion des Musées Nationaux/Art Resource, NY.

Figure 2.2 FPPH 224. Picasso Archives. Musée Picasso, Paris. Réunion des Musées Nationaux/Art Resource, NY.

Figure 2.3 The Metropolitan Museum of Art: Rogers Fund, 1956.

Figure 2.4 The Cleveland Museum of Art. Bequest of Leonard C. Hanna Jr. © 2001 Estate of Pablo Picasso/Artists Rights Society (ARS), New York.

Figure 3.1 A. Einstein Kantonsschule Arau, 1896.

Figure 3.3 Lotte Jacobi Archives Photographic Services, University of New Hampshire.

Figure 3.4 Schweizerisches Literaturarchiv—Bern.

Figure 3.11 © Lotte Jacobi.

Figure 3.12 Schweizerisches Literaturarchiv—Bern.

Figure 3.13 Besso Family, courtesy of AIP Emilio Segré Archives.

Figure 4.1 Acquired through the Lillie Bliss Request. Photograph © 1997. The Museum of Modern Art, New York. © 2001 Estate of Pablo Picasso/Artists Rights Society (ARS), New York.

Figure 4.2 Musée Picasso, Paris. Réunion des Musées Nationaux/Art Resource, NY.

Figure 4.3 Musée Picasso, Paris. Réunion des Musées Nationaux/Art Resource, NY.

Figure 4.4 Musée Picasso, Paris. © 2001 Estate of Pablo Picasso/Artists Rights Society (ARS), New York.

Figure 4.7 Musée Picasso, Paris. © 2001 Estate of Pablo Picasso/Artists Rights Society (ARS), New York.

Figure 4.8 Musée Picasso, Paris. Réunion des Musées Nationaux/Art Resource, NY.
Figure 4.9 Musée Picasso, Paris. Réunion des Musées Nationaux/Art Resource, NY.
Figure 4.10 Musée Picasso, Paris. Réunion des Musées Nationaux/Art Resource, NY.
Figure 4.11 Kunstmuseum, Basle. Private Collection, on loan. © 2001 Estate of Pablo Picasso/Artists Rights Society (ARS), New York.
Figure 4.12 Musée Picasso, Paris. Réunion des Musées Nationaux/Art Resource, NY.
Figure 4.14 Picasso Archives, Musée Picasso, Paris. Réunion des Musées Nationaux/Art Resource, NY.
Figure 4.15 Musée Picasso, Paris. Réunion des Musées Nationaux/Art Resource, NY.
Figure 4.16 National Gallery of Canada, Ottawa, gift of Dr. Robert Crook, Ottawa, 1982.
Figure 4.17 Archives de Cinémathèque Française, Paris.
Figure 4.18 Private Collection. © 2001 Estate of Pablo Picasso/Artists Rights Society (ARS), New York.

Figure 5.1 *Kunstmuseum,* Bern: Hermann und Margrit Rupf Stiftung. © 2001 Artists Rights Society (ARS), New York/ADAGP, Paris.
Figure 5.2 Private Collection. © 2001 Estate of Pablo Picasso/Artists Rights Society (ARS), New York.
Figure 5.3 Picasso Archives. Réunion des Musées Nationaux/Art Resource, NY.
Figure 5.4 Picasso Archives. Réunion des Musées Nationaux/Art Resource, NY.
Figure 5.6 Picasso Archives. Réunion des Musées Nationaux/Art Resource, NY.
Figure 5.7 Picasso Archives. Réunion des Musées Nationaux/Art Resource, NY.
Figure 5.8 Private Collection. © 2001 Estate of Pablo Picasso/Artists Rights Society (ARS), New York.
Figure 5.9 Private Collection. Réunion des Musées Nationaux/Art Resource, NY.
Figure 5.10 Private Collection. © 2001 Estate of Pablo Picasso/Artists Rights Society (ARS), New York.

Figure 5.11 Private Collection. © 2001 Estate of Pablo Picasso/Artists Rights Society (ARS), New York.

Figure 5.12 Private Collection. © 2001 Estate of Pablo Picasso/Artists Rights Society (ARS), New York.

Figure 5.13 Top left and bottom left, Musée Picasso. Réunion des Musées Nationaux/Art Resource, NY. Top right and bottom right, Private Collection. © 2001 Estate of Pablo Picasso/Artists Rights Society (ARS), New York.

Figure 5.14 Picasso Archives. Réunion des Musées Nationaux/Art Resource, NY.

Figure 5.15 Picasso Archives. © 2001 Estate of Pablo Picasso/Artists Rights Society (ARS), New York.

Figure 5.16 The Museum of Modern Art, New York. Nelson A. Rockefeller Bequest. Photograph © 2001 The Museum of Modern Art, New York.

Figure 5.17 Private Collection. © 2001 Estate of Pablo Picasso/Artists Rights Society (ARS), New York.

Figure 5.18 Private Collection. © 2001 Estate of Pablo Picasso/Artists Rights Society (ARS), New York.

Figure 5.19 Private Collection. © 2001 Estate of Pablo Picasso/Artists Rights Society (ARS), New York.

Figure 5.20 Private Collection. Réunion des Musées Nationaux/Art Resource, NY.

Figure 5.21 Private Collection. © 2001 Estate of Pablo Picasso/Artists Rights Society (ARS), New York.

Figure 5.22 Private Collection. © 2001 Estate of Pablo Picasso/Artists Rights Society (ARS), New York.

Figure 5.23 Collection Joseph Pulitzer, St. Louis. © 2001 Estate of Pablo Picasso/Artists Rights Society (ARS), New York.

Figure 5.24 Solomon R. Guggenheim Museum, New York. © 2001 Artists Rights Society (ARS), New York/ADAGP, Paris.

Figure 5.25 Albright-Knox Art Gallery, Buffalo, NY. Private Collection. © 2001 Estate of Pablo Picasso/Artists Rights Society (ARS), New York.

Figure 5.26 Musée National d'Art Moderne, Paris. © 2001 Artists Rights Society (ARS), New York/ADAGP, Paris.

Figure 5.27 Collection of Mrs. Gilbert W. Chapman, New York. © 2001 Estate of Pablo Picasso/Artists Rights Society (ARS), New York.

Figure 5.28 The Art Institute of Chicago. © 2001 Estate of Pablo Picasso/Artists Rights Society (ARS), New York.

Figure 5.29 Louise and Walter Arensberg Collection. Philadelphia Museum of Art. © 2001 Artists Rights Society (ARS), New York/ADAGP, Paris.

Figure I.1 Courtesy of Estate of Henri Poincaré.

Figure 6.1 Schweizerisches Literaturarchiv – Bern

Figure 7.2 Ullstein.

Figure 8.4 (a) Courtesy of Professor F.W. Bullock; (b) Acquired through the Lillie Bliss Request. Photograph © 1997. The Museum of Modern Art, New York. © 2001 Estate of Pablo Picasso/Artists Rights Society (ARS), New York. (d) Collection of Mrs. Gilbert W. Chapman, New York. © 2001 Estate of Pablo Picasso/Artists Rights Society (ARS), New York.

Figure 8.5 Instituts Internationaux de physique et de chimie Solvay, courtesy AIP Emilio Segré Visual Archives.

Figure 8.6 Réunion des Musées Nationaux/Art Resource, NY.

INDEX